T0281074

Computational
Organometallic
Chemistry

Computational Organometallic Chemistry

edited by

Thomas R. Cundari

The University of Memphis
Memphis, Tennessee

CRC Press
Taylor & Francis Group
Boca Raton London New York

CRC Press is an imprint of the
Taylor & Francis Group, an **informa** business

CRC Press
Taylor & Francis Group
6000 Broken Sound Parkway NW, Suite 300
Boca Raton, FL 33487-2742

First issued in paperback 2019

© 2001 by Taylor & Francis Group, LLC
CRC Press is an imprint of Taylor & Francis Group, an Informa business

No claim to original U.S. Government works

ISBN-13: 978-0-8247-0478-0 (hbk)
ISBN-13: 978-0-367-39749-4 (pbk)

Visit the Taylor & Francis Web site at
http://www.taylorandfrancis.com

and the CRC Press Web site at
http://www.crcpress.com

Preface

This book is intended to fill a gap in the literature by covering a broad range of topics in computational organometallic chemistry. Two objectives were foremost in putting together this volume. First, pedagogical aspects are emphasized throughout. The particular challenges inherent in reliable modeling (quantum or classical) in organometallic chemistry are discussed, and strategies for addressing these challenges are offered. Second, "how-to" aspects are complemented with applications-oriented material covering a wide spectrum of research areas, including catalysis, medicine, organic synthesis, actinide chemistry, and so forth. The first goal will assist those who may have limited experience in computational organometallic chemistry research upon entering this exciting and dynamic field. The second objective will provide motivation for undertaking such an intellectual journey.

Computational Organometallic Chemistry has been written to be accessible to a general scientific audience. These pages will provide upper-division undergraduate students and graduate students with useful lessons that can be employed in their future scientific endeavors, while the applications chapters will spark future research contributions. Similarly, senior researchers, academic and industrial, who may wish to bring their energies to bear on this field will find both motivation and suitable background to do so. To accomplish these ambitious goals, an internationally recognized group of experts has been assembled, each focusing on his or her particular area of expertise within this growing field of ·science.

Thomas R. Cundari

Contents

v

Contents

Contributors

Mohsen Abu-Khudeir, B.S. Department of Chemistry, The University of Memphis, Memphis, Tennessee

Brett M. Bode, Ph.D. Applied Mathematical Sciences, Ames Laboratory, Iowa State University, Ames, Iowa

Bruce E. Bursten, Ph.D. Department of Chemistry, The Ohio State University, Columbus, Ohio

Gyusung Chung, Ph.D. Department of Chemistry, Konyang University, Chungnam, Korea

Thomas R. Cundari, Ph.D. Department of Chemistry, Computational Research on Materials Institute, The University of Memphis, Memphis, Tennessee

Margaret Czerw, B.A. Department of Chemistry, Rutgers, The State University of New Jersey, New Brunswick, New Jersey

Freida S. Dale, B.S.   Department of Chemistry, The University of Alabama at Birmingham, Birmingham, Alabama

Michael Diedenhofen, Dipl.Chem.   Department of Chemistry, Philipps-Universität Marburg, Marburg, Germany

Warthen Douglass, M.S.   Department of Chemistry, University of North Carolina at Wilmington, Wilmington, North Carolina

Dmitri G. Fedorov, Ph.D.   Department of Applied Chemistry, School of Engineering, University of Tokyo, Tokyo, Japan

Gernot Frenking, Ph.D.   Department of Chemistry, Philipps-Universität Marburg, Marburg, Germany

Mark S. Gordon, Ph.D.   Department of Chemistry, Iowa State University, Ames, Iowa

Tracy P. Hamilton, Ph.D.   Department of Chemistry, The University of Alabama at Birmingham, Birmingham, Alabama

Jeremy Noel Harvey, D.Sc.   School of Chemistry, University of Bristol, Bristol, England

Amy Hirsh, B.S.   Department of Chemistry, The University of Memphis, Memphis, Tennessee

Angela Jolly, B.S.   Department of Chemistry, The University of Memphis, Memphis, Tennessee

Karsten Krogh-Jespersen, Ph.D.   Department of Chemistry, Rutgers, The State University of New Jersey, New Brunswick, New Jersey

Takako Kudo, Ph.D.   Department of Fundamental Studies, Faculty of Engineering, Gunma University, Kiryu, Japan

Ohyun Kwon, M.S.   Department of Chemistry, Auburn University, Auburn, Alabama

Pascual Lahuerta, Ph.D.   Department of Inorganic Chemistry, University of Valencia, Valencia, Spain

Jun Li, Ph.D. Department of Chemistry, The Ohio State University, Columbus, Ohio

James P. Louey, Ph.D. Department of Chemistry, Sacred Heart University, Fairfield, Connecticut

Scott C. Malcolm, Ph.D. Department of Chemistry and Chemical Biology, Harvard University, Cambridge, Massachusetts

Feliu Maseras, Ph.D. Division of Physical Chemistry, Department of Chemistry, Universitat Autònoma de Barcelona, Barcelona, Catalonia, Spain

Ashalla McGee, B.S. Department of Chemistry, The University of Alabama at Birmingham, Birmingham, Alabama

Michael L. McKee, Ph.D. Department of Chemistry, Auburn University, Auburn, Alabama

Robert P. Meagley, Ph.D. Fab Materials Operation, Intel Corporation, Hillsboro, Oregon

Jerzy Moc, Ph.D. Faculty of Chemistry, Wroclaw University of Technology, Wroclaw, Poland

Per-Ola Norrby, Ph.D.* Department of Medicinal Chemistry, Royal Danish School of Pharmacy, Copenhagen, Denmark

Abby L. Parrill, Ph.D. Department of Chemistry, The University of Memphis, Memphis, Tennessee

Kristine Pierloot, Ph.D. Department of Chemistry, Catholic University of Leuven, Leuven, Belgium

Gigi B. Ray, Ph.D. Department of Chemistry, The University of Memphis, Memphis, Tennessee

Salah-eddine Stiriba, Ph.D. Department of Inorganic Chemistry, University of Valencia, Valencia, Spain

*Current affiliation: Technical University of Denmark, Lyngby, Denmark.

Douglass F. Taber, Ph.D.   Department of Chemistry and Biochemistry, University of Delaware, Newark, Delaware

Thomas Wagener, Ph.D.   Department of Chemistry, Philipps-Universität Marburg, Marburg, Germany

Yanong Wang, Ph.D.   Division of Chemical Sciences, Wyeth-Ayerst Research, American Home Products, Pearl River, New York

Simon P. Webb, Ph.D.   Department of Chemistry, Pennsylvania State University, University Park, Pennsylvania

David P. White, Ph.D.   Department of Chemistry, University of North Carolina at Wilmington, Wilmington, North Carolina

Takeyce K. Whittingham, B.A.   Department of Chemistry, Rutgers, The State University of New Jersey, New Brunswick, New Jersey

Soon S. Yoon, B.S.   Department of Chemistry, The University of Alabama at Birmingham, Birmingham, Alabama

Kimberly K. You, Ph.D.   Plastics Application Center, BASF Corporation, Wyandotte, Michigan

Wei Zhang, M.S.   Process Research and Development, Bristol-Myers Squibb Pharmaceutical Research Institute, New Brunswick, New Jersey

1

Introduction

Thomas R. Cundari
The University of Memphis, Memphis, Tennessee

When I was invited to edit a volume on computational organometallic chemistry by the good folks at Marcel Dekker, I accepted with enthusiasm. My eagerness for this project sprang primarily from the fact that this monograph covers two types of chemistry that are near and dear to my heart—computational and organometallic. Additionally, after canvassing colleagues, experimental and computational, I felt that there would be sufficient interest in this undertaking from the scientific community. Perhaps most importantly, from these discussions there emerged a consensus that the time was ripe for just such a project.

The application of modern computational techniques to organometallic chemistry has truly undergone a renaissance in the past few years, as is more than evident from the breadth of methods and topics discussed in this book. Through the hard work and perseverance of numerous research groups around the globe, many of the challenges involved in modeling these species, particularly those concerning the reliable and efficient modeling of metallic elements, have been addressed. The computational chemist now has a much larger (not to mention more effective) arsenal in dealing with organometallic compounds than just a few short years ago. As is evident from the chapters in *Computational Organometallic Chemistry*, developments have occurred within the realm of quantum and classical techniques, as well as hybrid quantum-classical approaches.

1

Another major motivation for this volume is to organize in a single place much of the hard-won experience that speaks to the "how to" of computational organometallic chemistry. This monograph brings together experts in the field and is designed to combine instructional aspects with cutting-edge applications. The former are intended to introduce this exciting research field to those, experimentalists and theorists alike, who might wish to try their hand at computational organometallic chemistry, while the latter should provide motivation for embarking on the journey.

As we start the new millennium, we see that the face of scientific research has changed dramatically in just the past decade. Two of the most important trends are the growing importance of computers in all aspects of scientific research and the increasing interdisciplinary nature of the science being undertaken. These tendencies are well represented in the present volume. Computational chemistry and organometallic chemistry are, almost by definition, interdisciplinary endeavors. The latter exists at the interface between inorganic and organic chemistry, providing erstwhile inorganic chemists a chance to try their hand at making new organometallic compounds by manipulation of the metal and its environs. Closet organic chemists also play a major role in organometallic chemistry through their attention to the organic functionalities. Computational chemistry has also metamorphosed from its origins as a branch of physical chemistry to embracing all traditional and nontraditional chemical disciplines. Computational chemists now routinely tackle problems in organic, inorganic, analytical, materials, and biological chemistry, and the list goes on.

In many respects, progress in computational organometallic chemistry has traditionally lagged behind other areas, because it combines the inherent challenges of both organic and inorganic modeling. An organometallic compound, as the name implies, is made up of two chemical regions—a metallic "core" and an organic "coating"—if I might be allowed a little poetic license. The organic coating is often characterized by its large size, large in terms of the number of atoms, orbitals, and/or conformational possibilities. It takes very few t-butyl substituents before a calculation on an organometallic compound becomes onerous! For the metallic core, i.e., the metal (or metals) and its surrounding inner coordination sphere, the inherent challenges for the computational organometallic chemist are different. Metals, particularly those of the d- and f-block, typically give rise to three main challenges in their chemical modeling: the large number of orbitals (many of them core), the so-called electron correlation problem (which is exacerbated by the presence of low-energy excited states), and relativistic effects for the heaviest metals.

Two techniques for dealing with these challenges, effective core potentials (or pseudopotentials) and density functional theory, have quickly transformed themselves from marginal techniques, once primarily the domain of solid-state chemists and physicists, to almost de rigueur standards for the computational organometallic chemist. This is due in part to computational improvements but perhaps, more importantly, to the inclusion of these techniques into powerful, yet user-friendly, computational chemistry packages.

Another trend, and a very welcome one at that, in modern computational organometallic chemistry is in some respects a return to the roots of computational chemistry. In the Stone Age (at least according to some of the students who have worked in my research group), hardware and software limitations forced the utilization of less qualitative methodologies. This is best typified by the unparalleled work of Hoffmann and his colleagues employing extended Hückel methods. Much of this work spoke to the "how" and "why" of organometallic chemistry, with less concern for "how much." For a while, it seemed that the only trend in computational organometallic chemistry was to be more quantitative, particularly for nongeometric quantities, such as reaction energies. This increase toward what some have termed chemical accuracy was certainly needed for the field to realize its full promise, but in many cases quantitative concerns overshadowed qualitative insight. Chapter 5, by Pierloot, shows that fundamental chemical insight, and not just accurate energies and bond lengths, can be extracted from even the most high-level calculations. Likewise, Chapters 3 and 6, by White and Maseras, respectively, tackle an age-old problem in chemistry, quantification of steric effects, in the former using molecular mechanics techniques and in the latter with hybrid quantum mechanics/molecular mechanics approaches.

In putting together this volume, the overriding theme was diversity—diversity of methods, diversity of applications, and diversity of chemistry. The "something for everyone" approach is not only an attempt to attract the largest possible audience for this book, but is also meant to highlight the amazing breadth and depth of computational organometallic chemistry. Chapters 2, 3, and 10, respectively by Norrby, White, and White and Douglass, focus primarily on classical (molecular mechanics) descriptions of chemical bonding. Of course, quantum mechanical approaches receive attention. Diedenhofen et al. (Chap. 4) and Gordon et al. (Chap. 11) address the accuracy of different quantum chemical techniques.

At one extreme of quantum chemical methodology lie approximate methods. Such techniques (for example, semiempirical quantum mechanics) typically involve great latitude in the number and type of approximations made to the full Schrödinger treatment. Approximations generally involve either the replacement of difficult-to-calculate quantities with experimental or theoretical estimates or the neglect of interactions (typically between electrons) thought to be of less chemical importance. Hence, the tradeoff for approximate methods is one of computational efficiency versus accuracy. The balance between accuracy and speed can be quite problematic for semiempirical quantum calculations on organometallic compounds because of the challenges discussed earlier for modeling metal species. The development or extension of any approximate method (molecular mechanics included) has a prerequisite parameterization phase. In this process, one seeks to determine those parameters that maintain computational efficiency (not to mention realistic chemistry and physics) while maximizing the descriptive and predictive power of the model. Ideally, the parameterization process should take into account the full range of motifs that characterize a chemical family. One major issue in the parameterization of approximate methods for metal-containing species is therefore the development of a robust parameterization that can handle what our group has termed "chemical diversity." Progress has been made in this field, not only for the molecular mechanics approaches alluded to earlier (see, for example, Chapter 2, by Norrby), but also for semiempirical quantum mechanics, as typified by the chapters of Taber (Chaps. 8, 9).

Chemical diversity can be defined as the ability of metals to stabilize distinct bonding environments involving different bond (e.g., dative, single, and multiple bonds) and ligating-atom (e.g., hard and soft donors) types, spin and formal oxidation states, coordination numbers, and geometries. Chapter 12, by Harvey, is an excellent example of the challenges inherent in modeling organometallic species and processes in which "spin flips" occur. As has become apparent, as computational organometallic chemists have explored all regions of the periodic table, this chemical diversity is also part and parcel of elements other than those of the transition series. This is plainly evident in the contributions by Kwon and McKee (Chap. 16) and McGee et al. (Chap. 15) on main group chemistry and by Li and Bursten (Chap. 14) on organoactinides.

It can be argued that the tremendous growth in the popularity of research into organometallic chemistry, experimental and computational, is due in large part to their utility in industrial and academic applications. As the field of computational organometallic chemistry has matured it has become evident that it is the chemical diversity that characterizes these entities that gives rise to many of the challenges in their reliable and rapid modeling. One need only consider some of the myriad catalytic transformations involving organometallic species to appreciate the chemical gymnastics that alter oxidation states, coordination numbers, ligand types, etc. Thus, it is this very property of chemical diversity that makes organometallics so very interesting (and at times quite frustrating) as computational targets.

In putting together this volume, the traditional description of organometallics as entities with a metal–carbon bond has been expanded to include any entities with an organic and metallic functionality, whether they be joined by a direct metal–carbon bond or not. I have also tried to go beyond applications other than just those related to industrial catalysis, as admirably demonstrated by Czerw et al. in Chapter 13. Chapter 7, by Parrill and coworkers, with its biomedical bent, is a good demonstration of this philosophy, as are Chapters 8 and 9, by Taber et al., on the computer-aided design of organometallic catalysts for carrying out useful organic synthetic transformations.

I would like to conclude this introductory chapter by thanking the chapter authors, individually and as a group, for their good humor and spirit, particularly in dealing with the inadequacies of a first-time editor. I would also like to thank Anita Lekhwani (Acquisitions Editor), Moraima Suarez (Production Editor), and Jennifer Paizzi (Administrative Assistant) of Marcel Dekker for their encouragement and for answering my numerous questions. Much of the planning for *Computational Organometallic Chemistry* occurred while I was on a Professional Development Assignment (PDA), for which opportunity I am grateful to The University of Memphis College of Arts and Sciences and Chemistry Department. I'd also like to thank the Chemistry Department at Bristol University (UK), for providing a relaxing yet stimulating environment during this PDA, and the United States National Science Foundation Office of International Programs, for their support of travel between Memphis and Bristol. It would not have possible to become an "expert" (real or imagined) in computational organometallic chemistry without the hard work and dedication of a fabulous bunch of graduate and undergraduate research students at The University of Memphis. I thank the various agencies (American Chemical Society—Petroleum Research Foundation, Los Alamos National Laboratory, National Science Foundation, and U.S. Department of Energy) for their generous support of these students during their careers at The University of Memphis.

Saving the best for last, I would like to thank my lovely wife, Mary Anderson, for her support, suggestions, and spirited Texan ways. She has done more than help improve this monograph; she has improved my life in immeasurable ways. For these reasons, I dedicate this volume to her.

Finally, I take full responsibility for any errors of commission or omission that may exist in this volume.

2

Recipe for an Organometallic Force Field

Per-Ola Norrby*

Royal Danish School of Pharmacy, Copenhagen, Denmark

1. INTRODUCTION

The molecular mechanics (MM) method is well established in organic chemistry (1–4). For many types of molecules, reliable structures can be generated quickly and conformational energies can be calculated with a high degree of accuracy (5). Combination of force field methods with dynamic or stochastic schemes allows determination of thermodynamic and solvation properties (1–3). Force fields are routinely applied to large systems, consisting of several thousand atoms. It is also possible to perform exhaustive searches for low-energy conformations of molecules with 10–20 freely rotatable bonds (6). Compared to computational methods based on quantum mechanical (QM) calculations, force field methods are limited in scope, since only systems with identical bonding (i.e., conformers or diastereomers) can be directly compared. However, within this limitation, force fields are several orders of magnitude faster than any QM method. In addition, when high-quality parameters are available, the accuracy of force fields is competitive with standard QM methods, such as MP2 and B3LYP, and better than semiempirical schemes (5).

The situation is different for organometallic complexes. The tools and methods developed for organic systems are available, but application is hampered

---

*Current affiliation: Technical University of Denmark, Lyngby, Denmark.

by a lack of parameters. Metal systems are structurally more diverse than organic compounds (7). As an example, the C–O–C bond angle seldom deviates more than a few degrees from true tetrahedral, whereas observed P–Pd–P angles vary over a range of ca. 100° depending on coordination geometry and steric requirements. Thus, parameter transferability between different types of complexes is limited, and alternative functional forms may be required (vide infra) (8,9). Despite the apparent difficulties, several force fields exist that allow calculations to be performed for almost any type of complex (10). However, predictivity may well be low for complexes outside the set used in parameter generation (11).

An alternative approach, which will be pursued here, is to tailor a force field to one specific type of complex. For organometallic complexes, it is still possible to use existing parameters for the organic part of the system and to develop new parameters only for the coordination sphere. Many examples can be found in the literature (8,9,12), but the need to develop new parameters largely limits applications to force field experts, as opposed to the organic field, where practicing chemists can easily model the system with only basic computational experience. The goal of the current chapter is to simplify the process of producing a high-quality organometallic force field by providing a workable recipe for the procedure. Some examples from the literature are included, but the coverage is by no means complete.

1.1. Force Fields

A force field is essentially a relationship between the geometry of a molecule and the force on each atom. The force is a vector quantity, the derivative of the energy with respect to coordinates. To simplify the expressions, force fields are generally presented in the form of energy as a function of coordinates. The true zero of the energy is an unknown, different for each force field and molecule. Thus, the total energy calculated for any molecule cannot be interpreted in a physically meaningful way, and no special meaning should be attached to a calculated energy of zero (or a negative energy). However, when two energies are calculated from exactly the same functions (i.e., when the connectivities of two structures are identical), the unknown constants be considered identical, and the energies can be compared directly.*

One of the fundamental postulates of molecular mechanics is that the steric energy of a molecule can be separated into terms resulting from small, transferable moieties. For all bond lengths and angles, it is assumed that there exists an unstrained state with a steric energy of zero. All deviations from this "ideal"

---

*Formal heats of formation can be calculated from steric energies by adding geometry-independent terms for several structural features; see Ref. 4. By this method, structural isomers with different connectivity can also be compared.

$$E_{steric} = \sum_{bonds} k_s (l - l_0)^2 \qquad l = \text{bond length}$$

$$+ \sum_{angles} k_b (\theta - \theta_0)^2 \qquad \theta = \text{bond angle}$$

$$+ \sum_{torsions} v_n \cos n\omega \qquad \omega = \text{torsional angle}$$

$$+ \sum_{nonbonded} \left(\frac{q_i q_j}{\varepsilon r} + \frac{A}{r^{12}} - \frac{B}{r^6} \right) \qquad \begin{array}{l} r = \text{interatomic distance} \\ > 2 \text{ bond separation} \end{array}$$

FIGURE 1 A simple force field.

value will give rise to an energy increase.* It is generally impossible for all interactions to achieve their unstrained state in the same geometry, and thus the "ideal values" will never be directly observed, but in organic molecules, the deviations from the unstrained state are usually small. Other contributions to the total energy of the molecule come from rotations around bonds as well as non-bonded interactions. In order to reproduce strained structures or vibrational data, it has also been found necessary to employ cross-terms in the force field. An example is the stretch–bend interaction, which can be described as the change in a bond angle function when the constituent bonds are distorted. For trigonal atoms, it is also common to employ a term that differentiates between planar and pyramidal form (an out-of-plane or inversion term).

The functional form of a simple example force field is shown in Figure 1. Most current force fields are substantially more complicated, but the additions take many different forms and will not be covered here. For more detailed accounts, see, for example, Refs. 1 and 4.

The basic unit of a force field is the atom type. In general, there is at least one atom type for each element, more if several chemical environments are to be considered. For example, all force fields differentiate between sp^2 and sp^3 hybridized carbons, assigning a distinct atom type to each. For organometallic modeling, it is frequently necessary to add new metal atom types to existing force fields. Even when the metal atom types exist in the force field, there is seldom any differentiation based on, for example, oxidation state.† Atom types are used to classify other interactions. Any unique pair of connected atom types identifies a bond type; an angle type is labeled by a unique set of three connected atoms, etc. Each unique interaction type needs its own set of parameters. Many atom-

---

*The l_0 and θ_0 parameters are also called *reference values*. But to avoid confusion with bond and angle "reference data," the term *ideal values* will be used for these parameters throughout this chapter.

†One exception is the PCModel program, which allows at least a basic differentiation, see Ref. 10c.

type combinations will not have existing parameters; in particular the torsions would require determination of millions of parameters for a complete set.

1.2. Parameters

A complete force field consists of a functional form, as exemplified in Figure 1, and a set of parameters. For example, for each type of bond in the example force field, two parameters are needed: an ideal length l_0 (corresponding to the bond length in a hypothetical unstrained molecule), and a stretching force constant k_s. The latter can be seen as the relative stiffness of the bond, and determines how much the energy increases upon a certain distortion. Some parameters, such as the ideal bond length, correspond closely to observables. However, the optimum set of parameters can rarely be identified by observation.

Take the torsional parameters for the central bond in butane as an example (Fig. 2). There are several observable energies that are closely related to the v_n-parameter (Fig. 1), but each is also affected strongly by other parameters. The rotation barrier might be taken as the amplitude of a threefold cosine function (when $n = 3$, $v_3 = \Delta E^{\ddagger}_{rot}/2$ will give an energy difference of $\Delta E^{\ddagger}_{rot}$ between the lowest and highest point on the torsional profile). But in reality part of the barrier is due to van der Waals (vdW) repulsion, so v_3 should be less than half the observed barrier. Likewise, the conformational difference between gauche and anti forms is largely determined by vdW interactions, but the remaining error might be reproduced using an added v_1 parameter in the force field (the v_3 parameter has no influence on the relative energy of gauche and anti forms, because the contribution from $v_3 \cos 3\omega$ must always be equal at 60° and 180°).

From vibrational or microwave spectroscopy it is possible to obtain the curvature at the bottom of each well. Ignoring mixing with other structural elements, this corresponds to the second derivative of the energy with respect to the torsional angle, $\partial^2 E/\partial \omega^2$. Fitting to this observable may require either sacri-

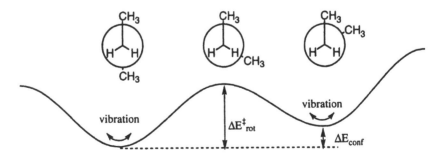

Figure 2 Butane torsional profile.

ficing some accuracy for other data points or adding more torsional terms (i.e., $n = 1$, 2, and 3, Fig. 1). When a substantial amount of data is used, no term in the final force field corresponds to only one type of observable. Instead, the optimal value for each parameter is that which, together with all other terms in the force field, gives the best overall fit to all observables. This concept will be defined more rigorously in Section 3.1. However, it should be clear that changing any parameter might lead to a shift in the optimum value for several others.

Despite what was said in the preceding paragraph, most parameters depend closely on some specific type of data. Good starting values for further refinement can therefore be obtained by manual fitting of one parameter at a time to small subsets of the reference data. The most intuitive example is the ideal bond lengths and bond angles (l_0 and θ_0, Fig. 1). Averages of observed values (possibly after removal of outliers) are good initial estimates for these parameters. Other examples are given in subsequent sections.

1.3. Parameterization

Defining new force fields has long been as much an art as a science. In the literature, there are two major schools on how to derive force field parameters, manually (4,13) and automatically (14). The manual method has the advantage of creating a deep familiarity with the force field and data, but it requires great expertise. Moreover, when the parameter set grows large, it becomes slow and tedious to ensure that fitting to new data retains consistency with all previously optimized sets.

An automated parameterization may be difficult to set up. But when this has been accomplished, the process is substantially faster than the manual method, and much larger bodies of data can be fitted simultaneously. The main drawback of the automated scheme is that errors may remain undetected more easily than in manual parameterization. Automated parameterization therefore requires substantial validation to identify outliers in the data set and deficiencies in the force field. Statistical tools should be used to verify that each parameter is well defined by the chosen set of reference data, and any ill-fitting data points can be rationalized on sound physical grounds.

The necessary steps in executing an automated parameterization for an organometallic complex are outlined in Figure 3 (15). Each step will be detailed in later sections. Selecting the basic force field is possibly the most critical step. The functional form of the basic force field must be flexible enough to accommodate the variability in metal complexes (7). In addition, the existing parameters for organic moieties will usually not be modified and will therefore limit the accuracy that can be obtained for organic ligands.

The target for an automated parameterization sequence is to enable the force field to reproduce a set of reference data, such as structures and relative

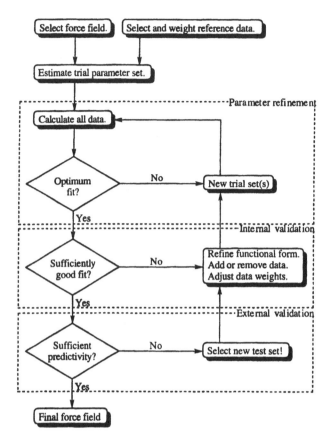

FIGURE 3   Parameterization flowchart.

energies. The quality of the reference data will therefore limit the attainable accuracy in the final force field. The accuracy can be improved by using large data sets, because random errors are expected to cancel to some extent. However, any systematic errors will be propagated into the final force field. It is also necessary to weight the reference data points, according to both quality and relevance to the intended use of the force field.

With the basic ingredients in hand, the next step is to set up a working force field. It is not necessary at this point to achieve a good fit, but the force field should allow calculations for all structures needed to reproduce the reference data set. This involves choosing functional forms for bonds and angles involving the metal and then guessing reasonable values for all previously undefined parameters. When all of this is accomplished, automatic procedures can vary the parameters and calculate all data points iteratively to obtain the best possible fit with the reference data. In the initial stages, it might be necessary to tether parameters and/or to divide them into subsets.

As a final step, the force field should be validated. In part, this is done by evaluating how well the reference data are reproduced and comparing that accuracy to the accuracy needed in the intended application of the force field. However, it is also advisable to apply the force field to an external test set, that is, data points that have not been used at any stage in the parameter refinement.

1.4. Force Fields for Catalysis

To predict reaction selectivities, a special type of force field is needed. Relative reactivities are determined in transition states, whereas most force fields are geared for calculating properties at energy minima. Only rarely have molecular mechanics methods been used for bond-breaking phenomena (16). However, an alternative method that has been successfully applied to selectivity predictions is to treat the transition state as an energy minimum and to develop a force field to reproduce the transition-state (TS) structure (17). This approach allows application of standard molecular mechanics tools such as conformational searching. In a recently developed method, the part of the potential energy surface (PES) perpendicular to the reaction path calculated by QM methods can be closely reproduced by force fields (18). The new method, dubbed "QM-guided molecular mechanics (Q2MM)," has been applied with good results to selectivity predictions in asymmetric synthesis (19) and catalysis (20).

1.5. Selecting a Force Field

There are many points to consider when selecting the program package and force field to be used as a basis for introduction of new organometallic moieties. The available modeling tools, the flexibility of the functional form, and the accuracy of the existing force field are all important. The intended use of the force field will dictate what tools must be present. Sometimes, all that is needed is the generation of good gas-phase structures. If so, the available tools need hardly be considered, since all existing MM packages allow energy minimization. In some situations, the graphical interface may be more important than performance, especially if the force field is to be used for visualization.

A very common use of force fields is to determine relative energies of isomeric forms, since most physical properties will depend on the relative energies of plausible isomers. In this case, it is very important that the underlying force field is already able to produce accurate energies (5). Prediction of thermodynamic properties, solvation, intermolecular interactions, etc. also requires that the basic force field already does well in calculating the particular property for organic molecules.

Finally, the selected force field must include functionality for describing the coordination environment. Bonds and angles around a metal atom do not behave like organic structures and can only rarely be described by the same functions. Several models for describing coordination angles have been implemented (8). In some types of complexes, the metal will exert only a weakly directing force. Such coordination can be implemented by replacing all metal-centered angles with nonbonded interactions while still retaining the metal–ligand bonds, as in the points-on-a-sphere (POS) model (21). Alternatively, even the metal–ligand bonds can be described by tailored nonbonded potentials (22), allowing also a variable coordination number.

For more rigid geometries, metal-centered angles are used. Depending on coordination geometry, it may be necessary to differentiate between, for example, cis and trans bond angles, with separate parameters for each. An alternative is to employ functional forms with multiple minima, such as trigonometric functions (23) or more complicated forms (24). More intricate problems are posed by π-ligands, in particular if rotation barriers around the metal–ligand axis are to be reproduced. This type of problem has frequently been addressed by bonding the ligand to the metal through a pseudoatom (8,25).

Coordination complexes frequently display trans-induction and Jahn–Teller distortions, which can been handled by specialized functional forms (9,26,27). In simple cases modified ligand–ligand interactions may suffice (28).

2. REFERENCE DATA

Molecular mechanics is essentially an interpolation method. Reliable predictions for a class of compounds usually require that the force field has been fitted to data of a similar type. When the functional form is physically sound and has been carefully parameterized, limited extrapolation can be successful, but generally only to new combinations of known structural moieties. Thus, an accurate and varied set of reference data is necessary for determination of a good force field. The exact selection depends on the intended use of the force field. Production of rough structures is easily accomplished, but selection of the most favored conformer requires accurate energetics. For prediction of vibrational frequencies, or strongly distorted structures, the shape of the local potential energy surface (PES) around minima must be well described. Solvation and docking requires a good set of nonbonded parameters. For each application, appropriate data must be included in the reference set.

2.1. Structures

The basis for all force field calculations is the generation of sound structures. Without consistent structures, no other properties can be reliably predicted. Thus,

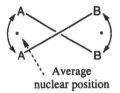

Average
nuclear position

FIGURE 4   Thermally induced oscillation resulting in offset nuclear positions.

the reference set *must* contain structural information. Depending on the intended application, it might also be necessary to consider the effect on bond lengths from differences in structural determination paradigms (4,29). Most computational methods will determine minima on the PES, that is, nuclear positions at zero Kelvin, without consideration of anharmonic vibrations.* All experimental determinations will take place at higher temperature and thus will include contributions from thermal vibrations, in effect lengthening most bonds slightly.

By far the most common source of structural data, particularly in organometallic chemistry, is X-ray crystallography. It must be noted here that atomic positions determined by X-ray are points of maximum electron density, not nuclear positions. For most atoms, it is a good approximation to consider that the electrons are centered on the nucleus. However, this is never true for hydrogens: the electron of a hydrogen atom always participates in bonding, and is thus offset from the nucleus. This is the major reason why X-ray structures should *never* be used for determining bond lengths to hydrogens. Other sources, such as neutron diffraction and QM structures, must be used for hydrogen positions.

A large majority of computational structures are determined in vacuo, corresponding most closely to experimental gas-phase structures. A fundamental difference between crystallographic and gas-phase structures is illustrated in Figure 4. Gas-phase methods generally determine bond lengths, whereas crystallographic methods find average atomic positions. Assuming that bonds are stiff and vary little in length, oscillations of rigid moieties in crystals can yield average positions that are closer together than any instantaneous bond length. This behavior is rather common, for example, in flexible chelate rings or in freely rotating phenyl groups, where the apparent $C_{ipso}-C_{ortho}$ bond is shortened. In extreme cases, the reported structures may even be averages of several cocrystalizing conformations, resulting in bond length errors exceeding 0.1 Å (30).

If possible, crystal structure reference data should be compared to calculations in a crystal environment (31). However, many packages do not include

---

*Exceptions include MM2 and MM3; see Ref. 4.

the necessary tools for solid-state calculations, necessitating the use of isolated structures in parameterization. When comparing in vacuo calculations with crystal structures, it should also be realized that the crystal structure need not be an energy minimum for the isolated molecule. Crystal packing can have a strong influence on torsions in particular, but also on any long interatomic distances. It is generally safe to compare lists of bond lengths and angles, for these interactions are strong compared to crystal packing. However, structural overlays or complete lists of interatomic distances should *not* be used as measures of force field quality.

As a validation tool, one may also measure the crystal distortion energy using the force field (11,28,30). This energy must be low, and certainly much lower than the total contribution from the packing forces. If the calculated energy of an error-free, nonionic crystal structure is high relative to the global minimum, the force field may be deficient.

2.2. Energies

Having obtained a good structure, the most important property to be calculated is the energy of the molecule. Except for completely rigid molecules, the structures are distributed among conformational forms where the population depends directly on the energy. Thus, to calculate any property, one must first know the relative energies of all conformers. It follows then that energies must be included in the reference set for any force fields that are not designed solely to yield crude geometries.

Comparing experimental and computational energies is not always straightforward. Molecular mechanics energies are "steric" or potential energies for a single fixed geometry. All experimental energies contain, at the very least, vibrational contributions and are therefore sensitive to the shape of the PES around the minimum. For relative energies it is frequently assumed that vibrational contributions cancel, allowing a direct comparison of MM potential energies with experimental enthalpy differences. A cruder but still common practice is to compare steric energies directly to experimental free energies, ignoring the effects of entropy and usually also of condensed-phase contributions.

Accurate comparisons to experimental enthalpies and free energies can be achieved in molecular mechanics by application of normal mode analysis, solvation models, solid-state calculations, and/or dynamic averaging over large ensembles. Such methods are time consuming and therefore are not easily implemented in a parameter refinement, where each data point is calculated multiple times with different trial force fields. However, the full calculation may be performed once, to derive a correction term allowing the use of the simple potential energy in further calculations. The correction term can be iteratively updated whenever the force field has changed substantially, allowing the use of rapid calculations in the parameter refinement.

2.3. PES data

The exact shape of the PES influences many properties, such as vibrational frequencies and the magnitude of distortions in strained structures. An exact representation of the PES will implicitly allow reproduction of structures and energies. It has been shown that a force field can be successfully derived from PES data alone (32).

Experimental information about the local PES around minima can be derived from vibrational spectroscopy. Employing a harmonic approximation, the vibrational modes and frequencies can be calculated by diagonalization of the mass-weighted Hessian (the matrix of Cartesian second derivatives of the energy) of a structure at an energy minimum. Unfortunately, it is by no means as easy to back-transform experimental frequencies to a Hessian. Experiments do not give any direct information about the vibrational modes. An exact assignment of all frequencies requires spectra of several isotopomers and extensive iterative fitting. Experimental frequencies are valuable in validation, but for parameterization of MM force fields, it is more efficient to find other types of PES data. A very attractive alternative is to use data from QM calculations. At correlated levels (e.g., MP2 or B3LYP), QM frequencies are close to the experimental results (33). Quantum mechanical methods also allow PES determinations at nonstationary points. Furthermore, both QM and MM methods determine structures as energy minima of nuclei on a PES, alleviating the need for conversion of bond-length types (4,29). Finally, the parameter refinement can be performed without time-consuming energy minimizations, because the energies and energy derivatives are calculated at fixed geometries (32).

2.4. Electrostatics

The largest difference between force fields is probably how they handle electrostatics. Each force field uses its own definition of what functions and data should be used. The well-known MM2 force field describes all electrostatic interactions by bond dipoles (4), but most other force fields utilize atomic point charges. The charges may in turn be obtained from fragment matching (34), from bond-type-dependent charge flux (35), or from more complex schemes that can also respond to the environment (36).

Neither atomic charges nor bond dipoles are observables. About the only experimental data for isolated molecules that can be used as parameterization reference are molecular dipoles and higher multipole moments. Substantial effort has also been expended to find electrostatic schemes that can rationalize the behavior of condensed phases (37). However, electrostatic data may be more conveniently obtained from QM calculations. Several schemes exist for partitioning the electron density into atomic charges (38). In general, methods that reproduce the QM-calculated electrostatic field outside the molecular surface are preferred, for the most important task of the electrostatic function is to reproduce intermolecular interactions.

The electrostatic parameters, together with other nonbonded parameters, should generally be set according to the specific force field rules ahead of actual refinement, and then kept fixed, at least in the initial refinement stages. However, when the new parameter set has stabilized with respect to the data, it is advantageous to allow a slight variation in the nonbonded parameters to improve the overall performance.

2.5. van der Waals Data

The types of data just listed will generally suffice for isolated molecules, if no new element types are introduced. However, applications to condensed phases, or sets incorporating unusual elements, may require additional data to fit vdW parameters. If QM data are to be used (39), very high levels of theory are required, because HF and most DFT methods do not incorporate London dispersion. Most correlated methods require huge basis sets if the vdW interactions are to be distinguishable from basis set deficiency errors. Experimental sources of data for nonbonded parameters include crystal cell constants and heats of sublimation. In all cases, the balance between vdW and electrostatic parameters is very important. However, if the metal atom is buried deep enough in coordinating ligands, direct attractive interactions may be unimportant. The repulsive vdW component may sometimes be determined from obvious strain in bulky ligands (30), but only if the force constants in the deformed moieties are known from other sources.

2.6. Quantum Mechanical Data

Quantum mechanical data can be very efficiently included in parameterizations, because no data conversion is necessary, and properties can be calculated for any point on the PES. However, it is important to realize that the goal of most force fields is to reproduce experiments, not QM results. The chosen QM level puts a limit on the attainable accuracy of the force field. Most systematic errors in the QM method will be reproduced by the force field. In particular for metal systems, it is necessary to use correlated levels, with reasonably flexible basis sets. Some DFT-based methods have proven to give excellent cost/performance ratios. Suitable theoretical levels are discussed more thoroughly by Diedenhofen et al. in Chapter 4. In all the examples given in later sections, the QM data have been obtained at the B3LYP (40) level, using an ECP (41) for the metal and at least valence double-ζ quality basis sets for all atoms.

2.7. Transition-State Data for Q2MM

In the Q2MM method, force fields describing transition states as minima are developed from QM data. Structures, charges, and relative energies of stationary points can be used as is in the parameterization. However, QM-derived Hessians must be modified, because they implicitly define the curvature to be negative in the direction of the reaction coordinate. The modification involves determination of normal modes, replacement of the negative eigenvalue with a positive value, and reformation of a new Hessian from the modified eigensystem (18). After modification, the curvature is positive in all directions, thus fulfilling all criteria for a regular force field. From this point on, the derivation and use of the force field is analogous to a regular ground-state force field, except that calculated steric energies will now correspond to relative activation energies for the reaction under investigation (19,20).

3. DERIVING PARAMETERS

With all the necessary ingredients in place, the task is now to derive a reliable force field. In an automated refinement, the first step is to define in machine-readable form what constitutes a good force field. Following that, the parameters are varied, randomly or systematically (15,42). For each new parameter set, the entire data set is recalculated, to yield the quality of the new force field. The best force field so far is retained and used as the basis for new trial parameter sets. The task is a standard one in nonlinear numerical optimization; many efficient procedures exist for selection of the optimum search direction (43). Only one recipe will be covered here, a combination of Newton–Raphson and Simplex methods that has been successfully employed in several recent parameterization efforts (11,19,20,28,44).

Parameter refinement is in many ways reminiscent of geometry optimization. The same problems apply—finding a minimum, and preferably the global minimum, of a function of many variables. Progress is not as easily visualized with a parameter set as with a set of coordinates, but the main implementation difference comes from the fact that gradients are not easily available in parameterization. For data that are calculated for minima on the PES (e.g., conformational energies), analytic gradients of the data with respect to the parameters cannot be determined. Thus, optimization must rely on numerical differentiation, approximate analytical derivatives (45), or methods that don't employ gradients. An alternative is to employ only reference data for which analytic gradients are available (32).

In simple geometry optimization, the result is sensitive to the starting geometry. A very distorted starting structure may lead to a strained high-energy optimum. A similar problem plagues automated parameter refinement. However, the problem is most serious in the initial phase of the refinement. Special techniques and frequent manual intervention may be needed until the force field has stabilized on track to the desired optimum.

$$\chi^2 = \sum t_i c_i \left(\frac{\hat{y}_i - y_i^\circ}{\sigma_i}\right)^2 = \sum w_i^2 \left(\hat{y}_i - y_i^\circ\right)^2 \qquad \text{penalty function}$$

$$\frac{\partial \chi^2}{\partial p_a} = 2 \sum w_i^2 \left(\hat{y}_i - y_i^\circ\right) \frac{\partial \hat{y}_i}{\partial p_a}$$

$$\frac{\partial^2 \chi^2}{\partial p_a \partial p_b} = 2 \sum w_i^2 \left[\frac{\partial \hat{y}_i}{\partial p_a} \frac{\partial \hat{y}_i}{\partial p_b} + \left(\hat{y}_i - y_i^\circ\right) \frac{\partial^2 \hat{y}_i}{\partial p_a \partial p_b}\right] \approx$$

$$\approx 2 \sum w_i^2 \frac{\partial \hat{y}_i}{\partial p_a} \frac{\partial \hat{y}_i}{\partial p_b} \qquad \text{approximate form}$$

FIGURE 5   Penalty function and derivatives with respect to parameters.

3.1. Defining the Goal

The goal of parameter refinement may be defined simply as minimizing the deviation of all calculated data points from the corresponding reference values. This is generally done in a least squares sense, employing the penalty function depicted in Figure 5,* where y_i and y_i° are the calculated and reference data points, respectively, σ_i corrects for the quality of the reference data, c_i corrects for different units of measure, and t_i is the relative importance of reproducing a specific type of data. The latter three, being constants for each data point, are conveniently combined into a weighting factor w_i (42). The weight factor must be set for each data point. At the very least, different types of data must be converted to a common unit of measure. If not, an error of $1°$ in an angle might have the same impact as an error of $1 \overset{\circ}{A}$ in a bond length!

An intuitive method for defining data weights is simply to use the inverse of the acceptable error. This can be either the acceptable error for one type of data in the final force field or the expected average error in a group of input data. Say, for example, that it is sufficient that the final force field reproduces bond lengths to within $0.01 \overset{\circ}{A}$, angles to $0.5°$, and torsions to $1°$. Suitable weight factors would then be $100 \overset{\circ}{A}{}^{-1}$, 2 degree^{-1}, and 1 degree^{-1}, respectively. If a low-quality structure with bond-length errors around $0.02 \overset{\circ}{A}$ is included in the refinement, bond lengths in that particular structure could be given a lower weight, $50 \overset{\circ}{A}{}^{-1}$. Weights for other types of data have been exemplified in the literature (11,15,19,20,28,44).

The balance between different types of data may be modified by further adjustment of the weight factors. In schemes employing QM-calculated energy derivatives, an extreme number of data points can be obtained with little effort.

---

*In earlier literature, the term *merit function* has been used (cf. Ref. 15). But because an increased value corresponds to a worse force field, *penalty function* is more appropriate.

It is then recommended that the weight of such data be reduced to avoid swamping the remaining reference data by sheer numbers. On the other hand, electrostatic parameters have a strong influence on conformational energies and may therefore be unduly adjusted by the automatic procedure in lieu of other, more relevant parameters. It can therefore be prudent to increase the weight of true electrostatic data (such as QM charges), especially in the initial stages of the refinement.

3.2. Initial Parameter Estimates

Setting up the initial force field is still largely a manual task. In particular when the reference set contains properties of energy minima, it is important that structures be reasonably accurate already before the parameter refinement is initiated. To achieve this, bond and angle ideal values and nonbonded parameters must be well estimated, whereas force constants, most torsional parameters, and cross-terms can be entrusted to the automated refinement. Initial values must be set also for these, but it may be sufficient to use "similar" values from the existing force field. It is usually best to err on the high side with force constants, to minimize deviations from the reference values, and on the low side with torsional parameters and cross-terms, to avoid introduction of physically unrealistic distortions.

Electrostatic parameters can be set directly from QM-calculated charges. With some force fields, the charges are fixed to QM values at the outset and not refined further (34). Other nonbonded parameters (vdW constants) are not easily elucidated directly from any type of input data. However, parameters for most of the periodic table are available in the literature (10,46). Different force fields do not use the same absolute parameter values, but the scales usually correlate. Thus, it is possible to fit the existing parameters in the force field of interest to any complete set and to obtain the missing parameters from the correlation.

Initial ideal bond lengths and angles can be obtained from averages of observed values in the reference data. However, if strained structures are included in the reference set, an improved procedure is available. After all the parameters have been given initial values, calculated bond lengths of one type can be correlated with the corresponding reference values. It may be postulated that for observed structures with small distortions, the "real" bond energy will follow the Hooke's law expression in Figure 1 reasonably well. It can also be assumed that for small parameter changes, the force from the surrounding structure acting upon one bond is constant. For one bond, the calculated force should thus equal the real force (Fig. 6). If the observed bond length l_{obs} is plotted against the calculated deviation from the estimated ideal length ($l_{calc} - l_{0,est}$), a better estimate of the ideal bond length is obtained as the intercept, and an improved force constant can be obtained by dividing the initial estimate with the regression slope. Note

$$k_{real}\left(l_{obs} - l_{0,real}\right) \approx k_{est}\left(l_{calc} - l_{0,est}\right)$$

$$l_{obs} \approx \frac{k_{est}}{k_{real}}\left(l_{calc} - l_{0,est}\right) + l_{0,real}$$

FIGURE 6   Regression of real and calculated bond lengths.

that the ideal length will always be improved by this procedure, but the force constant may not; the new value should be accepted only if it is physically reasonable and improves the fit. The equations in Figure 6 can easily be extended to more complicated bond functions, but higher terms may be ignored if the deviations are small. The procedure can be applied iteratively, but not too far, because all data except the observed bond lengths are ignored. Final parameters should be obtained by optimizing the full penalty function (Fig. 5).

The most important assumption in Figure 6 is that the forces in the surrounding structure will be unaffected by parameter changes for the bond under observation. This is true only if the same parameters are not used in proximal bonds. For example, the assumption may break down if two bonds of the same type are present in a small ring. For this reason, the procedure is also less useful for angles, which are frequently redundant and linearly dependent (vide infra). As a simple example, the procedure would fail completely for the H–C–H angle in methane, where the calculated bond angle will be 109.471° for all reasonable (and many unreasonable) ideal bond angles. In situations like this, the ideal bond angle should simply be set to the observed average.

Exact torsional parameters are important for conformational energies but frequently not for gross structural agreement. If the v_2 term for conjugated bonds is set to any large value, the remaining torsional parameters can usually be zeroed or set to values of "similar" torsions in the initial force field. However, this rule has many exceptions. When torsional parameters are important, reference data for the entire range of the rotational profile should be included. To avoid mixing with other force field terms, it is favorable to parameterize torsional parameters using QM data for rigid scans (47). In addition to the major advantage of avoiding mixing with other force field parameters, the QM calculations for rigid scans are also substantially cheaper than relaxed scans.

3.3. Refining Parameters

With the initial parameter set available, all data points can be calculated and compared to the reference data with suitable weighting (Fig. 5). The problem is then simply to vary the parameters in such a way that the penalty function decreases to a minimum. This is a very common task in all types of model development, and many numerical procedures are available (43). Here, we will focus on a joint application of two complementary techniques, Simplex and Newton–Raphson optimizations (15).

Simplex

The simplex optimization is a very simple and robust technique for optimizing any function of a moderate number of variables. Only the function values for different variable sets are needed. In this case, the function to be optimized is the penalty function, and the variables to vary are the force field parameters. To initialize a simplex optimization of N parameters, one must first select $N + 1$ linearly independent trial sets. A very simple way to achieve this is to start with the initial parameter estimate and then to vary each parameter in turn by a small amount, yielding N new trial sets. This is illustrated for a two-parameter case in Fig. 7. With two parameters, the shape of the simplex is a triangle, with three parameters a tetrahedron, and so on.

The penalty function is evaluated for all sets. The worst point is then selected and reflected through the centroid of the remaining points, yielding a new simplex. If the new point yields an even better result than the previous best point, an expansion is attempted. The expansion is accepted only if the result is better than for the simple reflection. If, on the other hand, the new point would be the worst in the new simplex, one of two possible contraction points is selected instead. Note that the contraction must always be accepted; if not, the simplex will just oscillate between two bad points.

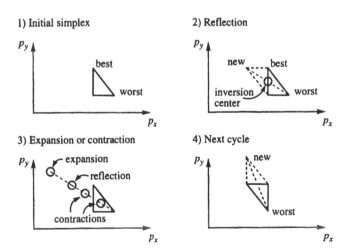

FIGURE 7　Simplex optimization of two parameters.

The simplex optimizations only run for a specified number of cycles, typically $10N$ to $30N$. If the same point is best for $3N$ cycles, the optimization is also terminated. The method is very robust, and boundary conditions for parameters are easily implemented (for example, ideal bond lengths should always be positive). However, convergence is slow when too many parameters are included. As a rule of thumb, no more than 10 parameters should be included in a standard simplex optimization, but a recently introduced biasing procedure where the inversion point is offset toward the best points can make the method competitive for up to 30–40 parameters (15).

Newton–Raphson

The Newton–Raphson method is a very efficient method for finding roots of well-behaved functions (43). A step for one variable is obtained by dividing the function value by the gradient. Finding minima is equivalent to finding points where the first derivative of the function with respect to the variable is zero (the root of the gradient). The distance to this point from the initial variable value can then be estimated by dividing the first derivative by the second. However, convergence can become problematic when the second derivative goes to zero or becomes negative.

The multidimensional version of the Newton–Raphson minimization is employed for functions of many variables. The matrix of second derivatives is inverted and multiplied by the first derivatives to obtain the optimum step for all parameters (Fig. 8). Again, convergence can be problematic if the curvature is negative or close to zero. Note that the matrix is positive definite if the approximate form of the second derivative is used (last equation, Fig. 5). Thus, only eigenvalues close to zero can give problems. However, parameters are frequently

Standard form:
$$\begin{pmatrix} \Delta p_1 \\ \Delta p_2 \\ \vdots \end{pmatrix} = -\begin{pmatrix} \dfrac{\partial^2 \chi^2}{\partial p_1^2} & \dfrac{\partial^2 \chi^2}{\partial p_2 \partial p_1} & \cdots \\ \dfrac{\partial^2 \chi^2}{\partial p_1 \partial p_2} & \dfrac{\partial^2 \chi^2}{\partial p_2^2} & \cdots \\ \vdots & \vdots & \ddots \end{pmatrix}^{-1} \begin{pmatrix} \dfrac{\partial \chi^2}{\partial p_1} \\ \dfrac{\partial \chi^2}{\partial p_2} \\ \vdots \end{pmatrix}$$

With Lagrange multipliers:
$$\begin{pmatrix} \Delta p_1 \\ \Delta p_2 \\ \vdots \end{pmatrix} = -\begin{pmatrix} \dfrac{\partial^2 \chi^2}{\partial p_1^2} + \gamma & \dfrac{\partial^2 \chi^2}{\partial p_2 \partial p_1} & \cdots \\ \dfrac{\partial^2 \chi^2}{\partial p_1 \partial p_2} & \dfrac{\partial^2 \chi^2}{\partial p_2^2} + \gamma & \cdots \\ \vdots & \vdots & \ddots \end{pmatrix}^{-1} \begin{pmatrix} \dfrac{\partial \chi^2}{\partial p_1} \\ \dfrac{\partial \chi^2}{\partial p_2} \\ \vdots \end{pmatrix}$$

FIGURE 8 Multidimensional Newton–Raphson.

interdependent, and therefore singularities are common.* The effect of a low eigenvalue is a very long step, possibly outside of the region where the quadratic approximation is valid. A very simple fix is to introduce Lagrange multipliers, in effect increasing the curvature by adding a constant to each diagonal element (15). Another method is to follow only search directions with a strong curvature, employing singular value decomposition (SVD) (15). Each of these methods has the undesired effect that parameters with a weak curvature are not optimized.

The Newton–Raphson method requires differentiation of all data points with respect to the parameters. For fixed-geometry properties (like energy derivatives), the force field derivatives can be obtained analytically (32). For other types of properties, an approximate analytical solution can be obtained by assuming that the shift in geometry is small upon parameter change (45). However, the most general and safest method is to obtain the derivatives numerically (15). The drawback is that the method is substantially slower than calculating analytical derivatives.

Alternating Between Methods

The Newton–Raphson method shows good convergence for parameters that display a strong penalty function curvature and are not too strongly interdependent. However, there are usually some parameters that will not be well converged by the method. This problem has been alleviated in some recent parameterization efforts (20,44) by alternating between optimization methods. In numerical schemes, the absolute second derivative of the penalty function with respect to each parameter is available from direct differentiation. It is assumed that a parameter will be badly determined by a Newton–Raphson step if this value is very low or negative. The 10–20 worst parameters are selected and subjected to a separate simplex optimization.

A complete automated refinement cycle is detailed shortly. It is assumed that the penalty function can be determined for each parameter set by automatic calculation of all data points. Further required input is a list of parameters to be refined and a numerical differentiation step size for each parameter. The number of parameters to be refined is denoted N. The value for data point i calculated with parameter set k is shown as $y_i(\mathbf{p}_k)$, and the total penalty function for the same parameter set is $\chi^2(\mathbf{p}_k)$. The initial parameter set is denoted $\mathbf{p}_0$, whereas a parameter set where parameter j has been differentiated is shown as $(\mathbf{p}_0 + \delta p_j)$.

1. Create $2N + 1$ trial parameter sets for central differentiation, by subtracting and adding a numerical differentiation step δp_j to each parameter in turn. Calculate all data points for all trial parameter sets.

---

*In the author's experience, the curvature matrix will always become singular, at least at some points in the parameter refinement.

$$dy_i/dp_j \approx \left[y_i\left(\mathbf{p_0} + \delta p_j\right) - y_i\left(\mathbf{p_0} - \delta p_j\right) \right]/2\delta p_j$$

$$d\chi^2/dp_j \approx \left[\chi^2\left(\mathbf{p_0} + \delta p_j\right) - \chi^2\left(\mathbf{p_0} - \delta p_j\right) \right]/2\delta p_j$$

$$d^2\chi^2/dp_j^2 \approx \left[\chi^2\left(\mathbf{p_0} + \delta p_j\right) + \chi^2\left(\mathbf{p_0} - \delta p_j\right) - 2\chi^2(\mathbf{p_0}) \right]/\delta p_j^2$$

FIGURE 9   Numerical differentiation.

2. Calculate data point derivatives by central differentiation as shown in Figure 9. The numerical derivatives are then used to calculate the penalty function derivatives according to Figure 5 ($\partial y/\partial p \approx dy/dp$). Several new trial parameter sets are calculated from the last equation in Figure 8, using different Lagrange multipliers γ. Additional trial sets can be obtained from SVD solutions, by varying the threshold for acceptable singular values (15). Very small steps are discarded, whereas very long steps can be either discarded or reduced to an acceptable size (a trust radius). The penalty function is calculated for all trial parameter sets, the best is selected, and all others are discarded.

3. Using data from step 1, the maximum variation in penalty function in response to each parameter is calculated from $\chi^2(\mathbf{p_0} + \delta p_j)$, $\chi^2(\mathbf{p_0})$, and $\chi^2(\mathbf{p_0} - \delta p_j)$. This value is used to balance the differentiation steps for the next iteration. As an example using arbitrary limits, if the variation is less than 1, δp_j is doubled, whereas if it is larger than 100, δp_j is halved.

4. Calculate the absolute derivatives of the penalty function with respect to each parameter by numerical differentiation as shown in Figure 9. Any parameters that result in a negative second derivative and as many as possible of those where the second derivative is small compared to the first (up to a maximum of 20–40 parameters) are selected for simplex optimization. The starting simplex is derived from the best parameter set (in step 2) by shifting each parameter in turn, using the updated step lengths from step 3.

5. The best parameter set after the simplex is compared to the initial parameter set. If improvement is lower than 0.1%, the refinement cycle is terminated.

3.4. Frequently Encountered Problems

The initially estimated force field will usually give very large errors for some data points. The automatic procedure will respond by large parameter changes, but not always in parameters that a chemist would consider natural. For example, energy second derivatives (the Hessian) are usually connected with force constants, but for any force field employing more complex bonding terms than the simple Hooke's law expression in Figure 1, Hessian elements are also affected by ideal bond lengths. Thus, large errors in the Hessian, which would naturally be corrected only by modification of force constants, might in an automatic procedure result in distorted ideal bond lengths. Many other types of parameters, in particular electrostatic parameters and cross terms, are sensitive to this type of "unnatural" correction.

Erroneous data may give strange effects in automated parameterization schemes. Since all deviations are squared, a single large error may totally dominate the refinement. For example, extreme bond-length shortenings of the type illustrated in Figure 4 are quite common in crystal structures, especially if the crystallographer has failed to take notice of cocrystallizing rotameric forms. Such errors must be identified and removed from the data set. Some low-quality data may have to be included in order to define all parameters, but should then be given low weight factors. It is also important that any errors be small and randomly distributed.

For metal complexes, specific problems may also arise from the coordination model. Angles around the metal are frequently soft, so geometries are easily distorted. Small parameter changes may lead to large distortions and sometimes to qualitatively wrong coordinations. In the initial stages, it may be safest to assign specific, relatively stiff angle interactions. Any scheme that dynamically updates the parameter values in the energy minimization is hazardous in a parameter refinement. If the chosen model uses no angle parameters, it may even be necessary to use weak restraints on the atomic coordinates to put a limit on the maximum error and avoiding falling into an erroneous geometry. If no precautions of this type can be taken, it is particularly important that each iteration start from one set of starting geometries, not the resulting geometries of the previous iteration.

Most parameterization problems arise because the parameters are not uniquely defined by the data. Molecular mechanics parameters are to some extent redundant and will therefore frequently show linear dependencies in the refinement. Ideal bond angles are good examples of this. Compare, for example, the simple molecules water and methane. The H–O–H ideal angle will be well defined if a water structure is included in the parameterization. Any change in the ideal bond angle will be immediately reflected in the calculated structure. For methane, on the other hand, any ideal bond angle larger than the standard tetrahedral angle of $109.471°$ will give a perfectly tetrahedral structure. Say, for example, that the ideal angle is set to the chemically unreasonable value of $130°$. The structures will be strained, but strain does not cause any increase in the penalty function, for the sum of forces on all atoms will still be zero. Vibrations will be affected, but a lowering of the force constant will have the same effect as lowering the ideal angle. In a more realistic parameterization also including ethane, the "erroneous" H–C–H ideal angle would be noticeable as a decrease in the C–C–H angle, but an automated procedure is just as likely to "correct" by increasing the C–C–H ideal angle as well.

What then is the solution to this problem? Nothing inherently says that the H–C–H ideal angle should be close to tetrahedral. It might very well be possible to set all ideal angles in a force field to 180° and to reproduce the entire data set by fitting the force constants. However, it has been seen that the predictive ability is enhanced if ideal values are set close to perceived "unstrained" states, and it is definitely more pleasing to the chemist. Therefore, the penalty function should be modified to favor the unstrained state, possibly by tethering.

3.5. Parameter Tethering

It is possible to bias the parameterization to specific parameter values by tethering. In essence, tethering is a way of telling the refinement "I know what this parameter should be, don't deviate too much from it." A "preferred" value for the parameter is set before refinement. The squared deviation of the parameter from the preferred value is then added to the penalty function. Another way to look at it (and the simplest way to implement tethering) is simply to see the "preferred" values of the tethered parameters as reference data and the actual values as the corresponding calculated data. The "weight factor" is set, according to the rules outlined previously, to the inverse of the acceptable deviation from the "preferred" values and then increased somewhat to compensate for the fact that only a few "data points" are included.

A weak tethering is generally beneficial for most parameters. If the parameter is well determined by the reference data, the effect of tethering will be negligible, as it should be. On the other hand, if the parameter is very badly determined, even a weak tethering potential will suffice to keep it close to a pleasing value. Ideal bond angles in particular should be tethered to perceived "unstrained" values (see Sec. 3.4). Torsional parameters and cross-terms may also be tethered, usually to a value of zero. Inclusion of QM charges as reference data may in a way be seen as a tethering of the electrostatic parameters, because many other data errors might otherwise have been "corrected" automatically by introduction of physically unrealistic charges. Tethering is especially valuable for achieving a balance in the initial stages of the refinement.

3.6. Strategies for Initial Refinement

Automated adjustments leading to "unnatural" parameter values are most frequently observed in the initial stages of the refinement, when errors are still large. Several techniques are available for minimizing unwanted deviations, including tethering, subset refinement, and analysis of outliers.

Tethering is a logical correction procedure for the "unnatural" parameter deviations observed in the initial refinement stages. The only problem is then to find "preferred" values for all parameters. Hopefully, any preferred values have been set in the initial force field, so the entire parameter set can be tethered to these values, with weight factors set from the confidence one has in each parameter value. As the refinement progresses, tethering weights can be lowered, to avoid biasing the final force field. However, for reasons already discussed, bond angle tethering could be retained throughout.

Initial refinement can be made more efficient by dividing parameters and data into subsets. For example, electrostatic parameters could be adjusted as a group to QM electrostatic data only (and possibly excluded from further refinement altogether). If ideal bond lengths and angles have been set to reliable values initially, force constants could be refined in a group using a penalty function based solely on QM Hessian data. Torsional parameters could be grouped with the force constants, but then the penalty function should be extended to include also conformational energies. When the force-related parameters have been balanced, ideal bond lengths and angles could be refined using structural data only. Unless a very large body of PES data is available, cross-terms should be given low values and left out of the refinement until a late stage.

Error detection should be attempted after a few refinement cycles. Erroneous reference data can frequently be detected by a complete failure of a partially refined force field to fit the data point. On the other hand, such failures are even more frequently a result of deficiencies in the functional form. In any case, the error must be corrected before refinement is finalized, either by removal of the offending reference data or by changes in the functional form of the force field. One can easily identify the data points that are most likely to affect the refinement adversely as the largest weighted contributions to the penalty function. All large deviations must be manually scrutinized to elucidate whether they result from data errors or force field deficiencies. If the latter, it should also be estimated whether a failure to reproduce the data point will adversely affect the intended use of the force field (this is not always the case). If so, the force field setup must be modified before continuing; otherwise the data point might simply be removed.

4. VALIDATION

New parameter sets should always be validated before use. The simple fact that no further improvement can be found is not a sufficient condition for accepting the force field. To verify that the new force field is accurate and predictive, several tests should be performed. First, it should be verified that all the reference data are indeed reproduced with an acceptable accuracy. Second, the precision of each parameter determination should be checked. If a large range of parameter values can give the same results, the parameter is not well determined by the data set. Finally, the predictivity of the force field should be tested against an external data set.

4.1. Internal Validation

The first step in validation is simply to verify that the remaining errors in the reproduction of the reference data are acceptably small. If the weight factors have been set, as suggested earlier, to the inverse of the acceptable error for each data type, the test is particularly simple. If the final penalty function is lower than the number of data points, the root mean square (rms) error will automatically fall within the acceptable range. The data should also be divided by type and retested, to make sure that the proper balance has been obtained. As before, outliers should be carefully scrutinized. Any errors in the reference data or deficiencies in the functional form are most easily detected at this stage. Plots of calculated vs. reference data can also give valuable information on trends in remaining deviations and possible systematic errors (20).

The penalty function derivatives calculated in Figure 9 give information about how well a parameter is determined by the available data. First of all, it should be verified that the second derivative of the penalty function with respect to each parameter is positive. If not, a lowering of the penalty function can always be obtained by a slight change in the parameter—indeed, one of the test points must have been better for the numerical derivative to be negative. The expected response of the penalty function to a small parameter change can now be calculated from a truncated Taylor expansion (Fig. 10). The parameter change that would be needed to effect a given penalty function change is available from solving the second-order equation.

At this point, it is necessary to decide a maximum "allowed" change in the penalty function, $\Delta\chi^2$. This choice is necessarily arbitrary, but should reflect a change in the data that corresponds to either the expected input error or the largest deviation that could be accepted in the intended use of the force field (28). Using this value with the equation in Figure 10, it is now possible to calculate the

$$\chi^2\left(\mathbf{p}_0 + \Delta p_j\right) \approx \chi^2\left(\mathbf{p}_0\right) + \frac{d\chi^2}{dp_j}\Delta p_j + \frac{d^2\chi^2}{dp_j^2}\frac{\Delta p_j^2}{2}$$

$$\Delta p_j \approx -\frac{d\chi^2/dp_j}{d^2\chi^2/dp_j^2} \pm \sqrt{\left(\frac{d\chi^2/dp_j}{d^2\chi^2/dp_j^2}\right)^2 + \frac{2\Delta\chi^2}{d^2\chi^2/dp_j^2}}$$

FIGURE 10   Relationship between small changes in a parameter and the penalty function.

maximum change in the parameter that could be accommodated without significant deterioration of the fit. This range could be reported together with the final parameter value as an indication of the quality in parameter determination. Notice that the equations in Figure 10 do not take account of possible errors in the data (except possibly in the choice of $\Delta\chi^2$). Thus, the calculated range is a measure not of the accuracy of the parameter but of the precision in the determination (11). In particular, if the reference data contain systematic errors, the "real" value of the parameter may well fall outside the calculated range.

4.2. External Validation

The final test for each force field is how well it reproduces data that have not been included in the reference set. The test set should reflect the intended future use of the force field. In most cases, the goal is to predict experimental properties. Thus, QM data are not needed in the test set, even if they are used extensively in parameter refinement. On the other hand, it is now possible to include data points that cannot practically be included in the parameterization. Examples are experimental IR spectra, where peak assignment is a problem with preliminary force fields, or equilibria, which must be calculated with time-consuming dynamic methods.

In a general treatise, it is not possible to state what discrepancies can be accepted for the final force field. However, if the deviations are substantially worse than the experimentally achievable accuracy, the usefulness of the force field will be very limited. As always, large deviations should be identified, and if possible the underlying causes should be rationalized.

It will sometimes happen that as a result of the final test a redefinition of the functional form with subsequent reparameterization becomes necessary. If the test data that pinpointed the failure are of a type that can be used in refinement, they should be transferred to the reference set. In any case, the test set should be discarded and a new test set selected. The reason for this is that the force field is no longer independent of the initial test set, since it has been used to influence a design decision. One can therefore argue that a reapplication of the same set tests for only internal, not external, predictivity.

5. EXAMPLES

The procedures described herein have been implemented as a package of small C programs and Unix scripts (15).* This implementation has principally been

---

*Updated versions of the programs and scripts are available from the author on request. Several force fields and example structures can be found at *http://compchem.dfh.dk/PeO/*.

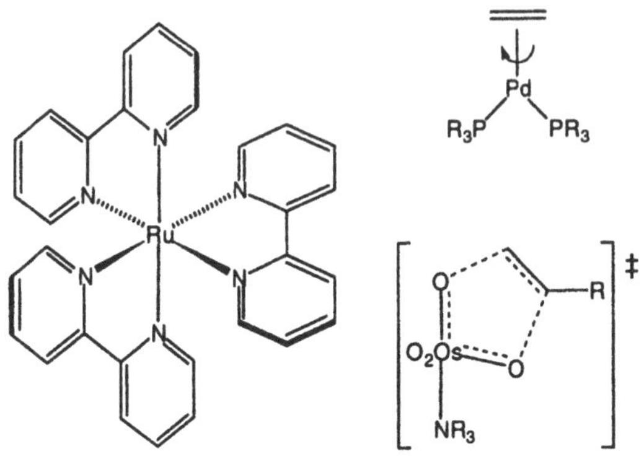

FIGURE 11   The three types of structures that have been parameterized in the examples.

developed in conjunction with the MacroModel package (35), but all procedures that are specific to one force field have been collected in a few scripts that are easily modified to accommodate other formats. Only minor modifications have been necessary for the procedures to work with other packages. Following are a few recently published examples, reflecting various development stages of the parameterization methods and several types of complexes. Typical structures of each type are depicted in Figure 11.

5.1. Rigid Octahedral Geometry: Ru(II)(bipy)₃

Octahedral geometry is among the easiest coordination modes to handle with force field methods, as evidenced by the large numbers of force fields that have been published (8,12). The functional forms of standard organic force fields must be extended to allow differentiation between cis and trans bond angles about the metal center, but the structures are generally rigid enough that standard bond and angle functions can be used. A point worth noticing is that the ideal angle parameter for the trans angle should be exactly 180°, to avoid creating a cusp. Observed deviations from this value should, as far as possible, be reproduced by a lowering of the force constant. The ideal cis angle should be close to 90°, but because there is no discontinuity in the function derivatives here, small variations are allowed. Alternative functional forms that could reproduce the octahedral geometry include POS (since the octahedron minimizes steric repulsion for six ligands) and trigonometric functions with minima at both 90° and 180° (23).

    The current implementation used the standard bond and angle functions available in MacroModel MM3*, with cis and trans parameters assigned by a test of the actual angle in the input geometry (in rare cases with strongly distorted structures, the test may erroneously assign one ligand as being trans to either two or no other ligands). A further refinement included a fourfold torsional term to describe rotation about the metal–ligand bond. The reference data consisted of several X-ray structures, together with QM normal modes and CHelpG charges (48) determined for one small model system. It could be shown that all parameters were well determined by the included data (28).

Several observed complexes with terpyridine display a trans induction. The bond from Ru to the central N is very short, resulting in an elongation of the trans Ru–N bond. This effect could possibly be described as a stretch–stretch cross-term in force field modeling, but no such function is available in Macro-Model. It was found that the observed distortions could be reproduced by a direct interaction between two trans ligands. This solution should not be given any physical significance, but should be considered only a working model for reproducing observed geometries (28).

5.2. Flexible Points-on-a-Sphere Model: $(\eta^2$-alkene)Pd(0)L$_2$

Alkene coordination to Pd(0) is rather loose, with the preferred in-plane geometry easily distorted by modest steric interactions. As a consequence, the barrier to rotation about the Pd–alkene axis is also low. The C–Pd–L angle is thus highly variable and cannot be well represented by a standard angle-bending function. The two most frequently used force field models for this type of coordination are the dummy atom approach and the POS model. The latter was used in the current model. The QM-calculated barrier to rotation could be reproduced by addition of a nonphysical dihedral angle parameter including all ligands but not Pd.

Very few relevant structures could be found: the reference set of four X-ray structures was therefore augmented with QM structures in addition to the QM normal modes and CHelpG charges. The final force field, although fully converged, could not fully reproduce the reference structures. A few bond angles deviated by several degrees, due mainly to the inadequacies of the POS model. The model could probably have been improved by treating the L–Pd–L angle by a standard function, but this type of differentiation was not easily achieved. On the positive side, the external predictivity was similar to the internal. The final model was judged adequate for the intended application, determination of relative energies of intermediates in the palladium-assisted alkylation reaction. Comparison to other available methods showed that the accuracy in structure determination was similar to the semiempirical method PM3(tm), albeit with different systematic errors, and substantially better than two force fields based on general metal parameters (11).

5.3. Transition-State Model: Os-Catalyzed Asymmetric Dihydroxylation (AD)

The Sharpless AD reaction is an almost ideal test case for a Q2MM study. The selectivity is determined in one well-defined step, which has been well characterized by a combination of high-level QM methods and isotope studies (49). Experimentally, the reaction is not overly sensitive to reaction conditions, tolerating a wide range of solvents, from toluene to water/alcohol mixtures (50). Selectivity data are available in the literature for a wide range of ligand–substrate combinations. Transition-state structures were obtained at the B3LYP level for 59 small model structures. Hessians and CHelpG data for three of the structures and relative energies of several distorted structures were also included in the parameterization, but no experimental data were employed. The final force field was tested on a range of substrates by extensive conformational searches for all low-energy reaction paths. The selectivities could then be calculated from the Boltzmann populations of diastereomeric structures and compared to experimental enantioselectivities. The results were very good, with most deviations below 2 kJ/mol (20).

6. SUMMARY

Force fields for organometallic complexes can be derived using a fast and consistent parameterization method. Structures, conformational energies, QM-derived vibrational modes, and charges should be used as basic reference data. Force fields derived from such data are useful in the prediction of structures and relative energies. For other types of property predictions, specific reference data may be added. The final parameter set should be tested for precision and internal and external predictivity. The method is complementary to the increasingly popular QM/MM methodology (see Chap. 6), in that extensive conformational searches can be performed rapidly, but the response to drastic electronic changes cannot be reliably predicted.

ABBREVIATIONS

B3LYP   DFT method, Becke 3-parameter hybrid exchange with the Lee–Yang–Parr correlation functional
DFT   density functional theory
CHelpG   a method for fitting atomic charges to a QM-derived electrostatic potential
ECP   effective core potential
HF   Hartree–Fock theory
MM   molecular mechanics

MP2 second-order Møller–Plesset perturbation theory
PES potential energy surface
PM3(tm) a modification of the PM3 semiempirical method developed for transition metal complexes
POS points-on-a-sphere coordination model
Q2MM QM-guided MM, a TS force field model based on QM data
QM quantum mechanics
QM/MM hybrid method with a QM core and MM environment
SVD singular value decomposition
TS transition state
vdW van der Waals; nonelectrostatic nonbonded interactions

REFERENCES

1. F Jensen. Introduction to Computational Chemistry. Chichester, UK: Wiley, 1999.
2. A Hinchliffe. Chemical Modeling from Atoms to Liquid. Chichester, UK: Wiley, 1999.
3. J Goodman. Chemical Applications of Molecular Modelling. London: Royal Society of Chemistry, 1998.
4. U Burkert, NL Allinger. Molecular Mechanics, ACS Monograph 177. Washington, DC: ACS, 1982.
5. (a) K Gundertofte, T Liljefors, PO Norrby, I Pettersson. J Comput Chem 17:429–449, 1996. (b) I Pettersson, T Liljefors. In: KB Lipkowitz, DB Boyd, eds. Reviews in Computational Chemistry. Vol. 9. New York: VCH, 1996, pp 167–189.
6. (a) JM Goodman, WC Still. J Comput Chem 12:1110–1117, 1991. (b) I Kolossvary, WC Guida. J Am Chem Soc 118:5011–5019, 1996.
7. (a) TR Cundari. J Chem Soc, Dalton Trans 2771–2776, 1998. (b) P Comba, M Zimmer. J Chem Educ 73:108–110, 1996.
8. CR Landis, DM Root, T Cleveland. In: KB Lipkowitz, DB Boyd, eds. Reviews in Computational Chemistry. Vol. 6. New York: VCH, 1995, pp 73–148.
9. P Comba. Coord Chem Rev 182:343–371, 1999.
10. (a) AK Rappé, CJ Casewit. Molecular Mechanics Across Chemistry. Sausalito, CA: University Science Books, 1997. (b) AK Rappé, CJ Casewit, KS Colwell, WA Goddard III, WM Skiff. J Am Chem Soc 114:10024–10035, 1992. (c) JJ Gajewski, KE Gilbert, TW Kreek. J Comput Chem 19 1167–1178, 1998.
11. H Hagelin, B Åkermark, M Svensson, PO Norrby. Organometallics 18:4574–4583, 1999.
12. (a) BP Hay. Coord Chem Rev 126:177–236, 1993. (b) P Comba. Comments Inorg Chem 16:133–151, 1994. (c) M Zimmer. Chem Rev 95:2629–2649, 1995.
13. JP Bowen, NL Allinger. In: KB Lipkowitz, DB Boyd, eds. Reviews in Computational Chemistry. Vol. 2. New York: VCH, 1991, pp 81–97.
14. U Dinur, AT Hagler. In: KB Lipkowitz, DB Boyd, eds. Reviews in Computational Chemistry. Vol. 2. New York: VCH, 1991, pp 99–164.
15. PO Norrby, T Liljefors. J Comput Chem 19:1146–1166, 1998.

16. (a) J Åqvist, A Warshel. Chem Rev 93:2523–2544, 1993. (b) F Jensen. J Comput Chem 15:1199–1216, 1994.
17. JE Eksterowicz, KN Houk. Chem Rev 93:2439–2461, 1993.
18. (a) PO Norrby. In: DG Truhlar, K Morokuma eds. Transition State Modeling for Catalysis. ACS Symposium Series No. 721, 1999, pp. 163–172. (b) PO Norrby. J Mol Struct (THEOCHEM) 506:9–16, 2000.
19. PO Norrby, P Brandt, T Rein. J Org Chem 64:5845–5852, 1999.
20. PO Norrby, T Rasmussen, J Haller, T Strassner, KN Houk. J Am Chem Soc 121: 10186–10192, 1999.
21. DL Kepert. Inorganic Stereochemistry. Berlin: Springer, 1982.
22. TV Timofeeva, JH Lii, NL Allinger. J Am Chem Soc 117:7452–7459, 1995.
23. (a) VS Allured, CM Kelly, CR Landis. J Am Chem Soc 113:1–12, 1991. (b) P Comba, TW Hambley, M Ströhle. Helv Chim Acta 78:2042–2047, 1995.
24. CR Landis, T Cleveland, TK Firman. J Am Chem Soc 120:2641–2649, 1998.
25. TN Doman, CR Landis, B Bosnich. J Am Chem Soc 114:7264–7272, 1992.
26. P Comba, M Zimmer. Inorg Chem 33:5368–5369, 1994.
27. VJ Burton, RJ Deeth, CM Kemp, PJ Gilbert. J Am Chem Soc 117:8407–8415, 1995.
28. P Brandt, T Norrby, B Åkermark, PO Norrby. Inorg Chem 37:4120–4127, 1998.
29. SB Engelsen, J Fabricius, K Rasmussen. Acta Chem Scand 48:548–552, 1994.
30. PO Norrby, B Åkermark, F Hæffner, S Hansson, M Blomberg. J Am Chem Soc 115:4859–4867, 1993.
31. J Sabolovic, K Rasmussen. Inorg Chem 34: 1221–1232, 1995.
32. JR Maple, MJ Hwang, TP Stockfisch, U Dinur, M Waldman, CS Ewig, AT Hagler. J Comput Chem 15:162–182, 1994.
33. AP Scott, L Radom. J Phys Chem 100:16502–16513, 1996.
34. WD Cornell, P Cieplak, CI Bayly, IR Gould, KM Merz Jr, DM Ferguson, DC Spellmeyer, T Fox, JW Caldwell, PA Kollman. J Am Chem Soc 117:5179–5197, 1995.
35. F Mohamadi, NGJ Richards, WC Guida, R Liskamp, M Lipton, C Caulfield, G Chang, T Hendrickson, WC Still. J Comput Chem 11:440–467, 1990.
36. (a) AK Rappé, WA Goddard III. J Phys Chem 95:3358–3363, 1991. (b) U Dinur, AT Hagler. J Comput Chem 16:154–170, 1995. (c) JW Caldwell, PA Kollman. J Am Chem Soc 117:4177–4178, 1995.
37. WL Jorgensen, J Tirado-Rives. J Am Chem Soc 110:1657–1666, 1988.
38. E Sigfridsson, U Ryde. J Comput Chem 19:377–395, 1998, and references cited therein.
39. TA Halgren. J Am Chem Soc 114:7827–7843, 1992, and references cited therein.
40. (a) AD Becke. J Chem Phys 98:5648–5652, 1993. (b) C Lee, W Yang, RG Parr. Phys Rev B 37:785–789, 1988.
41. (a) G Frenking, I Antes, M Böhme, S Dapprich, AW Ehlers, V Jonas, A Neuhaus, M Otto, R Stegmann, A Veldkamp, S Vyboishchikov. In: KB Lipkowitz, DB Body, eds. Reviews in Computational Chemistry. Vol. 8. New York: VCH, 1996, pp 63–144. (b) TR Cundari, MT Benson, ML Lutz, SO Sommerer. In: KB Lipkowitz, DB Boyd, eds. Reviews in Computational Chemistry. Vol. 8. New York: VCH, 1996, pp 145–202.
42. PO Norrby, P Brandt. Coord Chem Rev. In press.

43. WH Press, SA Teukolsky, WT Vetterling, BP Flannery. Numerical Recipes in C. 2nd ed. New York: Cambridge University Press, 1992.
44. H Hagelin, B Åkermark, PO Norrby. Organometallics 18:2884–2895, 1999.
45. (a) S Lifson, A Warshel. J Chem Phys 49:5116–5129, 1968. (b) JLM Dillen. J Comput Chem 13:257–267, 1992.
46. (a) A Bondi. J Phys Chem 68:441–451, 1964. (b) NL Allinger, X Zhou, J Bergsma. J Mol Struct (THEOCHEM) 312:69–83, 1994.
47. PO Norrby, K Wärnmark, B Åkermark, C Moberg. J Comput Chem 16:620–627, 1995.
48. CM Breneman, KB Wiberg. J Comput Chem 11:361–373, 1990.
49. AJ DelMonte, J Haller, KN Houk, KB Sharpless, DA Singleton, T Strassner, AA Thomas. J Am Chem Soc 119:9907–9908, 1997, and references cited therein.
50. HC Kolb, MS VanNieuwenzhe, KB Sharpless. Chem Rev 94:2483–2547, 1994.

3

Computational Approaches to the Quantification of Steric Effects

David P. White
University of North Carolina at Wilmington, Wilmington, North Carolina

1. INTRODUCTION

Many workers use quantitative steric and electronic parameters in linear free-energy relationships, LFERs, in which kinetic or thermodynamic properties are correlated with the steric and electronic parameters (Eq. 1) (1–3):

$$\text{Property} = aS + bE + c \tag{1}$$

where a, b, and c are constants and S and E are quantitative measures of steric and electronic effects, respectively. Each time an LFER is implemented, several pertinent questions arise. One of the most important is: Which steric and electronic parameter is most appropriate for my data set? In this chapter, we focus attention on the steric measure. There has been considerable attention given to the nature of quantitative steric measures in linear free-energy relationships. Recently, sections of inorganic and kinetics texts have been devoted to a discussion of the quantitative steric measures (4,5). We will examine the theory behind the quantification of steric effects in organometallic chemistry. Each of the measures presented has certain advantages that make it more appropriate for specific types of linear free-energy relationships. If the assumptions behind the steric measure are thoroughly understood, then deviations from linearity in the LFER are more easily understood, and the rationalizations for such deviations are more soundly based.

Steric effects have been recognized as important in organic chemistry since 1872 (6). However, it was not until the 1950s that a meaningful quantification of this steric effect appeared (7). Taft defined a steric parameter, E_S, as the average relative rate of acid-catalyzed ester hydrolysis:

$$E_S = \log\left(\frac{k}{k_0}\right) \tag{2}$$

where k is the observed rate for the acid-catalyzed ester hydrolysis and k_0 is the rate of methyl ester hydrolysis. Original values of E_S were averaged over four kinetic measurements. Since then, there have been several changes to the experiments upon which the E_S parameter is based, but the essential nature of the parameter has remained unchanged (1). There are several advantages to a steric measure based on kinetic data (1).

To successfully correlate thermodynamic or kinetic parameters with stereo-electronic effects, there must be a clean separation of the electronic from the steric effects (Eq. 1). Since there is no necessary reason for an experimentally based steric parameter to be free of electronic effects, workers have turned to applications of computational chemistry to achieve a quantitative measure of pure steric influence of a ligand.

In the early 1970s molecular mechanics was used to define the steric energy of a molecule (8). Molecular mechanics steric energies are the sum of all energies that cause a molecule to distort from an ideal, strain-free geometry. Thus, steric energy was defined as the sum of all bonded and nonbonded energies within a molecule. In 1992 Brown noted that the nonbonded interactions that express the steric requirements of a ligand are the nonbonded *repulsive* interactions between ligand and the binding site within a complex (9). Therefore, Brown defined a new parameter, the ligand repulsive energy, E_R, as a quantitative measure of the steric influence of a ligand.

Traditionally, the steric influence of a ligand in organometallic chemistry is in some way related to the physical size of that ligand. Hence, the cone and solid-angle methodologies have been used as quantitative measures of steric effects in organometallic chemistry (1–3).

We divide this chapter into two parts: the use of molecular mechanics in *defining* the steric requirement of a ligand and the use of mathematical *models* to quantify steric size. In the first part of the chapter, we present the original definitions of steric energy and Brown's extensions to formulate the ligand repulsive energy parameter. In the second part, we present Tolman's cone-angle methodology and its modifications leading to various ligand profiles. Also, we present the solid-angle methodology and its use in profiles. Finally, we conclude this chapter by presenting steric measures based upon Boltzmann-weighted averaging over the conformational space of the ligand.

2. STERIC EFFECTS IN MOLECULAR MECHANICS

2.1. Allinger's Molecular Mechanics Programs

In the original molecular mechanics work, a steric energy, E, for a molecule was defined as the sum of the potentials for bond stretch, E_s, angle bend, E_θ, torsional strain, E_ω, nonbonded interactions, E_{vdW}, and other terms, such as Urey–Bradley terms, cross-interaction terms, and electrostatic terms, (8).

$$E = \sum E_s + \sum E_\theta + \sum E_\omega + \sum E_{vdW} + \sum E_{other} \tag{3}$$

In the original force fields, Allinger and others used Hooke's-law harmonic potentials for a diagonal force field (Eq. 4). If l_0, θ_0, and ω_0 are the strain-free bond distances, angles, and torsion angles and $f_{l,i}$, $f_{\theta,k}$ and $f_{\alpha,m}$ are the relevant force constants, then the potential function for the molecule is given by

$$E = \frac{1}{2} \sum_i f_{l,i}(l_i - l_{0i})^2 + \frac{1}{2} \sum_k f_{\theta k}(\theta_k - \theta_{0k})^2 \tag{4}$$
$$+ \frac{1}{2} \sum_m f_{\alpha m}(\omega_m - \omega_{0m})^2$$

Because of their small magnitudes, all off-diagonal force constants are ignored. The Urey–Bradley force field (10) takes into account 1–3 nonbonded interactions, Eq. (5). If f, f', and f'' are harmonic force constants, then the Urey–Bradley potential is given by

$$E = \frac{1}{2} \sum_i f_{l,i}(l_i - l_{0i})^2 + \frac{1}{2} \sum_k f_{\theta,k}(\theta_k - \theta_{0k})^2$$
$$+ \frac{1}{2} \sum_m f_{\alpha,m}(\omega_m - \omega_{0m})^2 + \sum_{i,j=1}^{3n} f' l_{i,j}^2 + \sum_{i,j=1}^{3n} f'' l_{i,j}^2 + \sum_i f''_{l,i}(l_i - l_{0i})$$
$$+ \sum_k f''_{\theta,k}(\theta_k - \theta_{0k}) + \sum_m f''_{\alpha,m}(\omega_m - \omega_{0m}) \tag{5}$$

The original MM2 force field (represented by Eq. 4) was modified in 1989 (11–13). The MM3 force field employed a more refined set of functions to model structural, thermodynamic, and spectroscopic properties of molecules (for example, Eqs. 6–8; k and V represent force constants).

$$E_s = 71.94k_s(l - l_0)\left\{1 - 2.55(l - l_0) + \left(\frac{7}{12}\right)2.55(l - l_0)^2\right\} \tag{6}$$

$$E_\theta = 0.021914k_{\theta(\theta - \theta_0)})^2[1 - 0.014(\theta - \theta_0) + 5.6 \times 10^{-5}(\theta - \theta_0)^2$$
$$- 7.0 \times 10^{-7}(\theta - \theta_0)^3 + 9.0 \times 10^{-10}(\theta - \theta_0)^4] \tag{7}$$

$$E_\omega = \frac{V_1}{2}(1 + \cos \omega) + \frac{V_2}{2}(1 - \cos 2\omega) + \frac{V_3}{2}(1 + \cos 3\omega) \tag{8}$$

In Eqs. (6)–(8), energies and torsional constants are given in kcal/mol, stretching force constants in mdyn/Å, and bending force constants in mdyn · Å/rad². In 1996 Allinger improved on MM3 with MM4 (14,15). The stretch, bend, torsional, van der Waals, dipole, one-center bend–bend, and stretch–bend terms from MM3 were retained. The improper torsion* and torsion stretch terms from MM3 were modified. A number of new terms were added: stretch–stretch, torsion–bend, bend–torsion–bend, torsion–torsion, torsion–improper torsion, and improper torsion–improper torsion. The goal of MM4 is to begin to add terms to the force field that take into account chemical effects, such as electronegativity and hyperconjugation (14,15).

It has been recognized that nonbonded interactions have two parts: a short-range repulsive part and a longer-range attractive part (8). Both these parts tend asymptotically to zero with distance. The van der Waals function used to describe the behavior of noble gases has often been used as the nonbonded potential. A plot of van der Waals energy versus distance has the characteristic shape shown in Figure 1. There are three characteristic features to the shape of the van der Waals potential: (1) the minimum energy distance, r_0, (2) the depth of the well, ε, which is related to atom polarizabilities, and (3) the steepness of the repulsive part of the potential, which is related to atom hardness.

Second-order perturbation theory gives the form of the attractive part of the van der Waals potential:

$$V_{vdW} = -\frac{c_6}{r^6} - \frac{c_8}{r^8} - \frac{c_{10}}{r^{10}} - \Lambda \tag{9}$$

The first term, r^{-6}, arises from the instantaneous dipole/induced dipole energy of the interaction. If this coefficient, c_6, is adjusted empirically, then the higher terms may be ignored.

For neutral, nonpolar molecules or atoms, the Lennard–Jones potential can be used as the nonbonded potential:

---

*Improper torsion angles arise in considering angles about planar centers, for example, sp^2 carbon atoms.

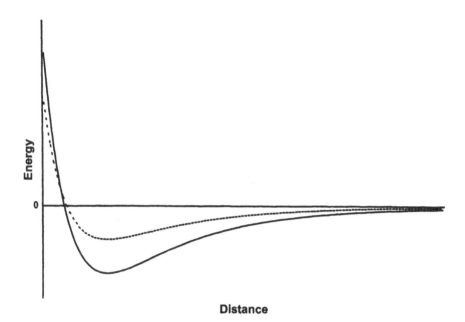

FIGURE 1   Plot of potential energy versus interatomic distance for a diatomic molecule. The solid curve is plotted with the Lennard–Jones potential (Eq. 11) and the dashed curve with the Buckingham potential (Eq. 12).

$$E_{vdW} = \frac{n\varepsilon}{n-m}\left[\frac{m}{n}\left(\frac{r_0}{r}\right)^n - \left(\frac{r_0}{r}\right)^m\right] \qquad (10)$$

In Eq. (10), ε is the potential well depth (Fig. 1). The Lennard–Jones potential can be used as the nonbonded potential because it contains both an attractive (if $n = 6$) and a repulsive part. In theory, $n > 6$. But when $n = 12$, the potential arranges to the simple Lennard–Jones 6/12 potential often used in molecular mechanics:

$$E_{vdW} = \varepsilon\left[\left(\frac{r_0}{r}\right)^{12} - 2\left(\frac{r_0}{r}\right)^6\right] \qquad (11)$$

Alternatively, the Buckingham potential can be used instead of the Lennard–Jones potential:

$$E_{vdW} = \varepsilon\left\{\left[\left(\frac{6}{\alpha-6}\right)e^{\alpha(1-r/r_0)}\right] - \left(\frac{\alpha}{\alpha-6}\right)\left(\frac{r_0}{r}\right)^6\right\} \qquad (12)$$

In order to get the steepness of the potential wells plotted from Eqs. (11) and (12) to agree, α is usually set to 12 or 12.5 (11,13,15,16). In MM3 and MM4 the Buckingham potential is favored as the van der Waals term.

Although the sum of all potential energies was used to define the steric energy of a molecule, Brown found a very poor correlation between this quantity and kinetic or thermodynamic parameters known to be under steric control (9). Thus, Brown defined a new quantitative measure of the steric influence of a ligand, called the ligand repulsive energy, E_R.

2.2. Brown's Ligand Repulsive Energy

Definition of Ligand Repulsive Energy

Brown reasoned that the steric influence of a ligand is more complicated than simply the physical size of that ligand (9). The steric effect of a ligand should optimally be considered to arise from the nonbonded repulsion between the ligand and its molecular environment. The geometry of an organometallic complex is determined by electronic as well as steric effects. To isolate the steric part, Brown reasoned that nonbonded *repulsive* interactions need to be considered. The attractive part of the van der Waals potential is derived from dispersion forces between nonbonded atoms in a molecule (Eq. 9). Since the dispersion forces affect electronic energy levels (17), inclusion of the attractive part of the van der Waals potential in a pure steric measure is inappropriate (9). These conclusions also support not using the total steric energy (defined in Eq. 3) as a quantitative measure of a ligand's steric requirement.

Since we are interested only in the steric influence exerted by the ligand on its environment, the van der Waals repulsion within the ligand needs to be excluded. Consider, for example, the calculation of the steric influence of a ligand, L, attached to a prototypical organometallic group such as $Cr(CO)_5$. (The choice of this prototypical fragment will be discussed in the upcoming section on "Generality of the Approach.") To exclude intraligand repulsion, Brown calculated the van der Waals repulsive energy, $E_{vdW,R}$, as a function of the Cr–L bond distance:

$$E_{vdW,R} = \varepsilon \exp\left\{\gamma\left(1 - \frac{r}{r_0}\right)\right\} \tag{13}$$

As the ligand moves toward the fragment in a direction perpendicular to the basal plane of the carbonyl groups, all intraligand repulsion is held constant. Thus, only the repulsion between ligand and organometallic group to which it is bound, $Cr(CO)_5$, will change when the Cr–L bond distance is varied. The slope of the plot of van der Waals repulsive energy versus distance gives dimensions of energy over distance. To give dimensions of energy, Brown scaled this slope by the equilibrium metal–ligand distance r_e. Thus, ligand repulsive energy, E_R, is defined as the slope of the plot of van der Waals repulsive energy versus distance scaled by r_e (the negative sign in the following equation ensures that the sign of the ligand repulsive energy is positive):

$$E_R = -r_e\left(\frac{\partial E_{vdW,R}}{\partial r}\right) \tag{14}$$

To date, ligand repulsive energies have been computed for a variety of different P-, As- (9), N-(18), O-, and S-donor ligands (19), η^2-coordinated olefins (20), and alkyl groups (21). In addition to the $Cr(CO)_5$ fragment, Brown and others have computed ligand repulsive energies, E'_R, using $[(\eta^5\text{-}C_5H_5)Rh(CO)]$ (19,20,22), CH_3, and CH_2COOH (21) fragments.*

Calculation Algorithm

In the original papers, ligand repulsive energies were computed manually (9,18–22). Recently, we developed a program, ERCODE, to compute ligand repulsive energies using the methodology developed by Brown (23). Ligand repulsive energies are computed as follows.

1. The $Cr(CO)_5L$ complex is built using molecular modeling software, and energy-minimized using a modified MMP2 (24–26) or the universal force field (27). (These modifications are listed in Tables 1–4.)
2. A conformational search is carried out to determine the best representation of the lowest-energy structure. Typically, 2000 conformers are generated using a Monte Carlo algorithm in which the torsion angles of all rotatable bonds are simultaneously varied by randomly different amounts (23).
3. The lowest-energy structure found in step 2 is energy-minimized using tight termination criteria (typically of 0.0100 kcal/mol · Å).
4. The van der Waals function is changed from the Buckingham potential (Eq. 12) to the pure repulsive form (Eq. 13).
5. With all other internal coordinates frozen, the metal–ligand bond is varied by small amounts. Typically, seven distances are used: one is the equilibrium distance, r_e, three are shorter than r_e (each by 0.01 Å), and three are longer than r_e (each by 0.01 Å).
6. A plot of van der Waals repulsive energy, $E_{vdW,R}$, versus distance, r, is constructed and the slope calculated. In practice, this plot is linear over

---

*The label E_R has been reserved for the $Cr(CO)_5$ fragment. Ligand repulsive energies computed with other fragments are called E'_R(fragment).

TABLE 1   Bond-Stretching Parameters Added to the MMP2 Force Field to Enable Modeling of $Cr(CO)_5L$ Complexes

Bond type	Stretching force constant, mdyn/Å	Strain-free bond length, Å
Cr–P	2.000	2.350 (phosphine)
		2.298 (phosphite)
Cr–N	1.500	2.140
Cr–O(sp^3)	1.600	2.08
Cr–S(sp^3)	1.600	2.40
Cr–C$_{centroid}$(olefin)	1.26	1.79
Cr–C(sp) basal	2.100	1.880
		1.895 (phosphite)
Cr–C(sp) axial	2.100	1.850
		1.861 (phosphite)
P–C(sp^3)	2.910	1.810
P–O	2.900	1.615
C(sp)–O(sp) basal	17.040	1.120
	17.029 (phosphite)	1.131 (phosphite)
C(sp)–O(sp) axial	17.040	1.150
	17.029 (phosphite)	1.135 (phosphite)
O(sp)–lone pair	4.600	0.600
C$_{centroid}$(olefin)–C(sp^2)	10.4	0.728

Source: Refs. 18–20, 25, and 26.

the small range of the distances by which the metal–ligand bond distance is varied.

7. The negative of the slope of the $E_{vdW,R}$ versus r plot from step 6 is multiplied by r_e to give E_R (Eq. 14).

Generality of the Approach

Brown chose the $Cr(CO)_5$ fragment for several practical reasons. The vibrational spectra of $Cr(CO)_6$ are well known, and force constants could be extracted (25,26), which allowed for the parameterization of the MMP2 force field. In addition, several crystal structures of $Cr(CO)_5L$ (L = P-donor ligand) were available, so computed and observed structures could be compared. In addition to parameterization arguments, Brown used the $Cr(CO)_5$ fragment for geometrical reasons. The $Cr(CO)_5$ fragment has a fourfold axis of symmetry, which, in the $Cr(CO)_5L$ complex, is collinear with the *pseudo* rotational axis of the ligand (threefold in the case of P-, As-, N-, and C-donor ligands, and twofold in the case of O- and S-donor ligands and olefins). These collinear axes of symmetry simplify the parameterization of the torsional part of the force field: to allow free

TABLE 2 Bond-Angle Deformation Parameters Added to the MMP2 Force Field to Enable Modeling of $Cr(CO)_5L$ Complexes

Bond type	Bending force constant, mdyn · Å/rad²	Strain-free bond angle, degrees
$C_{basal}-Cr-C_{basal}$	0.550	90.0
$C_{basal}-Cr-C_{basal}$	0.000	180.0
$C_{basal}-Cr-C_{ax}$	0.550	90.0
$C_{basal}-Cr-P$	0.500	90.0
$C_{ax}-Cr-P$	0.000	180.0
$C_{basal}-Cr-N$	0.500	90.0
$C_{ax}-Cr-N$	0.000	180.0
$C_{basal}-Cr-O(sp^3)$	0.500	90.0
$C_{ax}-Cr-O(sp^3)$	0.000	180.0
$C_{basal}-Cr-S(sp^3)$	0.500	90.0
$C_{ax}-Cr-S(sp^3)$	0.000	180.0
$C_{centroid}(olefin)-Cr-CO$	0.278	90.0
$Cr-C(sp)-O(sp)$	0.500	180.0
$Cr-P-O$	0.300	118.0
$Cr-P-C(sp^3)$	0.209	112.0
$Cr-N-C(sp^3)$	0.210	115.0
$Cr-N-H$	0.210	105.0
$Cr-O(sp^3)-C(sp^3)$	0.170	130.0
$Cr-O(sp^3)-H$	0.170	126.25
$Cr-S(sp^3)-C(sp^3)$	0.170	94.3
$Cr-O(sp^3)-lone$ pair	0.350	105.16
$C(sp)-O(sp)-lone$ pair	0.521	180.0
$P-C(sp^3)-H$	0.360	111.0
$P-C(sp^3)-C(sp^3)$	0.480	111.5
$C(sp^3)-P-C(sp^3)$	0.576	100.0
$C(sp^3)-O(sp^3)-C(sp^3)$	0.770	111.0
$H-O(sp^3)-H$	4.170	107.5
$C(sp^2)-C_{centroid}-C(sp^2)$	6.95	176.7
$C_{centroid}-C(sp^2)-C(sp^2)$	6.95	1.637
$C_{centroid}-C(sp^2)-H$	0.243	118.4
$C_{centroid}-C(sp^2)-C$	0.550	119.8

Source: Refs. 18–20, 25, and 26.

TABLE 3 Bond-Stretching Parameters Added to the MMP2 Force Field to
Enable Modeling of CpRh(CO)(L) Complexes

Bond type	Stretching force constant, mdyn/Å	Strain-free bond length, Å
Cp–Cp	2.780	1.42
Cp–H	2.606	1.08
Cp–Cp$_{centroid}$	10.425	1.21
Rh–Cp$_{centroid}$	3.000	1.90
Rh–C	2.100	1.81
Rh–P	2.085	2.25
Rh–C$_{centroid}$(olefin)	3.00	2.03

Source: Refs. 19, 20, and 22.

TABLE 4 Bond-Angle Deformation Parameters Added to the MMP2 Force
Field to Enable Modeling of CpRh(CO)(L) Complexes

Bond type	Bending force constant, mdyn · Å/rad^2	Strain-free bond angle, degrees
Cp–Cp–Cp	0.695	108
Cp–Cp–H	0.208	126
Cp$_{centroid}$–Cp–H	0.208	180
Cp–Cp$_{centroid}$–Cp	0.000	72
Cp–Cp–Cp$_{centroid}$	0.000	54
Cp–Cp$_{centroid}$–Rh	0.348	90
Cp$_{centroid}$–Rh–C	0.500	135
Cp$_{centroid}$–Rh–P	0.500	135
Cp$_{centroid}$–Rh–C$_{centroid}$(olefin)	0.500	135
P–Rh–C	0.500	90
C–Rh–C	0.500	90
C$_{centroid}$(olefin)–Rh–CO	0.500	90
Rh–C–O	0.500	180
Rh–P–C(sp^3)	0.209	112
Rh–P–C(sp^2)	0.209	112
Rh–P–O	0.278	118

Source: Refs. 19, 20, and 22.

rotation about the Cr–L bond, all L–Cr–CO$_{basal}$ torsion force constants were set to zero (see Tables 2 and 4). Finally, the basal CO groups of the Cr(CO)$_5$ fragment provide a relatively rigid structure from which the ligand is repelled.

To test the general applicability of E_R as a measure of the steric influence of a ligand, Brown computed ligand repulsive energies with a fragment of very different geometry: CpRh(CO) (22).* Ligand repulsive energy values generated from these two different fragments are highly correlated ($r = 0.95$; Ref. 22). Subsequent to this work, all other reported ligand repulsive energies have been computed with the Cr(CO)$_5$ fragments in addition to other fragments, e.g., CH$_3$ and CH$_2$COOH (21). In all cases, E_R is highly correlated with E'_R. Thus, ligand repulsive energy is a robust measure of the steric influence of a ligand in any prototypical environment. Brown has demonstrated that E_R can be used in linear free-energy relationships, giving generally superior correlation coefficients than the other steric measures traditionally used in organometallic chemistry (discussed in Sec. 3) (9).

3. MATHEMATICAL MODELING OF STERIC EFFECTS

3.1. Tolman's Cone Angle

Although not computational in nature, the cone angle, θ, is historically important as a steric measure in organometallic chemistry (28). Tolman reasoned that when a ligand binds to a metal, it will do so by adopting the least sterically demanding conformation to minimize any steric stress (28). To quantify the physical size of a ligand, Tolman built a CPK model of the ligand with a metal ligand distance typical of a Ni(CO)$_3$L complex. A right circular cone was placed around the ligand and the interior angle of the cone, θ, measured using a protractor (Fig. 2). Cone angles have been reported for P-, As- N-, and C-donor ligands (1–3).

In the late 1980s, workers replaced the CPK models with idealized structures generated from simple molecular modeling packages (1–3). (In all cases, no attempt was made to rigorously energy-minimize the structures.) Interior linear angles, equivalent to Tolman's cone angle, were calculated using measurement tools in the same molecular modeling package.

Recently, Coville and coworkers published a computational measure of cone angles (29). To exemplify the algorithm used, consider a PR$_3$ ligand attached to a metal. Cone angles were calculated as follows:

1. The ligand was divided into groups, with each group bonded to the donor atom, P. For example, for PHMePh the groups are H, Me, and Ph.

---

*Cp = η^5-C$_5$H$_5$

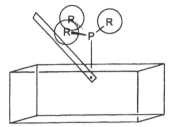

FIGURE 2   Measurement of the Tolman cone angle, θ. A protractor is used to measure the angle between the straight edge and the block.

2. For each atom in the group, a vector was defined from the metal to the center of that atom.

3. The angle a_x (Fig. 3) was defined as the angle between the M–P bond axis and the vector defined in step 2 (from the metal to the center of the atom).

4. The semivertex angle for the atom, α_x (Fig. 3), was defined as the angle between the vector from the metal to the center of the atom (defined in step 2) and the vector tangential to the van der Waals radius of the atom taken from the metal (Fig. 3).

5. The cone angle for that atom, γ_x, was as given by the following:

$$\gamma_x = 2(\alpha_x + a_x) \tag{15}$$

6. Since the largest cone angle for a ligand in a particular conformation was required, the vector resulting in the largest group cone angle, γ_i, was the one used for that group (i.e., γ_i is the largest of the γ_x values for a given group).

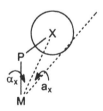

FIGURE 3   Definition of the angles a_x and α_x used to define the group cone angle, γ_x.

7. Finally, the Tolman cone angle, θ, for a ligand with n groups, each with group cone angle γ_i, was the average of all n group cone angles:

$$\theta = \frac{1}{n} \sum_{i=1}^{n} \gamma_i \qquad (16)$$

This algorithm is part of a program, *Steric*, published by Coville's group. At the time of writing, *Steric* was available by ftp to *hobbes.gh.wits.ac.za* in */pub/steric* (login as anonymous and use a full e-mail address as password). (Alternatively, the site may be accessed through: *ftp://hobbes.gh.wits.ac.za/pub/steric/*)

The simplicity of the cone-angle methodology has made it a very popular measure of steric size in organometallic chemistry and, recently, organic chemistry (1). In addition, Tolman's original choice of conformation for most of the ligands has resulted in a series of cone angles that are applicable to widely different reactions (2). However, any possible *attraction* between the ligand and its environment was ignored because the ligand is always placed in the conformation that gives rise to the smallest θ. Brown's ligand repulsive energies, by contrast, takes into account all attractions between the ligand and prototypical fragment prior to isolating the pure repulsive part of the van der Waals potential to define the steric influence of the ligand ("Calculation Algorithm" in Sec. 2.2).

Cone angles also ignore the finite spatial influence of a ligand on its molecular environment (30–33). For example, it is conceivable for a stable complex to form in which two ligands occupy adjacent coordination sites on a metal even though their cones overlap. One possible solution to the problem of interligand meshing is to generate a steric profile for each ligand.

3.2. Cone-Angle Profiles

Ligand Profiles

Cone-angle profiles were first introduced in 1977 by Alyea and Ferguson (34–36), Farrar and Payne (37), and Smith and Oliver (38). (Ligand profiles have also been used to define cone angles (39–42). This will be discussed more fully in Section 3.5.) In all cases, crystal structure data were used, hydrogen atoms (of radius 1.20 Å) were added and some sort of profile plotted. Usually, the semivertex angle, $\theta/2$, versus rotational angle, ϕ, was plotted in either Cartesian or polar coordinates (Fig. 4). These plots, called ligand profiles, were inspected for potential steric interaction between adjacent ligands. Workers have found ligand profiles useful in understanding coordination numbers of particularly bulky, conformationally flexible ligands, such as tricyclohexylphosphine, PCy_3. For example, Ferguson, Alyea, and coworkers found that the conformation of PCy_3 in $[Hg(OAc)_2PCy_3]_2$ was significantly different from its conformation in

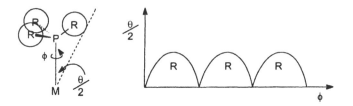

FIGURE 4   Generation of a ligand profile for a PR₃ ligand. As the ligand rotates about the M–P bond, ϕ, the half cone angle varies (shown on left). The plot of half cone angle versus ϕ is the ligand profile (shown on right).

[Hg(OAc)₂(PCy₃)₂] (36). This conformational difference was rationalized by examining the ligand profiles of PCy₃ in the two different environments.

Cone-Angle Radial Profiles

Coville and coworkers have attempted to generate a more general, and potentially more widely applicable, profile methodology (29). A cone-angle radial profile, CARP, is a plot of the variation of cone angle as a function of radial distance from the metal (Fig. 5). (This methodology was a modification of the solid-angle methodology presented in Sections 3.3 and 3.4.) As a sphere grows from the metal, it intersects different atoms in the ligand. At each radial distance, d, the

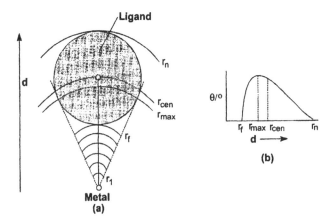

FIGURE 5   Generation of a cone-angle radial profile. (a) A sphere of radius d (a variable) is allowed to grow from the metal toward the ligand. (b) The plot of cone angle (or solid angle; see Sec. 3.4) versus radial distance, d, is called the cone-angle radial profile. (From Ref. 48.)

ligand exerts a different steric demand on its environment. For example, consider a simple sphere representing the ligand placed at some distance from the metal (Fig. 5). A second sphere of variable radius, r_1 to r_n, is allowed to grow from the metal. At r_f, the sphere growing from the metal firsts encounters the ligand (the ligand is also a sphere in this case). The growing sphere reaches a maximum at r_{max} when it intersects a cone enveloping the ligand (Fig. 5). (This is the point at which the Tolman cone angle would be defined.) At the center of the ligand sphere, the growing sphere has radius labeled r_{cen} (Fig. 5). At each radius, r_1 to r_n, the cone angle is calculated using *Steric* (see Sec. 3.1). The plot of cone angle, θ, versus the radius of the growing sphere, d, is called the cone-angle radial profile, CARP (Fig. 5).

Suppose at radius d the ligand has cone angle θ_{CARP}. Then a 3D plot can be generated using a circle of radius θ_{CARP} at each d (Fig. 6). Finally, Coville and coworkers generated a more meaningful 3D profile by plotting θ_{CARP} versus d versus ϕ_{CARP} (ϕ is as defined in Fig. 4), as shown for PH_3 in Figure 6.

Analyses of cone-angle radial profiles have been used to correlate multinuclear NMR data with spatial regions of steric overlap (43). Cone-angle radial profiles have the potential to be used to predict relative stabilities of cis and trans isomers and to predict coordination numbers of bulky ligands. However, cone-angle profiles are very sensitive to the conformation of a ligand.

3.3. Solid Angles

A different approach for taking into account interligand meshing is to mathematically remove all unoccupied space from the calculated ligand size. In other words, generate a computation of only the amount of space occupied by the atoms in a ligand (Fig. 7), called the solid angle, Ω (44)

The first attempt to apply solid angles to the calculation of steric sizes of ligands was by Immirzi and Musco, who placed a right circular cone around the ligand and calculated the solid angle of that cone (45). A solid angle can be thought of as the surface area of the shadow of the projection of a ligand into the inside of a unit sphere (Fig. 8). This means that the solid angle can provide shape-related quantification of the amount of space occupied by the ligand (Fig. 7). Unfortunately, by placing a cone around the ligand and calculating a solid cone angle, presumably because of the difficulty in obtaining an analytical solution to Eq. (17), the values reported by Immirzi and Musco are equivalent in concept to the cone angle.

In the 1980s and early 1990s several workers presented numerical solutions to Eq. (17) (1). In 1993, White et al. presented an analytical solution to the defining equation for a solid angle (46):

$$\Omega = \int_S \frac{\mathbf{r} \cdot \mathbf{dS}}{r^3} \tag{17}$$

Distance / A

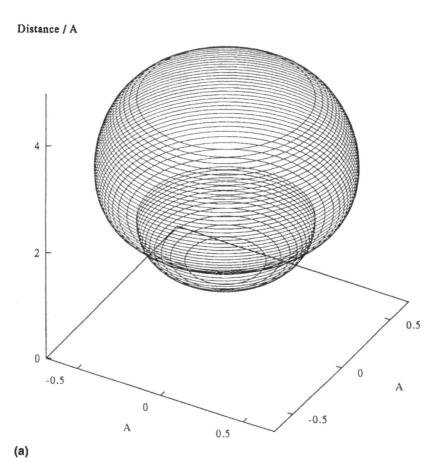

(a)

Figure 6 Three-dimensional cone-angle radial profiles. (a) A circle of radius θ_{CARP} is plotted at each distance, d (see Fig. 5). (b) At each distance d, a surface of θ_{CARP} versus ϕ (the M–P rotational angle; see Fig. 4) is plotted. (From Ref. 29.)

(In Eq. 17, **r** is the position vector of an element of surface with respect to an origin and r is the magnitude of **r**.) Since the solid angle is shape-dependent, it is very sensitive to the conformation of the ligand. In the limit of free rotation about the metal–ligand bond, which implies limited interligand meshing, the ligand occupies the same amount of space as a cone placed around it, so the solid-angle concept reduces to that of the cone angle.

For simplicity, let us consider the solution to Eq. (17) for a sphere of radius

Distance / A

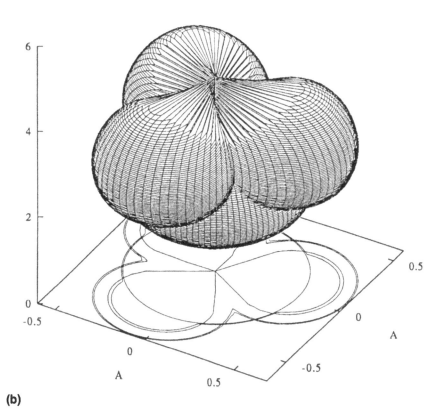

(b)

r_A a distance d_A from the origin enclosed in a right cone of semivertex angle α. The element of solid angle can be written as

$$d\Omega = \sin\theta\, d\theta\, d\phi \tag{18}$$

so that

$$\Omega = \int_0^{2\pi} d\phi \int_0^\alpha \sin\theta\, d\theta \tag{19}$$

and

$$\Omega = 2\pi\left[1 - \left\{1 - \left(\frac{r_A}{d_A}\right)^2\right\}^{1/2}\right] \tag{20}$$

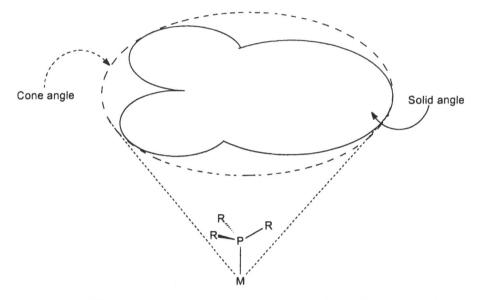

FIGURE 7 Difference between a cone angle and a solid angle. Cone angles completely envelop the ligand and include some unused space in the measure. Solid angles include only the space occupied by the ligand in the measure. (From Ref. 48.)

The key to solving Eq. (17) analytically is *not* attempting to work with an entire ligand at once. Thus, White et al. (46) considered only one pair of atoms in the ligand at a time, calculating the solid angle for that pair. As seen in Figure 9, some single atoms (spheres) will be counted too often. The solid angles for these atoms can be subtracted using the solid angle for a sphere (Eq. 20). To calculate the solid angle for a pair of atoms, a right circular cone was placed around them. The atoms (spheres) projected onto the base of the cone as ellipses. The solid angle of the cone can be calculated easily (45), and the solid angle of the unused space can be removed by integrating over the perimeter of the ellipses. The mathematical details of the integration are outside the scope of this chapter (approximately 46 equations form the basis of the algorithm), but can be obtained from Ref. 46.

As the spheres are projected, a certain amount of overlap between the ellipses occurs as a consequence of the projection. The original algorithm dealt only with overlap between two ellipses (46). It is possible that higher orders of overlap could result, and these have been removed in subsequent, numerical versions of the algorithm. (47)

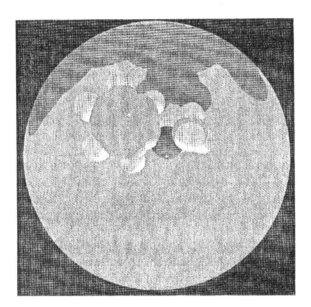

FIGURE 8 Solid angle, defined in Eq. (17), for PHMePh. A light source is placed at the center of a sphere and the ligand is projected onto the sphere. The area of the projection is the solid angle of the ligand. (From Ref. 49.)

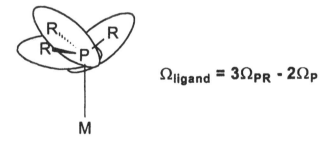

$$\Omega_{ligand} = 3\Omega_{PR} - 2\Omega_P$$

FIGURE 9 Calculation of the solid angle of PH_3 by traversing the molecule in a pairwise manner. Notice that the three pairs of P-H atoms would give the solid angle for P_3H_3, so the solid angle for two P atoms needs to be removed. (Redrawn from Ref. 46.)

3.4. Solid-Angle Profiles

Radial Profiles

In Section 3.3, the solid angle was defined as the surface area of the ligand projected onto the inside of a sphere (Eq. 17). White and Coville reasoned that if the sphere originates at the metal in an organometallic complex and is allowed to grow out toward the ligand, then a partial solid angle can be calculated at each radius of the growing sphere (48). The resulting plot of solid angle as a function of radial distance is called a solid-angle radial profile. (Solid-angle radial profiles are conceptually similar to cone-angle radial profiles—see Figure 5—but are easier to implement. The easier implementation arises because the solid angle is defined by the projection of the ligand onto the inside of a sphere. As the radial profile is computed, the sphere onto which the ligand is projected simply grows from the metal.)

Equation (17) reveals that the solid angle is inversely proportional to the square of distance. This means that the radial profile for a sphere is not symmetrical (Fig. 5). The shape of a radial profile can provide information about the relative orientation of atoms within a ligand (or, by extension, between ligands in an organometallic complex). For example, consider a sphere A placed at some distance from a metal. Now, place a second sphere, B, different distances from A, as illustrated in Fig. 10. The shape of the radial profile generated from the metal is different for each of the geometrical arrangements of the two spheres. The differences in appearance of the radial profiles have enabled Coville and coworkers to examine possible stearic interactions between adjacent ligands (Fig. 11) (31).

Perhaps more powerful than the solid-angle radial profile is the quantification of the amount of overlap between two adjacent ligands.

Angles of Overlap

If two ligands are arranged around a metal so that the van der Waals radii of their atoms overlap, then a severe stearic strain should develop. Quantification of the amount of overlap is useful in predicting whether a given geometrical arrangement of ligands could exist. For example, suppose there is a nominal amount of overlap between two adjacent ligands. Then it is possible for some bonds to stretch, angles to bend, etc. to relieve the stress caused by the overlap. As the amount of overlap increases, the complex should become less and less stable. Coville and coworkers introduced a quantification of interatom overlap using both solid-angle and linear or vertex-angle concepts (49).

Using the solid-angle concept, there are two different types of overlap: overlap as a consequence of projecting the atoms onto a sphere, and physical overlap between the van der Waals radii of the atoms (Fig. 12). Only the latter is related to steric congestion. By using the solid-angle radial profile methodol-

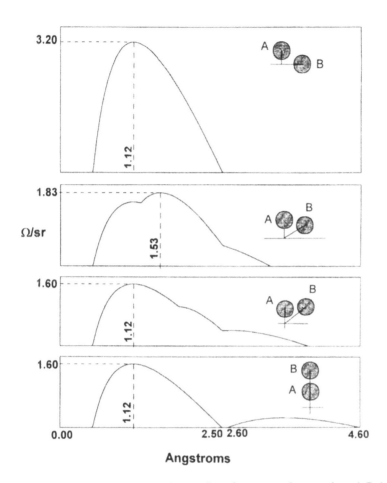

Figure 10 Solid-angle radial profiles for two spheres, A and B, in different geometrical arrangements. Sphere A is kept a constant distance of 1.5 Å from the metal, sphere B is placed at 1.5 Å, 2.2 Å, 2.8 Å, and 3.6 Å (top to bottom) from the metal. Notice that the shapes of the profiles, and the maximum solid angles, vary significantly as a function of the geometrical arrangement of the spheres. (From Ref. 48.)

ogy, all nonbonded overlap is eliminated from the computation. As the sphere grows from the metal in the solid-angle radial profile, the solid angle of only those atoms that intersect with the sphere is calculated. Thus, the possibility of overlap resulting from projecting the entire ligand onto a sphere is eliminated. For example, consider the case of the carbonyl ligand as viewed from a metal.

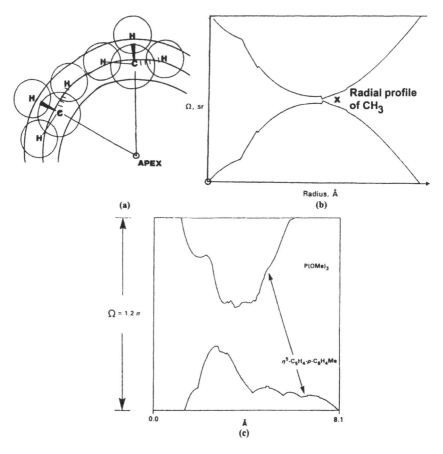

FIGURE 11 Use of solid-angle radial profiles. (a) Cis methyl groups give rise to the profiles indicated in (b). (b) The point marked **X** indicates the radial distance at which maximum steric interaction occurs. (From Ref. 48.) (c) Radial profiles for the cyclopentadienyl and P(OMe)$_3$ ligands in [(η^5-C$_5$H$_4$-p-C$_6$H$_4$Me)Fe(CO){P(OMe)$_3$}I]. Nuclear magnetic resonance spectroscopy reveals steric interaction between the ligands as indicated. (From Ref. 31.)

There is no physical overlap between the C and O atoms, yet in the projection of the ligand from the metal, the carbon atom eclipses oxygen. In the radial profile (Fig. 10), this overlap is eliminated. By using the solid-angle methodology, a quantitative measure of the amount of overlap can be attained (49).

The solid angle of overlap, Λ, is calculated using the original solid-angle algorithm (Sec. 3.3) as follows:

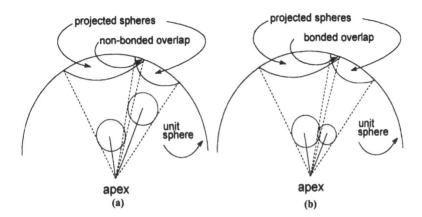

FIGURE 12 Different types of overlap encountered in the solid-angle measure. (a) In the process of projection, overlap that does not appear in the ligand appears in the projection. This overlap is called nonbonded overlap. (b) When two atoms physically overlap, the overlap in the projection is called bonded overlap. (From Ref. 48.)

1. A right circular cone is placed around two atoms.
2. The solid angle of the two intersecting atoms is calculated using the algorithm presented in Section 3.3.
3. The difference between the solid angle of each of the atoms (represented as spheres, obtained using Eq. 20) and the solid angle of the bonded atoms is the solid angle of overlap, Λ.

A semiquantitative measure of the amount of overlap between ligands can be attained using the semivertex angle of overlap, λ (49). In Figure 13, if the semivertex angle of atom A is α, for atom B is β, and the A-M-B bond angle is χ, then the vertex angle of overlap, λ, is

$$\lambda = (\alpha + \beta) - \chi \qquad (21)$$

It should be noted that the vertex angle of overlap, λ, is not additive, which limits its general utility. The solid angle of overlap, Λ, has been used to rationalize metal–ligand bond lengths in $Cr(CO)_5L$ complexes, the Mn–Re bond length in $MnRe(CO)_{10}$, and the conformational preferences in substituted cyclopentadienyl complexes of Mo and Ru (32,33,43,49).

3.5. Boltzmann-Weighted Steric Measures

Thus far in this chapter all steric measures presented have been based on a single conformation of the ligand. In the case of ligand repulsive energies, attempts

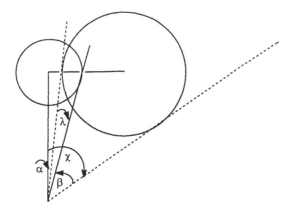

Figure 13 Definition of the vertex angle of overlap. The semivertex angle of one sphere, α, is indicated by the solid lines and the semivertex angle of the other sphere, β, indicated by dashed lines. The vertex angle of overlap is given by λ. (Redrawn from Ref. 49.)

were made to find a good representation of the lowest energy conformation prior to computation of the steric parameter (9,18–23). Mosbo and coworkers were the first to recognize that it would be appropriate to use a Boltzmann weighting factor to average cone angles over the entire conformational space of the ligand (39).

Mosbo began by plotting half cone angles $\theta/2$, versus M–P rotation axis, ϕ (Fig. 4), in 1° increments to obtain ligand profiles for each phosphines (see Sec. 3.2 for a description of "Ligand Profiles"). Four different definitions of the cone angle were proposed using these ligand profiles:

1. θ_I: twice the maximum in the ligand profile:

$$\theta_I = 2\left(\frac{\theta}{2}\right)_{max} \tag{22}$$

2. θ_{II}: twice the maximum for each group in the ligand profile averaged over the three groups attached to P. This is equivalent to the Tolman cone angle defined in Sec. 3.1 (2):

$$\theta_{II} = \frac{2}{3}\sum_{i=1}^{3}\left[\left(\frac{\theta}{2}\right)_{max_i}\right] \tag{23}$$

3. θ_{III}: twice the average maximum for each group in the ligand profile:

$$\theta_{III} = \frac{2}{3}\sum_{i=1}^{3}\left[\left\{\left(\frac{\theta}{2}\right)_{max}\right\}_{av}\right]_i \tag{24}$$

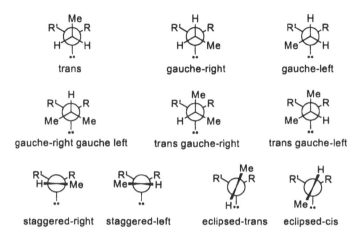

FIGURE 14 Classification of conformers of phosphine ligands used by Mosbo and coworkers. The metal is excluded from the molecular modeling calculations and is represented by a lone pair on phosphorus. (Redrawn from Ref. 39.)

4. θ_{IV}: twice the average of all 360 half cone angles:

$$\theta_{IV} = \frac{2}{360} \sum_{i=1}^{360} \left(\theta/2 \right)_i \tag{25}$$

When the ligand contains conformational degrees of freedom, each one angle, $\theta_I - \theta_{IV}$, was subjected to an energy-weighted averaging.

Several low-energy conformations for the phosphorus donor ligand were considered: trans, gauche-right, gauche-left, gauche-right gauche-left, trans gauche-right, and trans gauche-left (Fig. 14). For aryl-substituted phosphines, four additional conformers were considered: staggered-right, staggered-left, eclipsed-right, and eclipsed-left (Fig. 14). Symmetry-related conformers were considered degenerate. For ligands with conformational degrees of freedom, a Boltzmann-averaged cone angle, $\bar{\theta}$, was calculated:

$$\bar{\theta} = n_A\theta_A + n_B\theta_B + \cdots + n_i\theta_i \tag{26}$$

In Eq. (26), θ_i is the cone angle of conformer i with mole fraction n_i. Each mole fraction (Eq. 27) is obtained by calculating the heat of formation of the conformer using molecular modeling (discussed next):

$$n_A = \frac{g_A}{g_A + g_B e^{-E_{AB}/RT} + \cdots + g_i e^{-E_{Ai}/RT}} \tag{27}$$

TABLE 5   Bond Lengths and
Angles Used as Input Variables
for the Energy-Minimization of
Phosphines by Mosbo and
Coworkers

Bond	Bond length, Å, or angle, degrees
P–H	1.4
P–C	1.8
C–C$_{alkyl}$	1.5
C–C$_{aryl}$	1.4
C–H	1.1
R–P–R′	109.5
R–C–R′ (alkyl)	109.5
R–C–R′ (aryl)	120

Source: Refs. 39–42.

(In Eq. 27, g_i is the number of conformers with unique conformation i.)

Three different molecular modeling programs were used to obtain the heats of formation (or total molecular mechanics energy; see Eq. 3): MINDO/3 (39), MNDO (40), and MM2 (41). Similar methodologies were used in each case. The phosphorus ligand, in the absence of a metal, was placed in an appropriate low-energy conformation, identified in Figure. 14. Approximate bond lengths and angles were used as input variables (see Table 5). The ligand was allowed to energy-minimize, and the metal was placed 2.28 Å from the P atom perpendicular to the plane, illustrated in Figure 15. Finally, the ligand profile was plotted and

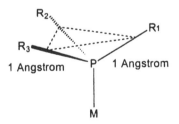

FIGURE 15   Placement of the metal after the conformation of the free phosphine has been energy-minimized. A plane is defined by marking a point 1.0 Å from the phosphorus along the P–C bond axis. The metal is placed 2.28 Å from the phosphorus normal to the plane. (Redrawn from Ref. 39.)

the cone angle calculated using customized code. In the case of the MM2 calculations, a slightly modified procedure was used for ligands containing conjugated π systems in which SCF (self-consistent field) were implemented in the final step. Additional conformers were also found by a slightly modified conformational search strategy. Consider PPh_3 as an example: One phenyl group was constrained and the dihedral driver of MM2 was used to construct a 2D-grid search of the conformational space of the other two rings (a 30° grid was used). For each of the minima found, the two varied phenyl rings were constrained and the ligand allowed to energy-minimize (thus generating a minimum for the third ring relative to the first two). Finally, all constraints were removed and the ligand was allowed to energy-minimize.

Cone angles generated from the preceding methodology, Eq. (22)–(25), were used in a linear free-energy relationship to rationalize cis:trans ratios in $W(CO)_4LL'$ complexes (42). In addition, several regression analyses were performed on this data set comparing Tolman's cone angle (2) to Mosbo's cone angles (39–41) to Brown's E_R values (9,41).

White and coworkers also computed Boltzmann-weighted solid angles (see Sec. 3.3) by implementing Eq. (28) (n_i is defined in Eq. (27) and Ω in Eq. 17) (50):

$$\overline{\Omega} = n_A\Omega_A + n_B\Omega_B + \Lambda + n_i\Omega_i \tag{28}$$

Instead of working with the isolated ligand, $Cr(CO)_5L$ complexes were built in a molecular modeling program and between 500 and 1000 conformers were generated per ligand. A random conformational search strategy was employed in which torsion angles for all rotatable bonds were simultaneously allowed to vary by randomly different amounts. (This is the same search strategy used to compute E_R (21).) Each conformer was energy-minimized using the MMP2 force field with modifications by Brown and coworkers (see Tables 1–4 for modifications) (9,25,26). Finally, all 500–1000 conformers were submitted to *Steric* (see Sec. 3.1) for energy-weighted solid-angle calculation (Eq. 28). These workers found that the energy-weighted solid angle correlated better with other steric parameters (θ, E_R) than the solid angle based on a single, low-energy conformer. In addition, the energy-weighted mean solid angle also performed better in linear free-energy relationships than the solid angle based only on the lowest-energy conformer (50).

4. SUMMARY

In essence there are three quantitative measures of steric size in organometallic chemistry: the cone angle, the solid angle, and the ligand repulsive energy. Cone angles have appeal because they are easy to visualize and because then perform well in linear free-energy relationships. However, cone angles may overstate the physical size of the ligand and cannot take into account ligand meshing. Solid angles can drastically understate the physical size of a ligand, but can be usefully applied to the problem of ligand meshing. The most appealing quantification of true steric demand is the ligand repulsive energy. E_R values are easy to measure and are robust. In addition, they are a sound quantitative measure of the steric influence of a ligand on its environment in the absence of electronic complications. The only current drawback of the measures is that current E_R values are based on a single, low-energy conformation. At the time of writing, Brown and White were in the process of deriving a set of Boltzmann-weighted ligand repulsive energies.

ACKNOWLEDGMENTS

The author would like to thank Professors Theodore L. Brown, University of Illinois at Urbana-Champaign, and S. Bart Jones and Michael Messina, University of North Carolina at Wilmington, for helpful comments in preparing this manuscript.

REFERENCES

1. D White, NJ Coville. Adv Organomet Chem 36:95–158, 1994.
2. CA Tolman. Chem Rev 77:313, 1977.
3. TL Brown, KJ Lee. Coord Chem Rev 128:89–116, 1993.
4. DMP Mingos. Essential Trends in Inorganic Chemistry. New York: Oxford University Press, 1998.
5. RB Jordan. Reaction Mechanisms of Inorganic and Organometallic Systems. 2nd ed. New York: Oxford University Press, 1998.
6. AW Hofmann. Chem. Ber. 5:704, 1872.
7. RW Taft. In: MS Newman, ed. Steric Effects in Organic Chemistry. New York: Wiley, 1956, pp 556.
8. U Burkert, NL Allinger. Molecular Mechanics. Washington, D.C.: ACS Monograph 177, 1982.
9. TL Brown. Inorg Chem 31:1286–1294, 1992.
10. HC Urey, CA Bradley Jr. Phys Rev 38:1969, 1931.
11. NL Allinger, YH Yuh, J-H Lii. J Am Chem Soc 111:8551-8566, 1989.
12. J-H Lii, NL Allinger. J Am Chem Soc 111:8566–8575, 1989.
13. J-H Lii, NL Allinger. J Am Chem Soc 111:8576–8582, 1989.
14. NL Allinger, K Chen, J-H Lii. J Comput Chem 17:642–668, 1996.
15. N Nevins, K Chen, NL Allinger. J Comput Chem 17:669–694, 1996.
16. NL Allinger, X Zhou, J Bergsma. J Mol Struct Theochem 312:69–83, 1994.
17. P Hobza, R Zahradnik. Intermolecular Complexes. Amsterdam: Elsevier, 1988.
18. M-G Choi, TL Brown. Inorg Chem 32:1548–1553, 1993.
19. M-G Choi, D White, TL Brown. Inorg Chem 33:5591–5594, 1994.

20. DP White, TL Brown. Inorg Chem 34:2718–2724, 1995.
21. DP White, JC Anthony, AO Oyefeso. J Org Chem 64:7707–7716, 1999.
22. M-G Choi, TL Brown. Inorg Chem 32:5603–5610, 1993.
23. RJ Bubel, W Douglass, DP White. J Comput Chem 21:239–246, 2000.
24. JT Sprague, JC Tai, Y Yuh, NL Allinger. J Comput Chem 8:581–603, 1987.
25. ML Caffery, TL Brown. Inorg Chem 30:3907–3914, 1991.
26. KJ Lee, TL Brown. Inorg Chem 31:289–294, 1992.
27. AK Rappé, CJ Casewit, KS Colwell, WA Goddard, WM Skiff. J Am Chem Soc 114: 10024, 1992.
28. CA Tolman. J Am Chem Soc. 92:2953, 1970.
29. JM Smith, BC Taverner, NJ Coville. J Organomet Chem 530:131–140, 1997.
30. D White, L Carlton, NJ Coville. J Organomet Chem 440:15–25, 1992.
31. D White, P Johnston, IA Levendis, JP Michael, NJ Coville. Inorg Chim Acta 215: 139–149, 1994.
32. JM Smith, DP White, NJ Coville. Polyhedron 15:4541–4554, 1996.
33. JM Smith, D White, NJ Coville. Bull Chem Soc Ethiop 10:1–8, 1996.
34. EC Alyea, SA Dias, G Ferguson, RJ Restivo. Inorg Chem 16:2329–2334, 1977.
35. G Ferguson, PJ Roberts, EC Alyea, M Khan. Inorg Chem 17:2965–2967, 1978.
36. EC Alyea, G Ferguson, A Somogyvari. Inorg Chem 21:1369–1371, 1982.
37. DH Farrar, NC Payne. Inorg Chem 20:821–828, 1981.
38. JD Smith, JD Oliver. Inorg Chem 17:2585–2589, 1978.
39. JT DeSanto, JA Mosbo, BN Storhoff, PL Bock, RE Bloss. Inorg Chem 19:3086–3092, 1980.
40. JA Mosbo, RK Atkins, PL Bock, BN Storhoff. Phosphorus Sulfur 11:11, 1981.
41. M Chin, GL Durst, SR Head, PL Bock, JA Mosbo. J Organomet Chem 470:73–85, 1994.
42. ML Boyles, DV Brown, DA Drake, CK Hostetler, CK Maves, JA Mosbo. Inorg Chem 24:3126–3131, 1985.
43. JM Smith, SC Pelly, NJ Coville. J Organomet Chem 525:159–166, 1996.
44. KF Riley. Mathematical Methods for the Physical Sciences. Cambridge, UK: Cambridge University Press, 1974, p 91.
45. A Immirzi, A Musco. Inorg Chim Acta 25: L41, 1977.
46. D White, BC Taverner, PGL Leach, NJ Coville. J Comput Chem 14:1042–1049, 1993.
47. BC Taverner. J Comput Chem 17:1612–1623, 1996.
48. D White, BC Taverner, PGL Leach, NJ Coville. J Organomet Chem 478: 205–211, 1994.
49. BC Taverner, JM Smith, DP White, NJ Coville. S Afr J Chem 50:59–66, 1997.
50. D White, BC Taverner, NJ Coville, PW Wade. J Organomet Chem 495:41–51, 1995.

4

The Accuracy of Quantum Chemical Methods for the Calculation of Transition Metal Compounds

Michael Diedenhofen,
Thomas Wagener, and Gernot Frenking
Philipps-Universität Marburg, Marburg, Germany

1. INTRODUCTION

The last decade has witnessed the establishment of quantum chemical methods as a standard tool for quantitative calculations of transition metal (TM) compounds, after numerous theoretical studies had proved that the calculated values are very accurate. The calculated data can be used to interpret experimental observations and to design new experiments and, thus, are very helpful for experimental chemistry. The theoretically predicted geometries, vibrational frequencies, bond dissociation energies, and other chemically important properties have become reliable enough to complement and sometimes even to challenge experimental data. This is particularly important for bond energies of TM compounds, which tend to be difficult to determine by experimental methods.

The situation at the end of the 1990s had dramatically changed from the 1980s, when only a small number of brave hearts of the theoretical chemistry community were tackling "the challenge of transition metals and coordination chemistry.'' This was the title of a NATO Advanced Study Institute that was held in Strasbourg in 1985 (1). The proceedings of the meeting reflected the cautious and reluctant opinion of most scientists about the accuracy of quantum chemical methods that might be achieved in the field in the near future. This reservation can still be found in the foreword of the editor of the special issue of *Chemical Reviews* about Theoretical Chemistry that was published in 1991: "The theory of transition-metal chemistry has lagged behind the quantum theory of organic chemistry because quantitative wave functions are more complicated" (2).

The enormous progress in quantum chemical methods for TM compounds is due mainly to quasi-relativistic effective core potentials (ECPs) and particularly to gradient-corrected (nonlocal) density functional theory (NL-DFT), which have become standard theoretical tools in computational chemistry. Pioneering work in method development and application of DFT methods in the field of TM chemistry has been carried out by Ziegler (3). Because computational chemistry has reached a status where available methods and programs are also used by scientists who are not specialists in the field, it is reasonable to give an overview of the accuracy that can be achieved with commonly used levels of theory. This has been done by us (4) and by Cundari et al. (5) in previous reviews, which summarize theoretical studies of TM compounds with ECPs in conjunction with classical ab initio methods at the HF, MP2, and CCSD(T) levels of theory. The two reviews, which were published in 1996, also give an overview of the available ECPs that have been optimized for TM elements. The same ECPs are usually employed in DFT calculations as well, although the ECP parameters have not been optimized in the framework of DFT but rather with respect to Hartree–Fock calculations or experimental results. Available are ECPs that have been generated from atomic DFT calculations (6,7). However, calculations of a representative set of TM complexes showed that ECPs generated from HF atomic calculations may be used with little loss of accuracy in DFT calculations as well (7).

There is general agreement in the theoretical community that gradient-corrected DFT methods are in most cases superior to classical ab initio methods at the HF and MP2 levels for the calculation of TM compounds, because the accuracy of the DFT results is similar or even better than the MP2 data, while the computational costs are less. For this reason most computational chemistry groups are now using DFT methods for TM compounds. It should be noted, however, that DFT methods are inferior to high-level ab initio methods such as CCSD(T) for very accurate energy calculations. We also want to point out that the statement about the superior results of DFT methods can at present be made only for the electronic ground states of diamagnetic (closed-shell) TM compounds. Density functional theory calculations of paramagnetic TM compounds have been carried out (22,23), but it seems that a standard DFT method for open-shell species has not yet been established.

In this chapter we want to give an overview of the scope and limitations of the presently available DFT methods commonly used for calculating TM compounds. For comparison, we also present in some cases results of ab initio calculations at the HF, MP2, and CCSD(T) levels of theory. The very large number of quantum chemical calculations of TM compounds published in the last decade makes it possible to estimate the accuracy of those DFT methods that can be considered as standard levels. The goal of this work is to serve as a guideline for nonspecialists who want to carry out DFT calculations of TM compounds. First, we will summarize the most important programs that can be employed for DFT calculations. We give an overview of the different functionals and ECPs commonly used for TM compounds. In the main part of the review we discuss selected topics and projects. These are not comprehensive but representative for the field of TM compounds. The cited references should be helpful for finding information about other fields of theoretical TM chemistry that are not discussed here.

2. QUANTUM CHEMICAL PROGRAMS, DENSITY FUNCTIONALS, ECPs, AND BASIS SETS FOR TRANSITION METALS

The most common quantum chemical programs—Gaussian (8), GAMESS (9), Turbomole (10), CADPAC (11), ACES II (12), MOLPRO (13), MOLCAS (14), and the newly developed TITAN (15)—are able to run pseudopotential calculations. Please note that CADPAC and MOLCAS can only use so-called ab initio model potentials (AIMPs) in pseudopotential calculations. Such AIMP differ from ECPs in the way that the valence orbitals of the former retain the correct nodal structure, while the lowest-lying valence orbital of an ECP is a nodeless function. Experience has shown that AIMPs do not give better results than ECPs, although the latter do not have the correct nodal behavior of the valence orbitals (16).

Most of the listed programs are also capable of running DFT calculations. In addition, there are some programs that have been developed specifically for DFT methods. The most common DFT programs are DMol (17), DGauss (18), DeMon (19), and ADF (20). The program ADF is unusual because it is the only widely distributed quantum chemical program that uses Slater orbitals as basis functions instead of the more common Gaussian functions. The use of Slater basis functions makes it a bit more difficult to compare the results of ADF with those of other programs that use Gaussian functions.

The central question for any DFT calculation concerns the choice of the exchange and correlation functionals for the energy expression. Numerous investigations have been carried out in order to examine the reliability of different mathematical expressions for the exchange and correlation functionals, and several studies were devoted to TM compounds (3,21–24). The following conclusions can be made from these investigations and from our work that will be discussed later. First, the geometries and particularly the energies become significantly improved when nonlocal (gradient-corrected) functionals $F(\rho, \nabla\rho)$ are employed rather than functionals that depend only on the electron density $F(\rho)$. Second, the nonlocal exchange functional suggested by Becke (B) in 1988 (25) has been established as a standard expression in NL-DFT calculations. Third, the choice of the best correlation functional, for which several mathematical expressions have been proposed, is less obvious than the choice of the exchange functional. The presently most popular correlation functionals are those of Perdew (P86) (26), Lee, Yang, and Parr (LYP) (27), Perdew and Wang (PW91) (28), and Vosko, Wilk, and Nuisar (VWN) (29).

The situation in choosing the proper combinations of exchange and correlation functionals became a bit confusing in the early 1990s when different functionals were combined and the resulting energy expression was given by a multiparameter fit of the functionals. The semiempirical weight factors were obtained from a fit to a set of well-established experimental values. The most commonly used functional combination of this type is the three-parameter fit of Becke (B3) (30). The original expression for the B3 hybrid functional is:

$$E_{XC} = 0.2(E_X^{HF}) + 0.8(E_X^{LDA}) + 0.72(E_X^B) + 1.0(E_C^{LDA}) + 0.81(E_C^{NL})$$

A widely used variant of the B3 hybrid functional termed B3LYP (31), which is slightly different from the original formulation of Becke, employs the LYP expression for the nonlocal exchange functional E_C^{NL}. It seems that the B3LYP hybrid functional is at present the most popular DFT method for calculating TM compounds. Other widely used functionals are BP86, which gives particularly good results for vibrational frequencies (32), BPW91, and BLYP. It is a wise idea to estimate the accuracy of a functional for the particular problem at the beginning of a research project by running some test calculations before the final choice of the functionals is made. The disadvantage of DFT compared with conventional ab initio methods is that the DFT calculations cannot systematically be improved toward better results by going to a higher level of theory.

We want to point out that the development of new functionals is at present a very active field in quantum chemistry. Promising new functionals have recently been proposed by Hamprecht et al. (33) and by Becke, who introduced multiparameter fits of functionals that involve first-order and second-order density gradients (34). The limits of gradient corrections in DFT were discussed by the same author (35). The accuracy of these functionals for TM compounds has not systematically been exploited yet, but it is possible that new functionals will soon be established as standard methods for TM compounds that surpass the already impressive reliability of the present methods.

The second crucial choice for a quantum chemical DFT calculation is the basis set. The valence shell of the TMs has s and d orbitals. As a minimum requirement for useful calculations it is necessary to have at least a double-zeta quality for the $n(s)$ and $(n - 1)d$ valence orbitals. The status of the lowest-lying empty $n(p)$ orbitals is at present controversial (36). However, it has been shown that the basis set should have at least one function that describes the empty $n(p)$ orbital of the TM (4,37). Extra f-type polarization functions improve the accuracy particularly of the calculated energies, but it seems that they are less important for the TMs than d-polarization functions for main-group elements.

Many DFT and ab initio calculations are carried out with the frozen-core approximation for the innermost electrons, or the core electrons are replaced by pseudopotentials, mostly in the form of an ECP but sometimes as an AIMP. It is generally recognized that the outermost $(n - 1)s^2$ and $(n - 1)p^6$ core electrons should *not* be replaced by an ECP, but should be retained in the calculations. Small-core ECPs are more reliable than large-core ECPs, where only the $(n)s$ and $(n - 1)d$ electrons of the TMs are calculated. Several groups developed valence basis sets in conjunction with small-core ECPs (38–41) and AIMPs (42) for the TMs. The ECP valence basis set suggested by the Stuttgart group (40) is very large and may be too big for calculations of larger molecules. It should be used for very accurate calculations. There is no report known to us that suggests that one of the other ECPs or AIMP is generally more accurate than the other.

An important theoretical aspect for calculating TM compounds concerns the effect of relativity. It is well known that relativistic effects must be considered in the calculation in order to obtain reliable geometries and energies of 2nd- and 3rd-TM-row molecules (43). Elements of the first TM row are little influenced by relativity, except for copper (43,44). The most convenient way to include relativistic effects in the calculations is the use of quasi-relativistic ECPs or AIMPs. The techniques of relativistic ECPs for molecules containing transition metals and other heavy atoms have recently been reviewed (45). Most ECPs and AIMPs have been derived from scalar-relativistic atom calculations, except the ECPs for the first TM row developed by Hay and Wadt (38). Note that the spin-orbit coupling term is not included in the scalar-relativistic ECPs. This seems to be not so important for the calculation of geometries, relative energies, and vibrational frequencies of closed-shell TM compounds, but spin-orbit interactions cannot be neglected for the calculations of NMR parameters of compounds of $5d$ TMs (see later). Various approximate treatments of relativistic effects in all-electron calculations have been suggested, and some of them have been implemented in computational chemistry programs (46). The status of relativistic DFT methods has recently been reviewed by van Wüllen (47). Most of the presently available implementations of relativistic all-electron DFT methods are based on different scalar-relativistic approximations; i.e. they are one-component approximations of the four-component Dirac equation (46,47). The results that have been published so far do not suggest that the scalar-relativistic all-electron methods are superior to quasi-relativistic ECP methods, except for the calculation of NMR parameters. The situation may become different in the future, when two-component methods that include spin-orbit effects become available and become more widely used for TM compounds. Work in this field is in progress (48).

3. RESULTS OF QUANTUM CHEMICAL CALCULATIONS OF TRANSITION METAL COMPOUNDS

In the following section we will discuss the results of quantum chemical calculations of TM compounds that may serve as a guideline for the search for a theoretical method. The data may also be used as an indicator of the accuracy that can be expected. The examples have been chosen to cover a large area of TM compounds that are not comprehensive, but representative of commonly used standard methods in the field.

3.1. Homoleptic Transition Metal Carbonyl Complexes

Carbonyl complexes are probably the theoretically best investigated class of TM complexes. Here we focus on the results reported in the last couple of years. Table 1 shows theoretical and experimental bond lengths and first bond dissociation energies (FBDEs) of the hexacarbonyls $TM(CO)_6$ (TM = Cr, Mo, W).

The following conclusions can be drawn from the calculated data. The DFT methods BP86 and B3LYP predict bond lengths that are in excellent agreement with experimental values that have been taken from gas-phase measurements. The calculated bond lengths at LDA are clearly inferior. The Mo–CO and W–CO bond lengths predicted at MP2 are very good, but the Cr–CO distance is too short. This is a general weakness of the MP2 method. Systematic studies have shown that MP2 gives good metal–ligand bond lengths for $4d$ and $5d$ metal, while bond lengths of $3d$ TMs are too short (4). The results for the FBDEs lead to a similar conclusion. BP86 and B3LYP give bond energies in good agreement with experiment. LDA and MP2 give bond energies that are in all cases too high. Previous calculations have shown that MP2 systematically overestimates the FBDE of TM complexes, particularly for the first TM row (4,49). CCSD(T) gives very accurate bond energies. Because all theoretical methods predict a higher value for the FBDE of $Cr(CO)_6$ than the experimental value, the accuracy of the latter has been questioned (50). Please note that the methods listed in Table 1 have been used in conjunction with different basis sets, and that some results were obtained from ECP calculations while others used all-electron basis sets. However, this does not affect the general conclusions about the methods.

TABLE 1 Bond Lengths r (Å) and First TM–CO Bond Dissociation Energies D_e (kcal/mol) for TM(CO)$_6$

Method	Cr(CO)$_6$			Mo(CO)$_6$			W(CO)$_6$		
	r (TM–C)	r (C–O)	D_e (TM–(CO))	r (TM–C)	r (C–O)	D_e (TM–(CO))	r (TM–C)	r (C–O)	D_e (TM–(CO))
MP2[a]	1.861	1.168	58.0	2.061	1.164	46.1	2.060	1.166	54.9
CCSD(T)[b]	1.938	1.172	42.7 (45.8)[a]			40.4[a]			48.0[a]
LDA[c]	1.866	1.145	62.1	2.035	1.144	52.7	2.060	1.144	48.4
BP86[c]	1.910	1.153	46.2	2.076	1.153	39.7	2.049	1.155	43.7
B3LYP[d]	1.921	1.155	40.7	2.068	1.155	40.1	2.078	1.156	44.8
Expt.[e]	1.918	1.141	36.8 ± 2	2.063	1.145	40.5 ± 2	2.058	1.148	46.0 ± 2

[a] Ref. 100 using the standard basis set II.
[b] Ref. 101.
[c] Ref. 102.
[d] Ref. 22.
[e] Ref. 103.

A comparison of the different methods—BP86, B3LYP, MP2, and CCSD(T)—with the same ECP/basis set combination is available from a study of the isoelectronic hexacarbonyls $TM^q(CO)_6$ with the third-TM-row elements $TM^q = Hf^{2-}$, Ta^-, W, Re^+, Os^{2+}, Ir^{3+} by Szilagyi and Frenking (51). Table 2 shows the calculated and experimental bond lengths obtained with a quasi-relativistic ECP and a valence basis set that has DZ + P quality. Table 3 gives the FBDEs of the complexes.

The calculated bond lengths obtained via B3LYP, BP86, and MP2 are in very good agreement with the experimental results. Please note that the gas-phase value for the W–CO bond (2.058 Å) is longer than the solid-state values (2.018–2.032 Å). Bond lengths of donor–acceptor bonds measured in the solid state are always *shorter* than in the gas phase (52). This must be considered when the theoretical and experimental TM–CO distances of the TM hexacarbonyl ions shall be compared. The calculated bond energies support the conclusion that BP86 and B3LYP give values that agree with the very accurate but expensive CCSD(T) method. MP2 gives bond energies that are too high. However, the trend that is predicted for the hexacarbonyls by MP2 agrees with the other methods. Note that the BP86 and B3LYP values for $W(CO)_6$ given in Table 3 are slightly different from the data shown in Table 1. The results were reported by different groups using different basis sets.

The vibrational spectra of TM carbonyls have also been calculated in numerous theoretical studies. Table 4 gives the theoretical and experimental stretching frequencies v_{CO} and force constants F_{CO} of the preceding series of isoelectronic hexacarbonyls. Figure 1 shows a plot of the t_{1u} mode of v_{CO}. It is obvious that the calculated trend of the force constants and vibrational frequencies is in accord with experiment. Please note that the calculations refer to harmonic fundamentals, while the experimental values are taken from the observed anharmonic modes. Systematic studies of the performance of BP86 with different ECPs for the vibrational spectra of many neutral and ionic TM carbonyls by Jonas and Thiel have shown that reliable harmonic force fields can be obtained at this level of theory (32).

Another theoretical study of TM carbonyls in which a comparison of different methods has been made was recently published by Lupinetti et al. (53). The focus of the paper was the analysis of the metal–CO bond in the series of homoleptic d^{10} carbonyls $TM^q(CO)_n$, with $TM = Cu^+$, Ag^+, Au^+, Zn^{2+}, Cd^{2+}, Hg^{2+}, where $n = 1–6$. In order to estimate the accuracy of the theoretical level the authors calculated the TM carbonyls of the group 11 elements Cu–Au with $n = 1–4$ at B3LYP, BP86, MP2, and CCSD(T) levels of theory. Table 5 shows the theoretically predicted and experimental FBDEs.

The results clearly indicate a limitation of the DFT method in the calculation of the TM d^{10} carbonyls. The CCSD(T) values are in very good agreement with the experimental results, except for $Cu(CO)^+$ and $Cu(CO)_2^+$, for which the

TABLE 2 Calculated and Experimental Bond Lengths (Å) of TM Hexacarbonyl Complexes and CO

Compound	TM–C	C–O	Method
$[Hf(CO)_6]^{2-}$ (O_h)	2.211	1.182	B3LYP
	2.206	1.196	BP86
	2.174(3); 2.179(3)	1.162(5); 1.165(4)	X-ray
	2.180(3)	1.162(4)	
$[Ta(CO)_6]^-$ (O_h)	2.124	1.166	B3LYP
	2.118	1.179	BP86
	2.113	1.180	MP2
	2.083(6)	1.149(8)	X-ray
$W(CO)_6$ (O_h)	2.074	1.151	B3LYP
	2.066	1.164	BP86
	2.060	1.166	MP2
	2.058	1.148	ED
	2.018; 2.025	1.130; 1.139	X-ray
	2.032; 2.033	1.152; 1.158	
$[Re(CO)_6]^+$ (O_h)	2.046	1.138	B3LYP
	2.036	1.151	BP86
	2.026	1.155	MP2
	1.98(3); 2.03(6)	1.14(4); 1.12(7)	X-ray
	2.02(6); 1.89(7)	1.16(8); 1.19(8)	
	2.07(7)	1.12(9)	
$[Os(CO)_6]^{2+}$ (O_h)	2.049	1.128	B3LYP
	2.038	1.141	BP86
	2.025	1.148	MP2
$[Ir(CO)_6]^{3+}$ (O_h)	2.068	1.121	B3LYP
	2.057	1.135	BP86
	2.041	1.144	MP2
	2.05(1); 2.01(1)	1.07(1); 1.08(2)	X-ray
	2.04(1); 2.00(2)	1.12(2)	
	2.02(2)		
CO ($C_{\infty v}$)		1.150	B3LYP
		1.138	BP86
		1.152	MP2
		1.143	exptl

Source: Ref. 51.

TABLE 3 Calculated and Experimental First Bond Dissociation Energies D_e (kcal/mol) of Isoelectronic TM Hexacarbonyls

Method	$[Hf(CO)_6]^{2-}$	$[Ta(CO)_6]^-$	$W(CO)_6$	$[Re(CO)_6]^+$	$[Os(CO)_6]^{2+}$	$[Ir(CO)_6]^{3+}$
B3LYP	51.40	47.81	45.93	48.22	58.20	74.94
	(49.61)	(45.95)	(43.84)	(45.99)	(55.87)	(72.59)
BP86	54.86	50.94	49.43	52.25	62.57	79.05
	(53.18)	(49.08)	(47.34)	(50.02)	(60.24)	(76.70)
MP2		53.08	54.76	58.21	69.90	85.71
		(51.21)	(52.67)	(55.98)	(67.27)	(83.36)
CCSD(T)		47.94	48.02	50.57	60.95	77.45
		(46.07)	(45.93)	(48.34)	(58.62)	(75.10)
Experimental			46.0 ± 2			

ZPE corrected values are given in parentheses.
Source: Ref. 51.

TABLE 4 Calculated and Experimental Carbonyl Stretching Frequencies v_{CO} (cm^{-1}) and Force Constants F_{CO} (mdyn/Å) of the Hexacarbonyl Complexes and CO

Compound	t_{1u} v_{CO}	F_{CO}	$E_g v_{CO}$	$A_{1g} v_{CO}$	Method
[Hf(CO)$_6$]$^{2-}$	1863.3	14.50	1873.8	1990.9	B3LYP
	1798.5	13.49	1805.4	1910.3	BP86
	1757				Experimental
[Ta(CO)$_6$]$^-$	1969.7	16.22	1988.7	2098.6	B3LYP
	1899.4	15.06	1914.6	2015.6	BP86
	1896.4	14.75	1882.9	2019.2	MP2
	1850				Experimental
W(CO)$_6$	2074.2	18.00	2097.7	2191.8	B3LYP
	1996.4	16.65	2017.3	2106.7	BP86
	1977.7	16.35	1998.5	2095.1	MP2
	1977	17.0^f	1998	2115	Experimental
[Re(CO)$_6$]$^+$	2176.8	19.83	2200.1	2271.9	B3LYP
	2088.8	18.24	2112.4	2184.2	BP86
	2053.1	17.64	2087.2	2148.1	MP2
	2085	18.1^f	2122	2197	Experimental
[Os(CO)$_6$]$^{2+}$	2267.5	21.50	2287.4	2333.1	B3LYP
	2165.5	19.60	2187.0	2237.3	BP86
	2113.9	18.67	2144.1	2172.7	MP2
	2190	19.8^f	2218	2259	Experimental
[Ir(CO)$_6$]$^{3+}$	2335.2	22.77	2349.6	2373.4	B3LYP
	2223.5	20.65	2240.4	2269.3	BP86
	2139.6	19.09	2163.2	2167.9	MP2
	2254	20.8^f	2276	2295	Experimental
CO	2211.6	20.21			B3LYP
	2117.6	18.79			BP86
	2118.9	18.81			MP2
	2143	18.9			Experimental

Source: Ref. 51.

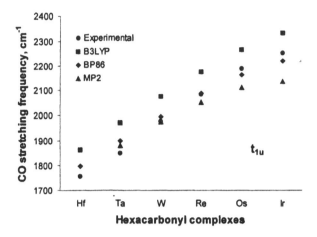

FIGURE 1 Trend of the calculated and experimental C-O stretching frequencies of the t_{1u} mode of $Hf(CO)_6^{2-}$, $Ta(CO)_6^-$, $W(CO)_6$, $Re(CO)_6^+$, $Os(CO)_6^{2+}$, $Ir(CO)_6^{3+}$. (From Ref. 51.)

theoretical values are higher than given by experiment. Lupinetti et al. investigated the discrepancy between the CCSD(T) values and the experimental results. They carried out additional calculations using very large basis sets (53). The calculated bond energies did not change very much. BP86 and B3LYP give bond energies for the mono- and dicarbonyls that are too high. The DFT methods also have problems with the relative FBDEs of $Cu(CO)^+$ and $Cu(CO)_2^+$. CCSD(T) and even MP2, which notoriously gives bond energies that are too high, agree with the experimental observation that the FBDE of $Cu(CO)_2^+$ is higher than that of $Cu(CO)^+$ (Table 5). The bond energies predicted at BP86 are even higher than the MP2 values that are notoriously too high. The failure of the DFT methods for the mono- and dicarbonyls of the group 11 metal ions is a warning against the indiscriminate use of DFT functionals without initial calibration calculations having been carried out.

The different approximations for all-electron relativistic calculations using one-component methods have recently been compared with each other and with relativistic ECP calculations of TM carbonyls by several workers (47,55). Table 6 shows the calculated bond lengths and FBDEs for the group 6 hexacarbonyls predicted when different relativistic methods are used. The results, which were obtained at the nonrelativistic DFT level, show the increase in the relativistic effects from $3d$ to $4d$ and $5d$ elements. It becomes obvious that the all-electron DFT calculations using the different relativistic approximations—scalar-relativistic (SR) zero-order regular approximation (ZORA), quasi-relativistic (QR) Pauli

TABLE 5 Calculated and Experimental TM–CO First Bond Dissociation Energies D_e (kcal/mol) for $[TM(CO)_n]^+$ Complexes (TM = Cu, Ag, Au; $n = 1$–4)

Compound	MP2/I// MP2/I	CCSD(T)/I// MP2/I	CCSD(T)/I// CCSD(T)/I	BP86/I// BP86/I	B3LYP/I// B3LYP/I	Experimental
$[Cu(CO)]^+$	38.1	32.3	32.9	51.7	43.3	37.4
$[Cu(CO)_2]^+$	43.1	36.2	36.7	47.1	42.6	42.9
$[Cu(CO)_3]^+$	23.4	18.6	19.6	25.0	20.5	19.3
$[Cu(CO)_4]^+$	22.8	16.5	18.0	21.7	17.3	14.8
$[Ag(CO)]^+$	23.3	21.8	22.0	35.2	29.5	22.0
$[Ag(CO)_2]^+$	28.6	26.4	26.6	38.4	33.1	27.5
$[Ag(CO)_3]^+$	13.9	12.6	12.8	16.4	13.8	13.6
$[Ag(CO)_4]^+$	12.3	11.1	11.3	13.0	11.2	11.7
$[Au(CO)]^+$	40.8	38.3	38.5	61.3	49.9	—
$[Au(CO)_2]^+$	51.0	47.0	47.3	57.2	51.9	—
$[Au(CO)_3]^+$	9.2	6.4	6.9	11.5	7.2	—
$[Au(CO)_4]^+$	9.3	6.7	7.3	10.1	7.1	—

Basis set I: 6–31G(d) for C, O; quasi-relativistic ECP with a valence basis set [311111/22111/411] for TM.
Source: Ref. 53.

TABLE 6 Theoretical and Experimental Bond Lengths (Å) and First Bond Dissociation Energies D_e (kcal/mol) for TM(CO)$_6$ (M = Cr, Mo, W)

Method	Cr(CO)$_6$		Mo(CO)$_6$		W(CO)$_6$	
	r (TM–C)	D_e (TM–(CO))	r (TM–C)	D_e (TM–(CO))	r (TM–C)	D_e (TM–(CO))
DFT n.r.	1.908	41.6	2.079	37.2	2.106	37.5
DFT (SR) ZORA	1.904	42.0	2.068	39.6	2.062	45.0
DFT QR	1.910	46.2	2.076	39.7	2.049	43.7
DFT DKH		43.7	2.068	39.3	2.063	46.9
DFT DPT	1.905	45.4	2.064	39.1	2.060	46.1
BP86[a]	1.903	58.0	2.072	42.0	2.066	49.4
MP2	1.861	45.8	2.061	46.1	2.060	54.9
CCSD(T)//MP2				40.4		48.0
Experimental	1.918	36.8 ± 2	2.063	40.5 ± 2	2.058	46.0 ± 2

[a] BP86 values taken from Ref. 104.
Source: Ref. 54b.

Hamiltonian (PH), Douglas-Kroll-Hess (DKH), and direct perturbation theory (DPT)—give similar results as the much cheaper, quasi-relativistic ECP calculations. There is no reason for not using ECPs for the calculation of geometries, energies, and vibrational spectra of TM compounds. We want to point out that relativistic all-electron methods become important for the calculation of NMR chemical shifts and coupling constants, because the effect of the core electrons is not negligible anymore. This is discussed in more detail later.

The results shown in Table 6 seem to indicate that the different approximations for relativistic effects in all-electron calculations have a comparable accuracy. This is not the case. It has been found that the QR method using the Pauli Hamiltonian can lead to significant errors in the bond energy (55). An example will be given in the following section, about substituted carbonyl complexes.

3.2. Substituted Transition Metal Carbonyl Complexes

The accuracy of DFT methods and ab initio calculations has also been investigated for substituted TM carbonyl complexes $TM(CO)_nL$. Two papers focused on group 6 and group 10 carbonyls with the formula $TM(CO)_5L$ (TM = Cr, Mo, W) and $TM(CO)_3L$ (TM = Ni, Pd, Pt), respectively (56,57). Table 7 shows a comparison of the calculated bond lengths and $(CO)_nTM-L$ BDEs of some complexes at the BP86, MP2, and CCSD(T) levels of theory. The BP86 calculations were carried out with all-electron basis sets and first-order relativistic corrections estimated by direct perturbation theory (57), while the MP2 and CCSD(T) results have been obtained using quasi-relativistic ECPs for Mo and W and nonrelativistic ECPs for Cu (56).

The results shown in Table 7 span a range between weakly (N_2) and very strongly (NO^+) bonded ligands. It becomes obvious that the bond lengths of the molybdenum and tungsten complexes calculated with BP86 and MP2 are very similar, while the BDEs predicted at MP2 are clearly higher than the BP86 values. Since the BP86 values for the bond energies are very similar to the data obtained at CCSD(T) it can be concluded that BP86 gives rather accurate bond lengths and bond energies for these systems.

It is frequently said that present DFT methods are not reliable enough to calculate weakly bonded systems. An important work by Ehlers et al. (58) about TM–noble gas complexes $TM(CO)_5NG$ (TM = Cr, Mo, W; NG = Ar, Kr, Xe) showed that the NL-DFT methods BP86 and PW91 have a comparable accuracy for calculating TM–NG bond energies as the CCSD(T) method. Table 8 shows the theoretical and experimental bond energies. The calculated values have been corrected for the basis set superposition error (BSSE). The large all-electron triple-zeta Slater-type basis set III augmented by polarization functions and diffuse functions was used in the DFT calculations. The basis set II for the CCSD(T) calculations employed DZ + P valence basis sets for the TMs and QZ + P

TABLE 7 Calculated Bond Lengths (Å) and Dissociation Energies D_e (kcal/mol) for TM(CO)$_5$L complexes

L	r(TM-(CO)$_{cis}$) MP2/II	BP86	r(TM-(CO)$_{trans}$) MP2/II	BP86	r(TM-L) MP2/II	BP86	D_e (TM-L) MP2/II	CCSD(T)/II[a]	BP86
Cr(CO)$_5$L									
N$_2$	1.870	1.907	1.803	1.877	1.936	1.961	33.9	24.8	22.5
CO	1.861	1.905	1.920	1.930	1.804	1.869	58.0	45.8	43.7
CS	1.860	1.907	2.055	1.994	1.761	1.749	84.9	65.8	59.5
NO$^+$	1.900	1.948					126.0	106.7	103.0
Mo(CO)$_5$L									
N$_2$	2.060	2.062	1.996	2.017	2.164	2.128	26.3	22.0	19.8
CO	2.061	2.064	2.119	2.095	1.985	2.024	46.1	40.4	39.0
CS	2.066	2.068	2.233	2.157	1.877	1.888	70.0	60.8	54.5
NO$^+$	2.119	2.102					123.6	104.4	101.8
W(CO)$_5$L									
N$_2$	2.057	2.059	2.013	2.022	2.126	2.099	32.6	26.4	25.6
CO	2.060	2.061	2.094	2.085	2.006	2.025	54.9	48.0	46.1
CS	2.063	2.063	2.178	2.144	1.891	1.887	80.3	70.7	63.1
NO$^+$	2.107	2.093					129.2	110.0	109.0

[a] Using MP2 optimized geometries.
Source: MP2 and CCSD(T) values are taken from Ref. 56; the BP86 data are taken from Ref. 57.

TABLE 8 Dissociation Energies and Enthalpies (kcal/mol) of the Noble Gas Complexes TM(CO)$_5$–NG (TM = Cr, Mo, W; NG = Ar, Kr, Xe)

| | CCSD(T)/II | | | BP86/III | | | PW91/III | | | Experimental |
	ΔE	ΔH_{298}	ΔH_{298} BSSE corr	ΔE	ΔH_{298}	ΔH_{298} BSSE corr	ΔE	ΔH_{298}	ΔH_{298} BSSE corr	ΔH_{298}
Cr(CO)$_5$Ar	4.9	6.3	3.5	1.9	3.3	3.0	3.8	5.2	4.8	
Cr(CO)$_5$Kr	6.2	7.5	4.7	3.0	4.3	4.0	5.0	6.3	5.9	
Cr(CO)$_5$Xe	7.2	8.5	5.0	5.4	6.7	6.4	7.6	8.9	8.5	9.0 ± 0.9
Mo(CO)$_5$Ar	5.4	6.8	2.2	2.7	4.1	3.6	4.0	5.4	4.9	
Mo(CO)$_5$Kr	6.9	8.2	4.4	3.9	5.2	4.7	5.1	6.4	5.9	
Mo(CO)$_5$Xe	8.2	9.5	4.7	7.0	8.3	7.9	8.6	9.9	9.4	8.0 ± 1.0
W(CO)$_5$Ar	8.0	9.4	4.3	3.6	5.0	4.6	5.2	6.6	6.1	≲~3
W(CO)$_5$Kr	10.0	11.3	6.7	5.1	6.4	6.0	7.0	8.3	7.8	≤6
W(CO)$_5$Xe	11.9	13.2	7.6	7.6	8.9	8.8	9.8	11.1	10.7	8.2 ± 1.0

Source: Ref. 58.

valence basis sets for the noble gas elements. The authors also calculated the electric polarizabilities of the noble gases, because the TM-NG bonding is mainly due to dipole-induced dipole interactions. Table 9 shows the calculated results. The data in Table 9 show that the larger basis sets used for the DFT calculations yield atomic polarizabilities in good agreement with experiment. The DFT values are even slightly too high. The polarizabilities predicted by the ab initio methods are clearly too low, which is caused by the significantly smaller basis sets. This is partly corrected by the contribution of the basis set superposition of the $TM(CO)_5$ fragment orbitals to the calculated polarizabilities, which leads to ab initio results that are 75–80% of the experimental values. Table 8 shows that the theoretically predicted TM-NG BDEs (CCSD(T)/II, BP86/III, and PW91/III) after BSSE correction are in reasonable agreement with experiment. The BSSE corrections at the DFT levels are very small. PW91 gives always larger bond energies than BP86. The error range of the experimental values is too large to discriminate among the methods. The CCSD(T)/II energy calculations used geometries optimized at MP2/II. The bond lengths calculated at MP2/II, BP86/III, and PW91/III were found to be very similar (58). The message of this study is that NL-DFT methods may also be used for TM complexes with weakly bonded ligands.

Another class of substituted carbonyl complexes that has been investigated to test the accuracy of theoretical methods are phosphine complexes $TM(CO)_5PR_3$. The theoretical studies focused on the results obtained when different approximations for the treatment of relativistic effects are used (47,55). Table 10 shows the W–P bond lengths and bond energies of the complexes $(CO)_5W–PR_3$ (R = H, CH_3, F, Cl) that have been calculated with the BP86 functional

TABLE 9 Calculated and Experimental Electric Polarizabilities of the Noble Gas Atoms (10^{-24} cm^3)

Method	Basis Set[a]	Ar	Kr	Xe
HF	II	0.716	1.086	2.352
MP2	II	0.723	1.064	2.340
CCSD(T)	II	0.730	1.068	2.364
HF	II + $W(CO)_5$ ghost functions	1.254	2.019	3.388
BP86	III	1.737	2.597	4.188
PW91	III	1.787	2.666	4.250
Experimental	Experimental	1.64	2.48	4.04

[a] Basis set II: ECP with [3111/3111/1] valence basis set. Basis set III: Triple-zeta Slater functions augmented by two s, p, d diffuse functions.
Source: Ref. 58.

TABLE 10 Calculated Bond Lengths r (Å) and Bond Dissociation Energies D_e (kcal/mol) of W–PR$_3$ Bonds of Octahedral W(CO)$_5$PR$_3$ Complexes Using Different Approximations for Relativistic Effects

	r(W-P)				D_e(W-P)			
R	BP86/II[a]	BP86 (QR)[b]	BP86 (ZORA)[c]	BP86 (DPT)[d]	BP86/II[a]	BP86 (QR)[b]	BP86 (ZORA)[c]	BP86 (DPT)[d]
H	2.515	2.496	—	2.518	35.3	37.2	—	33.0
F	2.409	2.374	—	2.399	43.3	39.6	—	35.1
Cl	2.442	2.431	—	2.442	32.6	33.3	—	29.2
Me	2.543	2.460	2.553	2.542	45.5	75.7	43.8	43.5

[a] Quasi-relativistic ECPs.
[b] Quasi-relativistic Pauli Hamiltonian.
[c] ZORA approximation.
[d] Direct perturbation theory.
Source: Ref. 55.

using relativistic ECPs and all-electron basis sets, where the relativistic effects have been estimated by direct perturbation theory (DPT) (59), by the quasi-relativistic approximation using the Pauli Hamiltonian (PH) (60), and by the zero-order regular approximation (ZORA) (61). A comparison with the experimental bond lengths shows that the ECPs perform equally well as the more expensive all-electron calculations. The three methods—DPT, PH, and ZORA—give very similar results, with one notable exception. The calculated $W-P(CH_3)_3$ bond length at BP86(PH) is too short, and the theoretical value for the $W-P(CH_3)_3$ BDE at this level is much higher (75.7 kcal/mol) than predicted by the other three methods (43.5–45.5 kcal/mol). It has been pointed out that the PH scheme is problematic from a theoretical point of view, because the Pauli operator is not bounded from below, and nonphysical low energies may result from the variational treatment (53).

The failure of the PH method is disturbing, because it occurs for only one molecule $[(CO)_5WP(CH_3)_3]$, while the PH result for the related compound $(CO)_5WPH_3$ is in agreement with the other methods. A related situation has recently been reported by van Lenthe et al. (54a), who calculated the geometries and bond energies of the TM carbonyls $W(CO)_6$, $Os(CO)_5$, and $Pt(CO)_4$ with the PH and ZORA approximations using different basis sets. It was found that the variational collapse of the PH method occurs when the basis set becomes very large. The FBDE of $Os(CO)_5$ was predicted with the Pauli Hamiltonian and a triple-ζ basis set for oxygen and carbon to be 42.9 kcal/mol, which agrees with the results given by other methods. The bond energy becomes unrealistically high (191.5 kcal/mol) when a quadruple-ζ basis set is employed (54a). The unpredictability of cases where significant errors may occur makes the PH approach unsuitable for reliable calculation of TM compounds. The ZORA approach (61), which includes higher-order relativistic effects than the Pauli Hamiltonian, is clearly the better method for relativistic calculations.

As a final example of substituted carbonyl complexes we want to mention TM complexes with group-13 diyl ligands $(CO)_nTM-ER$, where E is either B, Al, Ga, or In. This is a rather young class of compounds, for which the first examples of stable molecules have been synthesized only in the last couple of years (62). A review about theoretical work of group-13 diyl complexes has just been published (63). Table 11 shows calculated and experimental bond lengths and theoretically predicted bond dissociation energies at BP86. It becomes obvious that the theoretical bond lengths are in very good agreement with experimental values of related compounds. Table 11 shows also two examples of homoleptic group-13 diyl complexes that confirm the good performance of BP86. Reliable theoretical studies in this field are particularly important, because there is still not much known experimentally about the stabilities and properties of these compounds.

TABLE 11 Calculated Bond Lengths (Å) and Bond Dissociation Energies of the TM–E Bond D_e (kcal/mol) at BP86/II

E	TM–E	E–C$_{Cp/Ph/CH3}$/E–N	TM–CO$_{ax/trans}$	TM–CO$_{eq/cis}$	∠E–TM–CO$_{ax/trans}$	∠E–TM–CO$_{eq/cis}$	D_e
		(CO)$_4$–Fe–ECp (axial isomers)					
B	1.962 (2.010)	1.830–1.838 (1.811–1.817)	1.788 (1.793)	1.765 (1.774, 1.786)	179.6	84.7[a]	78.0
Al	2.242 (2.231)	2.240–2.243 (2.140–2.153)	1.768 (1.796)	1.772 (1.768)	179.6	85.2[a]	53.1
Ga	2.330 (2.273)	2.355–2.356 (2.226)	1.755 (1.781)	1.782 (1.789)	180.0	87.7[a]	32.9
		(CO)$_5$–W–EN(SiH$_3$)$_2$					
B	2.125 (2.152)	1.383 (1.339)	2.078	2.059	180.0	88.3[a]	75.1
		(CO)$_4$–Fe–EPh (axial isomer)					
Ga	2.263 (2.225)	1.983 (1.943)	1.771 (1.766)	1.780 (1.764)	179.6	86.8[a]	55.0
		TM(ECH$_3$)$_4$					
Ga	2.214 (2.170)	2.047 (2.014)					49.6
In	2.347 (2.310)	2.190 (2.195)					51.1

[a] Average over slightly different angles.
Source: Ref. 63. Experimental values are given in parentheses.

3.3. Complexes with TM≡N and TM≡P Triple Bonds

Transition-metal nitrido and phosphido complexes have been investigated by us in two theoretical studies (64,65). The results are interesting in the present context, because the calculated molecules have metal–ligand triple bonds. It is known that the MP2 method predicts bond lengths of multiple bonds between main group elements that are too long (66), and it is important to compare the performance of MP2 and DFT methods for calculating multiple bonds between a TM and a main group element. This has been done in Refs. 64 and 65.

Figure 2 shows the theoretically predicted structures of the rhenium nitrido complex $Cl_2(PH_3)_3ReN$ and the compounds where the nitrido ligand is bonded to different Lewis acids $Cl_2(PH_3)_3ReN-X$ (X = BH_3, BCl_3, BBr_3, AlH_3, $AlCl_3$, $AlBr_3$, GaH_3, $GaCl_3$, $GaBr_3$, O, S, Se, Te)(64). The geometries were optimized at the MP2 and B3LYP levels of theory using our standard basis set II (4). A comparison of the calculated interatomic distances shows that the MP2 values for all $Re-PH_3$ and most $Re-Cl$ bond lengths are slightly smaller than the B3LYP data, while MP2 calculates the TM–N multiple bond longer than at B3LYP. The MP2-calculated N–chalcogen bond lengths are shorter than for B3LYP, except for the N–O bond, which has the highest N–chalcogen double bond character according to the NBO analysis (64). It has been found by us in several investigations of TM compounds that MP2 tends to give slightly too long interatomic distances for TM–X shared-electron multiple bonds, while TM–X single bonds and donor–acceptor bonds are usually shorter than at the DFT (B3LYP and BP86) level (4). However, the differences are in most cases not very large, and both methods give geometries that are reasonably accurate.

Experimental values of three related compounds may be used to estimate the accuracy of the theoretical data given in Figure 2. The geometry of the parent nitrido complex, $Cl_2(PH_3)_3ReN$, can be compared with the X-ray structure analysis of $Cl_2(PMe_2Ph)_3ReN$ (67). The most important bond lengths at B3LYP/II are Re–N = 1.668 Å (exp. 1.660 Å), Re–P = 2.440 and 2.460 Å (exp. 2.42–2.46 Å), and Re–Cl_{cis} = 2.455 Å (2.442 Å). The MP2/II values are similar, but the Re–N bond length (1.703 Å) is slightly too long. An X-ray structure analysis has also been reported for $Cl_2(PMe_2Ph)_3ReN-GaCl_3$, which shows that the Re–N distance becomes a little longer (1.68 Å) than in the parent compounds (1.660 Å) (68). This is in agreement with the calculated data, which predict at both levels of theory that the Re–N bond of $Cl_2(PH_3)_3ReN-GaCl_3$ is ca. 0.02 Å longer than in the parent compound (Fig. 2). The calculations show that the complexation of $Cl_2(PH_3)_3ReN$ by $GaCl_3$ leads to a significant shortening of the Re–Cl bond trans to the nitrido ligand by ~0.13 Å. The experimentally observed Re–Cl_{trans} bond length of $Cl_2(PMe_2Ph)_3ReN-GaCl_3$ is 0.15 Å shorter than in the parent compound (68). The calculated RE–N–Ga bond angle is 162°, while the experimental value is 168°. A significant difference between theory and experiment is

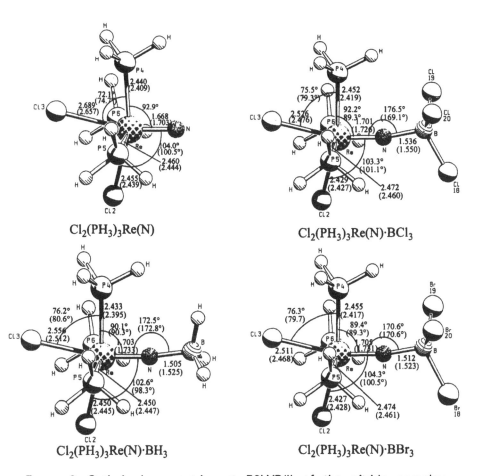

$Cl_2(PH_3)_3Re(N)$

$Cl_2(PH_3)_3Re(N)\cdot BCl_3$

$Cl_2(PH_3)_3Re(N)\cdot BH_3$

$Cl_2(PH_3)_3Re(N)\cdot BBr_3$

FIGURE 2 Optimized geometries at B3LYP/II of the nitrido complex $Cl_2(PH_3)_3ReN$ and the nitrido adducts with various Lewis acids and chalcogen atoms. MP2/II values are shown in parentheses. Bond lengths are given in angstroms, bond angles in degrees. (From Ref. 64.)

found only for the N–Ga bond length. The X-ray structure analysis gives a value of 1.97 Å, which is clearly shorter than the calculated values of 2.055 Å (B3LYP/ II) and 2.080 Å (MP2/II). The disagreement between experiment and theory is probably not caused by an insufficient level of the calculation, but rather by intermolecular forces that lead to a shortening of bonds between Lewis acids and Lewis bases (52). The calculated structures of the N–chalcogen complexes show that the Re–N distances are significantly longer (1.767–1.745 Å) than in the

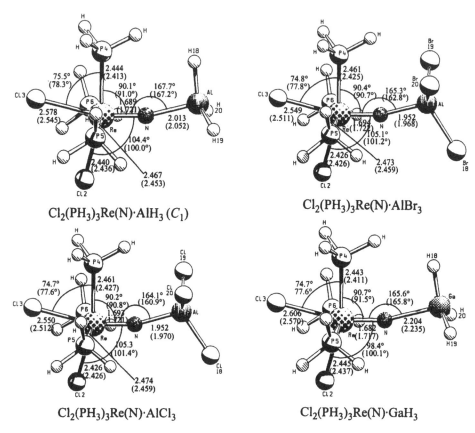

$Cl_2(PH_3)_3Re(N){\cdot}AlH_3$ (C_1)

$Cl_2(PH_3)_3Re(N){\cdot}AlBr_3$

$Cl_2(PH_3)_3Re(N){\cdot}AlCl_3$

$Cl_2(PH_3)_3Re(N){\cdot}GaH_3$

FIGURE 2   Continued

parent compound (1.668 Å, Fig. 2). This is in agreement with the measured Re–N bond lengths for $Cl(PMe_2Ph)_2(Et_2dtc)ReNS$ (1.72(1) and 1.795(9) Å), which are clearly longer than in parent nitrido complexes (69).

While the MP2 and B3LYP geometries of the complexes shown in Figure 2 are quite similar, there are larger differences between the theoretically predicted N–X bond dissociation energies. Table 12 gives the calculated results. The BDEs of the chalcogen complexes, where X = O, S, Se, Te calculated at MP2 and B3LYP, are nearly the same. The MP2 values of the BDEs of the nitrido–group 13 complexes $Cl_2(PH_3)_3ReN–AY_3$ are clearly higher than the B3LYP data, except for the BH_3 complex. MP2 predicts that BH_3 is slightly stronger bonded to the nitrido ligand than BCl_3, while B3LYP strongly favors BH_3 over BCl_3. The MP2

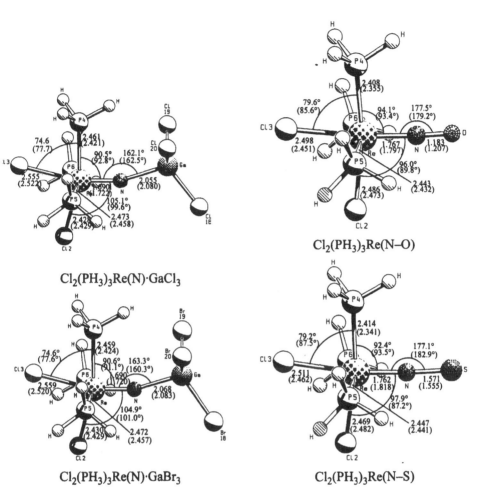

Cl$_2$(PH$_3$)$_3$Re(N–O)

Cl$_2$(PH$_3$)$_3$Re(N)·GaCl$_3$

Cl$_2$(PH$_3$)$_3$Re(N)·GaBr$_3$

Cl$_2$(PH$_3$)$_3$Re(N–S)

FIGURE 2 Continued

result is supported by the CCSD(T) calculations, which also give similar BDEs for the BH$_3$ and BCl$_3$ complexes.

Theoretical and experimental geometries of molybdenum and tungsten phosphido and phosphorous sulfide complexes are shown in Figure 3 (65). The geometry optimizations were carried out at the HF, MP2, and B3LYP levels of theory using our standard basis set II (4). Because the TM has the high oxidation state VI in the molecules it can be expected that the HF geometries should be in reasonable agreement with experiment. The results given in Figure 3 generally confirm the expectation. The HF, MP2, and B3LYP bond lengths and bond angles

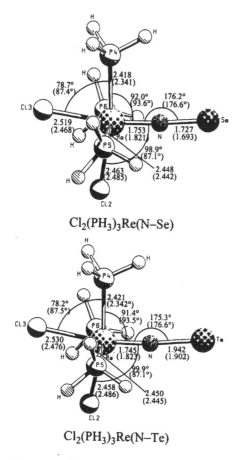

$$Cl_2(PH_3)_3Re(N-Se)$$

$$Cl_2(PH_3)_3Re(N-Te)$$

FIGURE 2 Continued

are in most cases not very different from each other. Please note, however, that the TM $\equiv$ P triple bond length is predicted to be too short at HF and too long at MP2, while the B3LYP value is in good agreement with the experimental values of the two compounds that have been reported (70,71).

Table 13 gives the calculated BDEs at HF, MP2, B3LYP, and CCSD(T) levels of the P–S bonds and the TM–NH$_3$ bonds trans to the TM–P(S) ligand. The CCSD(T) energies were calculated using B3LYP optimized geometries. Since experimental values for the BDEs are not available, the CCSD(T) results may be used as reference data. The B3LYP BDEs of the P–S bonds are 8–10 kcal/mol higher than the CCSD(T) values, but the trend is the same. MP2 and particu-

TABLE 12   Dissociation Energies D_e and Zero-Point Energy Corrected Values D_0 of the Rhenium Nitrido–Bridged Complexes $Cl_2(PH_3)_3ReN-X$ (kcal/mol) with Respect to $Cl_2(PH_3)_3ReN$ and X at the B3LYP and MP2 Levels Using Basis Set II

Complex	B3LYP/II		MP2/II		CCSD(T)/II[a]	
	D_e	D_0	D_e	D_0	D_e	D_0
$Cl_2(PH_3)_3ReN\cdot BH_3$	33.1	30.4	31.9	29.3	31.8	29.2
$Cl_2(PH_3)_3ReN\cdot BCl_3$	23.8	22.4	32.6	31.2	30.2	28.8
$Cl_2(PH_3)_3ReN\cdot BBr_3$	25.3	24.1	36.2	35.0	31.6	30.4
$Cl_2(PH_3)_3ReN\cdot AlH_3$	25.4	23.6	27.1	25.3		
$Cl_2(PH_3)_3ReN\cdot AlCl_3$	36.8	35.7	43.7	42.6		
$Cl_2(PH_3)_3ReN\cdot AlBr_3$	33.9	32.9	42.0	41.0		
$Cl_2(PH_3)_3ReN\cdot GaH_3$	14.9	13.4	18.5	17.0		
$Cl_2(PH_3)_3ReN\cdot GaCl_3$	28.3	26.8	35.0	33.5		
$Cl_2(PH_3)_3ReN\cdot GaBr_3$	23.3	22.0	33.2	31.9		
$Cl_2(PH_3)_3ReN\cdot O$	101.2	98.6	100.2	97.6		
$Cl_2(PH_3)_3ReN\cdot S$	65.1	63.6	66.5	65.0		
$Cl_2(PH_3)_3ReN\cdot Se$	47.2	46.3	47.3	46.4		
$Cl_2(PH_3)_3ReN\cdot Te$	36.3	35.8	35.2	34.7		

[a] Using B3LYP/II optimized geometries.
Source: Ref. 64. For the $Cl_2(PH_3)_3Re(N-X)$ chalcogen complexes the dissociation energies correspond to formation of $Cl_2(PH_3)_3ReN$ and X in its ^3P state.

larly HF give P–S bond energies that are too low. MP2 even gives the wrong trend between the Mo and W complexes. MP2 performs better for the bond energies of the TM–NH$_3$ donor–acceptor bond, where the calculated values exhibit good agreement with the CCSD(T) results. B3LYP gives similar TM–NH$_3$ bond energies to HF, which are clearly too low.

3.4. Reaction Energies of Transition Metal–Catalyzed Processes

Quantum chemical calculations have become important tools for elucidating the reaction mechanisms of TM-catalyzed processes (99). Most of the recent studies have been carried out at the NL-DFT level of theory, which has clearly become the standard method in the field. There are very few studies of reaction profiles of catalytic reactions, however, where the results of DFT methods are compared with data that are predicted by ab initio methods. In the following we will give three examples where NL-DFT methods are compared with CCSD(T) and MP2.

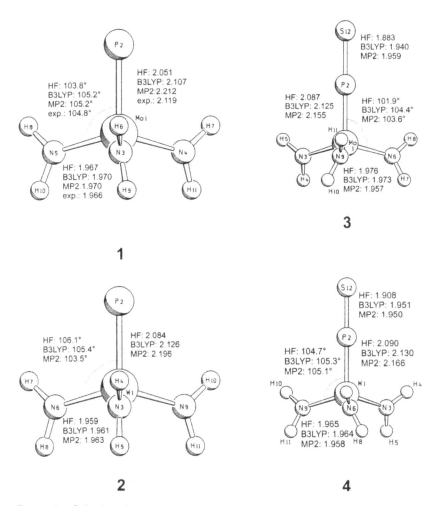

FIGURE 3 Calculated and experimental geometries of phosphido complexes and phosphido-sulfur adducts. The experimental values for **1** and **8** are taken from substituted analogs that have been reported in Refs. 70 and 71, respectively. (From Ref. 65.)

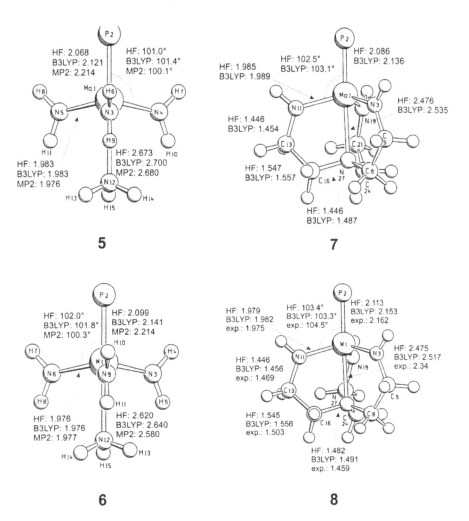

FIGURE 3 Continued

Water–Gas Shift Reaction

A recent theoretical work by Torrent et al. (72) investigated gas-phase reactions of Fe(CO)$_5$ with OH$^-$, which are relevant for the water–gas shift reaction (WGSR):

$$CO + H_2O \rightarrow CO_2 + H_2 \tag{1}$$

The WGSR can be catalyzed by TM compounds, and Fe(CO)$_5$ is a promising candidate for a mononuclear catalyst (73). The mechanism of the reaction is

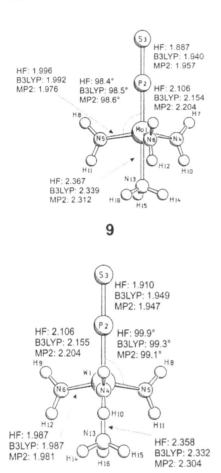

9

10

FIGURE 3   Continued

not precisely known. Sunderlin and Squires (74) recently carried out gas-phase experiments that made it possible to obtain thermochemical data for selected steps of the reaction cycle depicted in Figure 4. The published data are ideally suited for comparison with theoretical results, because the experimental values of the reaction energies and activation barriers refer to processes in the gas phase.

   Torrent et al. optimized the geometries of the intermediates and transition states sketched in Figure 4 at the B3LYP/II level of theory (72). Single-point energies were then calculated at B3LYP and CCSD(T) levels using the larger

TABLE 13 Theoretically Predicted Bond Dissociation Energies D_e (kcal/mol) of the P–S and TM–NH$_3$ Bonds

| Compound | No. | D_e(S) | | | | D_e (NH$_3$) | | | |
		HF	MP2	B3LYP	CCSD(T)[a]	HF	MP2	B3LYP	CCSD(T)[a]
[Mo(PS)(NH$_2$)$_3$]	3	28.2	34.6	57.9	47.6	—	—	—	—
[W(PS)(NH$_2$)$_3$]	4	19.8	37.4	50.3	42.2	—	—	—	—
[Mo(P)(NH$_2$)$_3$(NH$_3$)]	5	—	—	—	—	8.6	10.3	7.2	15.8
[W(P)(NH$_2$)$_3$(NH$_3$)]	6	—	—	—	—	10.5	16.0	8.9	17.4
[Mo(PS)(NH$_2$)$_3$(NH$_3$)]	9	41.2	53.9	71.6	62.9	21.6	29.6	22.8	31.2
[W(PS)(NH$_2$)$_3$(NH$_3$)]	10	33.2	56.0	66.4	58.5	24.0	32.4	24.9	33.7

[a] Using B3LYP/II optimized geometries.
Source: Ref. 65.

basis set II+ +, which has additional diffuse s and p functions on hydrogen, carbon, and oxygen. The diffuse functions turned out to be crucial for the accuracy of the theoretically predicted energies, because some of the calculated species are anions. Table 14 shows the calculated and experimental reaction energies and activation barriers.

The data listed in Table 14 show that the calculated reaction energies at 298 K at the CCSD(T)//II+ + level after zero-point energy (ZPE) and thermal corrections are in excellent agreement with the experimental values. The B3LYP/II+ + values show larger differences from the experimental values, but the *relative* reaction energies correspond with the observed data. Note the dramatic effect of the diffuse functions on the B3LYP reaction energies of reactions 2, 6, and 7. The theoretical values at B3LYP/II+ + for reactions 2 (−61.8 kcal/mol) and 7 (70.3 kcal/mol) are in perfect agreement with experiment, while the B3LYP/II values have errors of ~30 kcal/mol. The very large effect of the diffuse functions on the calculated reaction energies in reactions 2, 6, and 7 arises because the negative charge is constrained on OH⁻ on the adduct (reaction 2) or product (reactions 6 and 7) side of the reaction, while it is delocalized over the iron complex on the other side of the reaction. Note that the theoretically predicted activation barrier for the reaction 12a → 13 + CO$_2$, which is nearly the same at the three levels of theory, is also in reasonable agreement with experiment.

Olefin Addition to Transition Metal Oxides: [2 + 2] Versus [3 + 2] Cycloaddition

The question whether the addition of TM oxides to olefins proceeds as a [3 + 2] cycloaddition yielding a dioxylate or via a [2 + 2] addition forming a metallaoxetane, which then may rearrange to the dioxylate (Scheme 1), has been

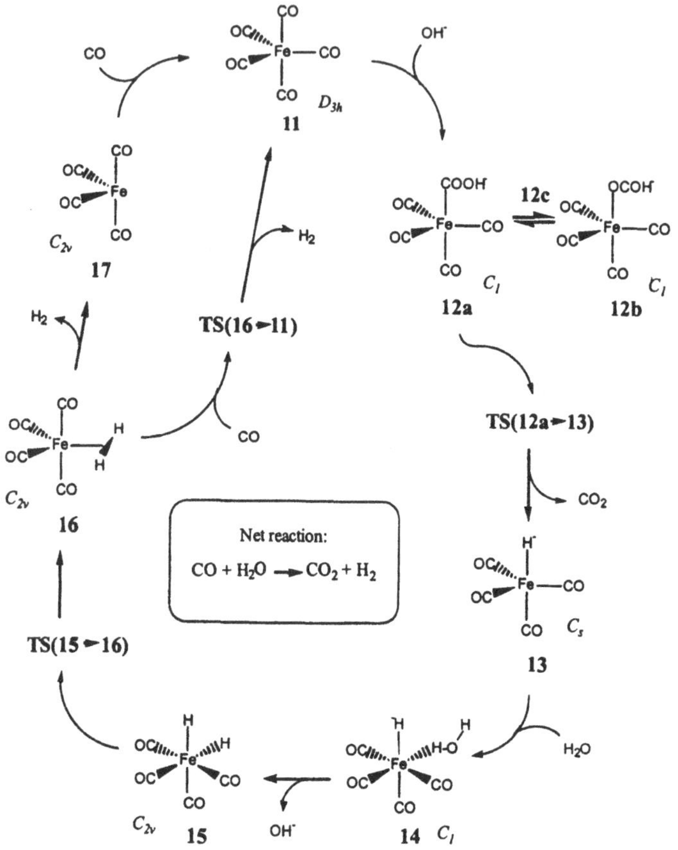

FIGURE 4   Calculated reaction steps in the water-gas shift reaction. (From Ref. 72.)

the topic of several theoretical studies. Most investigations focused on the osmy-lation reaction (74–80). The unequivocal answer to our question was that the barriers for the two-step mechanism are much higher than for the [3 + 2] addition (76–80). A recent theoretical study by Deubel and Frenking also investigated the addition of $LReO_3$ (L = O^-, Cl, Cp) to ethylene (81). The results were compared with theoretical data for the osmylation reaction using different basis sets.

Table 15 shows that the B3LYP energies with basis set III+ are in some cases considerably different from the B3LYP/II values. The largest differences are found for the relative energy of the dioxylate with respect to TM oxide and ethylene. The B3LYP/III+ values for the relative energy of the dioxylate are

TABLE 14 Calculated and Experimental Reaction Energies and Activation Barriers ΔE (in kcal/mol) for Some Reactions Related to the Catalytic Cycle of the WGSR

No.	Reaction	ΔE (B3LYP/II)	ΔE (B3LYP/II++)	ΔE (CCSD(T))/II++)	Experimental
1	$CO + H_2O \rightarrow CO_2 + H_2$	−19.0 (−21.9) *−24.9*	−11.3 (−14.2) *−17.2*	−3.7 (−6.6) *−9.6*	−9.8
2	$Fe(CO)_5(11) + OH^- \rightarrow (CO)_4FeCOOH^- (12a)$	−106.5 (−101.8) *−97.8*	−70.5 (−65.8) *−61.8*	−71.1 (−66.4) *−62.4*	−60.8 ± 3.4
3	$(CO)_4FeCOOH^- (12a) \rightarrow (CO)_4FeOCOH^- (12b)$	−3.4 (−4.4) *−5.0*	−3.4 (−4.4) *−5.0*	—	—
4	$(CO)_4FeCOOH^- (12a) \rightarrow (CO)_4FeH^- (13) + CO_2$	−5.2 (8.6) *−12.5*	−3.9 (−7.3) *−11.2*	6.4 (3.0) *−0.9*	−4 ± 7
4TS	TS $(12a \rightarrow 13 + CO_2)$	33.6 (29.3) *24.8*	33.7 (29.4) *24.9*	33.8 (29.5) *25.0*	18.9 ± 3
5	$(CO)_4FeH^- (13) + H_2O \rightarrow (CO)_4FeH^-H_2O (14)$	−8.5 (−6.9) *−4.8*	−7.0 (−5.4) *−3.3*	—	—
6	$(CO)_4FeH^-H_2O (14) \rightarrow (CO)_4FeH_2 (15) + OH^-$	115.9 (111.8) *106.9*	82.6 (78.5) *73.6*	—	—
7	$(CO)_4FeH^- (13) + H_2O \rightarrow (CO)_4FeH_2 (15) + OH^-$	107.4 (104.9) *102.1*	75.6 (73.1) *70.3*	71.9 (69.4) *66.6*	71.5
8	$(CO)_4FeH_2 (15) \rightarrow (CO)_4FeH_2(16)$	8.4 (8.2) *8.3*	7.9 (7.7) *7.8*	12.0 (11.8) *11.9*	—
9	$(CO)_4FeH_2 (16) \rightarrow Fe(CO)_4(17) + H_2$	17.1 (12.8) *9.7*	17.5 (13.2) *10.1*	24.4 (20.1) *17.0*	—
10	$(CO)_4FeH_2 (15) \rightarrow Fe(CO)_4 (17) + H_2$	25.5 (21.0) *18.0*	25.4 (20.9) *17.9*	36.4 (31.9) *28.9*	26 ± 2
11	$Fe(CO)_4 (17) + CO \rightarrow Fe(CO)_5(11)$	−40.2 (−37.4) *−34.7*	−37.9 (−35.1) *−32.4*	−47.3 (−44.5) *−41.9*	−41.5

Numbering of the species as in Fig. 4. Numbers in parentheses include the ZPE correction computed at B3LYP/II. Numbers in italics include ZPE + thermal corrections computed at B3LYP/II.

Source: Ref. 72.

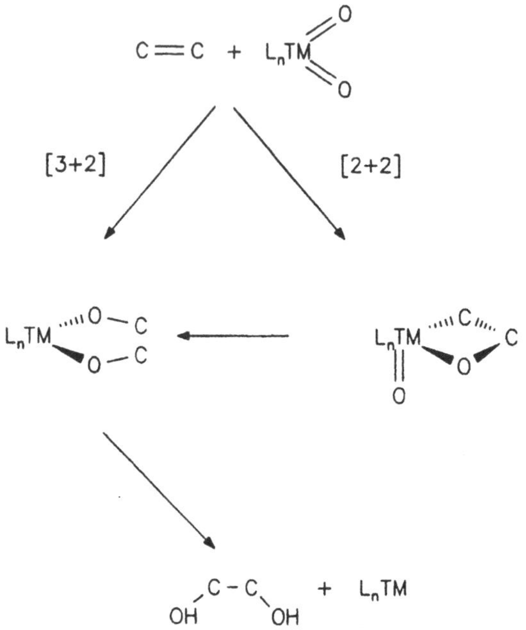

C $=$ C $+$ L$_n$TM (with two O attached)

[3+2]   [2+2]

L$_n$TM (with O—C, O—C) ← L$_n$TM (with C, O, O)

C—C with OH and OH $+$ L$_n$TM

SCHEME 1 Schematic representation of the one-step [3 + 2] addition and the two-step reaction via [2 + 2] addition of transition metal oxides to olefins.

uniformly 13–15 kcal/mol higher than the B3LYP/II data. Note that the formal oxidation state of the metal changes from +8 to +6 for Os and from +7 to +5 for Re when the dioxylate is formed. It is our general experience that the relative energies of TM compounds are particularly sensitive to the theoretical level of the calculation when the oxidation state or the coordination number of the molecule changes. This finding holds for DFT calculations and, to a less extent, also for CCSD(T) results. Table 15 shows that the CCSD(T) values for the relative energy of the osmium dioxylate using basis sets III+ and II differ only by 6 kcal/mol.

Reductive Elimination and Oxidative Addition

Reductive elimination and oxidative addition are ubiquitous reaction steps in many TM-catalyzed processes. A recent study by Beste and Frenking (82) may serve as example for the general finding that relative energies of TM complexes with different coordination numbers may be subject to systematic errors at the DFT level of theory. Table 16 shows calculated energies at the CCSD(T)/II level and at B3LYP using three different basis sets, II–IV, for platinum complexes

TABLE 15 Calculated (B3LYP/II, B3LYP/III+//B3LYP/II, CCSD(T)/II//B3LYP/II, and CCSD(T)/III+//B3LYP/II) Relative Energies E_{rel} (kcal/mol) for the Stationary Points of the [3 + 2] and [2 + 2] Additions of TM Oxides LTMO$_3$ to Ethylene

LTMO$_3$	Method	LTMO$_3$ + C$_2$H$_4$	TS [3 + 2]	Dioxylate	TS [2 + 2]	Oxetane	TS [rear]
OsO$_4$	B3LYP/II[a]	0.0	5.0	−32.3	44.0	5.0	36.2
OsO$_4$	B3LYP/III+//B3LYP/II	0.0	11.8	−19.1	47.9	12.7	45.4
OsO$_4$	CCSD(T)/II//B3LYP/II[a]	0.0	9.6	−21.2	44.7	11.1	41.8
OsO$_4$	CCSD(T)/III+//B3LYP/II	0.0	11.0	−15.2			
ReO$_4^-$	B3LYP/II	0.0	36.0	15.5	46.2	19.1	95.4
ReO$_4^-$	B3LYO/III+//B3LYP/II	0.0	44.8	30.8	50.6	27.2	108.1
ClReO$_3$	B3LYP/II	0.0	21.8	1.4	30.0; 35.2	5.9; 2.2	52.2[c]; 49.8
ClReO$_3$	B3LYP/III+//B3LYP/II	0.0	30.6	14.7	34.5; 39.1	13.1; 9.2	61.8; 60.0
CpReO$_3$	B3LYP/II	0.0	13.6	−21.0	25.4; 33.3	−2.2; −6.3	49.3[c]; 49.6[c]
CpReO$_3$	B3LYP/III+//B3LYP/II	0.0	20.7	−8.7	29.9; 37.8[b]	4.7; 0.9	58.3; 60.0
Cp*ReO$_3$	B3LYP/II	0.0	16.8		41.5[b]; 40.3[b]		
Cp*ReO$_3$	B3LYP/III+//B3LYP/II	0.0	23.0				

[a] In cases where cis and trans isomers are found, the cis isomer is given first and the trans isomer is given second.
[b] From a geometry taken from the analogous Cp system, only internal Cp* coordinates and Re–C(Cp*) distances were optimized.
[c] Instability of the restricted wave function. The unrestricted ansatz leads to an energy correction lower than 0.1 kcal/mol.
Source: Ref. 81.

TABLE 16 Relative Energies (kcal/mol) of Platinum Complexes with Different Coordination Numbers

Compound	No.	Coordination no.	E_{rel}			
			CCSD(T)/II	B3LYP/II	B3LYP/III	B3LYP/IV
cis-[Pt(CH$_3$)(SiH$_3$)(PH$_3$)$_2$] + C$_2$H$_2$	18	4	0.0	0.0	0.0	0.0
[Pt(CH$_3$)(SiH$_3$)(PH$_3$] + C$_2$H$_2$ + PH$_3$	19	3	21.1	14.1	14.4	13.5
[Pt(CH$_3$)(SiH$_3$)(PH$_3$)$_2$(C$_2$H$_2$)]	20	5	1.8	11.1	10.6	14.0
[Pt(CH$_3$)(SiH$_3$)(PH$_3$)(C$_2$H$_2$)PH$_3$	21	4	7.5	6.3	6.3	8.3
[Pt(PH$_3$)(C$_2$H$_2$)] + CH$_3$SiH$_3$ + PH$_3$	22	2	10.9	2.0	1.8	3.1
[Pt(PH$_3$)$_2$(C$_2$H$_2$)] + CH$_3$SiH$_3$	23	3	−16.0	−16.5	−16.9	−14.2

Source: Ref. 82.

with the formal coordination numbers 2–5. The calculations are part of a theoretical study of the reductive elimination of CH_3EH_3 from cis-[$Pt(CH_3)(EH_3)(PH_3)_2$] (E = Si, Ge) (82).

The energy of the four-coordinate complex **18** and free acetylene serves as a reference. Table 16 shows that the relative energy of the three-coordinate complex **19**, which gives the Pt-PH_3 BDE, is higher than that of **18** + C_2H_2. The energy difference at CCSD(T) is 21.1 kcal/mol, but it is only 14.1 kcal/mol at B3LYP/II. The B3LYP value hardly changes when the basis set becomes larger. The five-coordinate complex **20** is predicted at CCSD(T)/II to be only 1.8 kcal/mol higher in energy than **18** + C_2H_2, while B3LYP gives an energy difference of over 10 kcal/mol. Thus, the three-coordinate complex **19** is calculated too low in energy and the five-coordinate complex **20** is predicted too unstable at B3LYP. The relative energies of the four-coordinate complexes **18** and **21** at CCSD(T)/II and B3LYP/II-IV are very similar.

The same trend is found for the relative energies of **22** and **23** with respect to **18**. Complexes **22** and **23** are formally two-coordinate and three-coordinate species, respectively. However, the acetylene ligand may serve as a four-electron donor in electron-deficient species (83). Therefore, **22** should be considered as three-coordinate and **23** as four-coordinate. The relative energies given in Table 16 show that CCSD(T)/II and B3LYP/II-IV give similar values for **23**, while **22** is predicted too stable at the B3LYP level compared with CCSD(T)/II.

Epoxidation of Ethylene with Rheniumperoxide Complexes

Another TM-catalyzed reaction that has been studied at the DFT and ab initio levels of theory is the epoxidation of ethylene with rhenium peroxo complexes. Table 17 shows calculated reaction energies at the MP2, B3LYP, and CCSD(T) levels of theory using basis set II (84).

The energies of reactions 1–3 in Table 17 give the strength of the Lewis acidity of the Re(VII) complexes with respect to water. All methods agree that the water complexation energy of the diperoxo complex is clearly higher than the binding energies of the monoperoxo and trioxo complexes. Reactions 4 and 5 are peroxidation reactions. CCSD(T) and B3LYP agree that the second step of the peroxidation (reaction 5) is energetically less favorable than the first step (reaction 4), while MP2 gives the opposite result. Thus, MP2 may be qualitatively wrong in predicting reaction energies involving high-valent TM compounds. Reactions 6 and 7 are the analogous peroxidation reactions of the rhenium complexes with an additional water ligand. It is interesting that all three methods concur that the second peroxidation step (reaction 7) is thermodynamically more favored than the first reaction (reaction 6). B3LYP energies are not available for the energies of the epoxidation reactions 8 and 9, which are predicted by MP2 and CCSD(T) to be strongly exothermic.

TABLE 17 Reaction Energies (kcal/mol) for the Trioxorhenium-Catalyzed Oxidation of Olefins

Reaction	No.	ΔE_{MP2}	ΔE_{B3LYP}	$\Delta E_{CCSD(T)}$
	1	−11.1 (−8.9)	−9.4	−11.8 (−9.2)
	2	−12.7 (−10.1)	−8.5	−11.8 (−8.9)
	3	−16.0 (−13.2)	−16.3	−21.5 (−18.7)
	4	2.4 (1.7)	1.6	−2.4 (−3.1)
	5	−3.4 (−4.4)	5.2	2.0 (1.0)

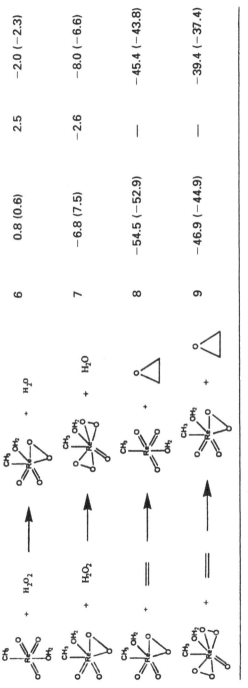

6	0.8 (0.6)	2.5	−2.0 (−2.3)
7	−6.8 (7.5)	−2.6	−8.0 (−6.6)
8	−54.5 (−52.9)	—	−45.4 (−43.8)
9	−46.9 (−44.9)	—	−39.4 (−37.4)

ZPE (MP2) corrected values in parentheses.
Source: Ref. 84.

3.5. Calculation of Nuclear Magnetic Resonance Chemical Shifts of Transition Metal Compounds

There has been a stunning progress in the development of quantum chemical methods for calculating NMR chemical shifts of TM compounds in the last decade. Pioneering work in the field was carried out by Nakatsuji, who used Hartree–Fock/finite perturbation theory (HF/FPT) to predict TM chemical shifts of tetrahedral and octahedral $3d$ and $4d$ TM compounds (85). The ab initio values were generally in reasonable agreement with experimental values. However, the use of a common gauge origin at the metal center and the neglect of correlation energy precluded the extension of the method as a reliable tool for predicting chemical shifts of more complicated compounds and for ligand atoms. A breakthrough towards more accurate theoretical methods for the calculation of NMR parameters of TM compounds came when several groups developed methods for calculating NMR chemical shifts that are based on DFT. Different variants of the uncoupled DFT method with the IGLO (individual gauge for localized orbitals) (86) approach and with the GIAO (gauge including atomic orbitals) (87) method have been suggested (88). The most common methods presently used for TM compounds are the modified sum-over-states (SOS) density-functional perturbation theory (DFPT) IGLO approach by Malkin et al. (89) and the DFT-GIAO work of Schreckenbach and Ziegler (90). Details about the methods can be found in the original publications. The scope and limitations of the methods have recently been reviewed in several papers (88,90,91). A review has also been published that focuses on the performance of DFT methods in predicting NMR parameters of TM compounds (92). The DFT-IGLO and DFT-GIAO methods are commonly used for TM compounds. While it seems that the GIAO approach is perhaps more robust than the IGLO method, it is too early to make a definite statement about the accuracy of the two methods. It is possible to obtain quite accurate results with IGLO and GIAO. The choice of the functionals, the basis set, and the method for calculating relativistic effects seems to be more important.

Although DFT methods for calculating NMR chemical shifts are still rather young, some standards have already been established. This became possible because systematic studies of the accuracy of the methods in the field have been made by Bühl (91,93), by Kaupp (92,94), by Oldfield and coworkers (95), and by Schreckenbach, Ziegler, et al. (90,96). The most important conclusions of some representative papers shall shortly be summarized. We want to point out, however, that the calculation of NMR parameters for TM compounds is not a trivial task and that it is wise to consult an expert prior to running calculations. Also, much progress can be expected in the next couple of years, which may soon outdate the present recommendations. Yet, it is helpful to see what accuracy can already be expected from present standard methods.

The choice of the method for calculating an NMR spectrum depends on whether one wants to know the chemical shifts of the ligand atoms or those of the metal atoms, which are usually more difficult to calculate. We will first focus on chemical shifts of the TMs. It is important to recognize that calculations of TM chemical shifts should not be carried out with ECPs for the metal, because the valence orbitals do not have the correct nodal behavior near the nucleus. This holds even when one is interested only in the relative shifts with respect to a standard compound, i.e., in the δ values of the chemical shifts for which some error cancellation may be expected. Thus, calculations of TM chemical shifts must be carried out with all-electron basis sets. It seems, however, that the effect of relativity on the predicted NMR chemical shifts can be neglected for 3d and perhaps even for 4d elements, but definitely not for the heaviest TM 5d elements. Table 18 shows calculated results of the absolute shielding and the chemical shifts of TM(CO)$_6$ (TM = Cr, Mo, W) at the nonrelativistic (NR) and quasi-relativistic (QR) levels, which includes scalar-relativistic effects but not spin-orbit coupling (95). It becomes obvious that relativity significantly influences the absolute values of the TM shielding. However, the chemical shifts at the NR and QR levels of Cr(CO)$_6$ and Mo(CO)$_6$ are very similar to each other. They are also in good agreement with experiment. The NR value for W(CO)$_6$ clearly deviates from experiment, while the QR value is in much better agreement. We want to point out that the relativistic calculations were carried out with the frozen-core approximation; i.e., only the valence electrons have been treated with the QR approximation in the calculation of the molecules.

A very important point concerns the choice of the exchange and correlations functionals for the NMR calculations. The functionals should, in principle, depend on the current density induced by the magnetic field. Calculations using approximate current-DFT that have been published so far suggest that the current-dependent contributions to the chemical shifts are probably negligible (97). Thus, standard NL-DFT methods that do not depend on the current density are commonly used for the calculation of chemical shifts. No general trend has been established about the accuracy of the various functionals, although the results obtained with different functionals may vary significantly. The only general conclusion that can be made is that local DFT is insufficient for chemical shifts (92).

The most common functionals used in conjunction with the IGLO and GIAO approaches are BPW91, BP86, and B3LYP. A good example that shows the importance of the choice of the functionals is the study by Bühl et al. (91) about the theoretically predicted [57]Fe chemical shifts of various organometallic compounds. Table 19 shows the chemical shifts calculated at different levels of theory. It is obvious that the B3LYP-GIAO results are clearly superior to the SOS-DFPT-BPW91 data using the IGLO approach in conjunction with different

TABLE 18　Calculated Metal Shieldings and Chemical Shifts (ppm) in Transition Metal Carbonyls $TM(CO)_6$ (TM = Cr, Mo, W) and in $[WO_4]^{2-}$ Using DFT-GIAO

Compound	Calculated metal shieldings		Calculated metal NMR chemical shifts[a]		
	Nonrelativistic	Relativistic	Nonrelativistic	Relativistic	Experimental
$Cr(CO)_6$	−509 (−507)	−451 (−449)	−1831 (−1866)	−1812 (−1846)	−1795
$Mo(CO)_6$	1431 (1452)	1704 (1720)	−1805 (−1814)	−1804 (−1804)	−1857
$W(CO)_6$	4900 (4892)	5834 (5767)	−4075 (−4050)	−3703 (−3615)	−3505
$[WO_4]^{2-}$	825 (841)	2131 (2152)	0(0)	0 (0)	0

[a]Shifts are taken relative to the metal oxides $[TMO_4]^{2-}$.
Source: Ref. 96a.

TABLE 19 ^{57}Fe Chemical Shifts Calculated with Various Density Functionals

Molecule	SOS-DFPT-PW91	SOS-DFPT-BP86	SOS-DFPT-DBP86	UDFT-GIAO-B3LYP	Experimental
Fe(CO)$_3$(cyclo-C$_4$H$_4$)	-445	-444	-424	-504	-538
Fe(CO)$_5$	0	0	0	0	0
Fe(CO)$_3$(H$_2$C=CHCH=CH$_2$)	-113	-123	-102	32	4
Fe(CO)$_4$(H$_2$C=CHCN)	102	90	104	210	303
Fe(CO)$_3$(H$_2$C=CHCH=O)	627	629	664	1237	1274
Fe(C$_5$H$_5$)$_2$	156	166	251	1485	1532

Source: Ref. 91.

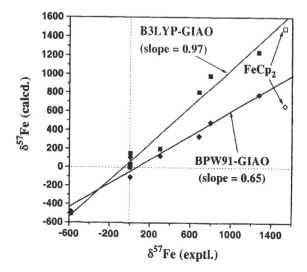

δ^{57}Fe (exptl.)

FIGURE 5 Correlation between calculated and experimental ^{57}Fe chemical shifts (ppm relative to Fe(CO)$_5$) for two different functionals. The regression lines do not include the FeCp$_2$ results. (From Ref. 92.)

TABLE 20 Comparison of Calculated and Experimental ^{13}CO Chemical Shifts (ppm) Relative to TMS for Some First-Row Transition Metal Carbonyl Complexes

Compound	Exp.	ECP	AE-DZVP	AE-ext.
CO	185.3–187.1			175.0
H$_2$CO	185.3			185.9
V(CO)$_6^-$	225.7	216.8 (219.2)	215.5	215.7
Cr(CO)$_6$	211.2–214.6	205.8 (207.7)	201.8	204.0 (205.1)
Mn(CO)$_5$H cis	211.4	209.6 (210.3)	205.4	206.2 (207.4)
Mn(CO)$_5$H trans	210.8	207.5 (207.9)	202.5	203.3 (204.0)
Ni(CO)$_4$	191.6–193.0	197.6 (198.8)	192.3	192.9 (192.2)

Calculated at the SOS-DFPT-IGLO level using ECPs and all-electron basis sets with DZVP quality and extended basis sets.
Source: Ref. 98b.

DFT options. In particular, the calculated IGLO values for ferrocene are dramatically in error by more than 1300 ppm, while the GIAO result concurs with experiment. However, the differences are caused mainly by the functionals used and not by the NMR methods. It has been shown that the GIAO results strongly depend on the functional being used. Figure 5 displays a correlation of experimental and calculated ^{57}Fe chemical shifts predicted using the B3LYP and BPW91 functionals in conjunction with GIAO. The regression lines have slopes of 0.65 for BPW91-GIAO and 0.97 for B3LYP-GIAO. Thus, there is a dramatic improvement of the theoretically predicted ^{57}Fe chemical shifts when the hybrid functional B3LYP is used. A comparison of the two methods for the calculation of ^{103}Rh chemical shifts in various rhodium compounds showed a similar situation, but the slopes of the regression lines were 0.90 at BPW91-GIAO and close to 1 at B3LYP-GIAO (93a).

The calculation of the NMR chemical shifts of ligand atoms of TM complexes is somewhat easier than the metal chemical shifts, because the use of quasi-relativistic small-core ECPs for the metal-core electrons leads to negligible errors for complexes of $3d$ and $4d$ elements. The errors for complexes of the $5d$ elements caused by neglecting spin-orbit effects may become larger, but trends are usually reproduced correctly. Thus, calculations of ^{1}H, ^{13}C, ^{15}N, ^{17}O, ^{31}P, and other chemical shifts of TM ligands can be carried out analogous to molecules composed of main group elements (94b,d,e,95d,96,98). Table 20 shows a comparison of theoretically predicted ^{13}C chemical shifts of carbonyl ligands using ECPs and all-electron (AE) basis sets (98b). The calculations were carried out at the SOS-DFPT-IGLO level. It becomes obvious that the AE basis set does not lead to better results than the ECP method. The same paper reports a large number of calculated ^{13}C chemical shifts of TM carbonyl and cyanide complexes, which demonstrates the good performance of the method (98b).

The quantum chemical methods may be used to predict not only chemical shifts but also the tensor components of the chemical shielding. Table 21 shows the experimental and calculated ^{13}C and ^{17}O shielding tensor components predicted by the NL-DFT-GIAO method of the transition metal hexacarbonyls of Cr, Mo, and W (96b). The experimental values have a rather large error bar, which makes it difficult to estimate the accuracy of the calculated absolute values. However, the trend of the observed data is well reproduced. Unusually large differences between the vector components of the chemical shifts have recently been reported for phosphido complexes (65). Table 22 gives experimental anisotropies and calculated values using GIAO and IGLO of model complexes with smaller substituents. Both methods predict the record-high differences between the tensor components quite well.

TABLE 21 Comparison Between Experimental and Calculated Absolute ^{13}C and [17]O Chemical Shielding Tensor Components (ppm) for the CO Molecule and Group 6 Transition Metal Carbonyls

	CO		Cr(CO)$_6$		Mo(CO)$_6$		W(CO)$_6$	
	[13]C	[17]O	[13]C	[17]O	[13]C	[17]O	[13]C	[17]O
σ_{xx}	−149.4 (−132.3)	−307.3 (−267.6 ± 26(srl))	−174.5 (−167.6 ± 15)	−302.4 (−307.1 ± 10−20)	−169.6 (−157.6 ± 15)	−295.6 (−277 ± 10−20)	−167.9[a] −157.9[b] (−138.6 ± 15)	−291.8[a] −268.7[b] (−259.1 ± 10−20)
σ_{yy}	−149.4 (−132.3)	−307.3 (−267.6 ± 26(srl))	−174.5 (−167.6 ± 15)	−302.4 (−271.1 ± 10−20)	−169.6 (−157.6 ± 15)	−295.6 (−248.1 ± 10−20)	−167.9[a] −157.9[b] (−138.6 ± 15)	−291.8[a] −268.7[b] (−228.1 ± 10−20)
σ_{zz}	273.6 (273.4)	410.6 (408.47 ± 26)	265.8 (255.4 ± 15)	374.3 (401.9 ± 10−20)	267.6 (260.4 ± 15)	362.9 (386.9 ± 10−20)	271.5[a] 267.5[b] (256.4 ± 15)	359.7[a] 351.9[b] (374.9 ± 10−20)
Anisotropy $\Delta\sigma$	423.0 (406(s) ± 1.4)	717.9 (676.1 ± 26(srl))	440.3 (423 ± 30)	676.7 (691 ± 10−20)	437.5 (417 ± 30)	658.5 (650 ± 10−20)	439.4[a] 425.4[b] (395 ± 30)	651.5[a] 620.6[b] (619 ± 10−20)
Isotropic shielding σ	−8.4 (1.0 (sr))	−67.9 (−42.7 ± 17.2)	−27.7 (−26.6 ± 15(s,l))	−76.8 (−59.1 ± 10−20)	−23.9 (−17.6 ± 15)	−76.1 (−46.1 ± 10−20)	−21.4[a] −16.1[b] (−6.6 ± 15(s,l))	−74.6[a] −61.8[b] (−40.1 ± 10−20)

Calculated values at BP86. Experimental values in parentheses. Experimental data are generally solid-state data (s) or, as indicated, liquid (l), gas (g), or a combination of spin-rotation constants plus standard diamagnetic shieldings (sr).
[a] Nonrelativistic NL-DFT calculation.
[b] Relativistic NL-DFT-QR calculation.
Source: Ref. 96b.

TABLE 22 Experimental and Theoretical Anisotropies of the Calculated ^{31}P Chemical Shifts (ppm)

Compound	IGLO			GIAO			Experimental		
	$\delta_\parallel$	$\delta_\perp$	$\delta_\parallel - \delta_\perp$	$\delta_\parallel$	$\delta_\perp$	$\delta_\parallel - \delta_\perp$	$\delta_\parallel$	$\delta_\perp$	$\delta_\parallel - \delta_\perp$
[Mo(P)(NH$_2$)$_3$]	−181.0	1676.6	−1857.6	−144.8	1707.6	−1852.4	−324	1987	−2311
[W(P)(NH$_2$)$_3$]	−211.7	1395.9	−1607.6	−187.9	1402.7	−1590.6	—	—	—
[Mo(PS)(NH$_2$)$_3$]	−292.6	769.8	1062.4	−268.4	711.7	−980.1	—	—	—
[W(PS)(NH$_2$)$_3$]	−267.3	667.1	−934.4	−247.8	613.1	−860.9	—	—	—
[Mo(P)(NH$_2$)$_3$(NH$_3$)]	−204.5	1755.9	−1960.4	−253.7	1948.5	−2202.2	—	—	—
[W(P)(NH$_2$)$_3$(NH$_3$)]	−295.9	1825.8	−2121.7	−240.1	1764.9	−2005.0	—	—	—
[Mo(P)(N$_3$N)]	nc^a	nc	nc	−96.0	1823.2	−1919.2	−267	2125	−2392
[W(P)(N$_3$N)]	nc	nc	nc	−136.3	1495.0	−1631.3	−280	1728	−2008
[Mo(PS)(NH$_2$)$_3$(NH$_3$)]	nc	nc	nc	−402.0	679.4	−1081.4	—	—	—
[W(PS)(NH$_2$)$_3$NH$_3$]	nc	nc	nc	−385.5	598.2	−983.7	—	—	—

^{a} No convergence.
Source: Ref. 65.

4. SUMMARY AND CONCLUSION

The progress that has been made in the development of quantum chemical methods for accurate calculations of TM compounds in the last decade is truly impressive. The account given here is not comprehensive, but the examples presented reflect the already high accuracy of presently available methods. The review focused on methods for calculating geometries, energies, vibrational frequencies, and NMR chemical shifts, because they are probably the most important properties of a molecule. Theoretical methods for other properties are also available but not reviewed here. Standard levels of theory have been established that complement experimental techniques that aim at gaining insight into the structure and reactivity of TM compounds. Chemistry is no longer a purely experimental science. This has already been accepted in recent decades by organic chemists. Inorganic chemistry, particularly TM chemistry, appeared to be a much more difficult challenge. New theoretical methods, particularly NL-DFT techniques, have helped to conquer the field. Method development in TM chemistry has not come to an end yet. In particular, methods for calculating relativistic effects are a field where further progress can be expected in the near future. Theoretical TM chemistry is a booming field with a bright future.

ACKNOWLEDGMENTS

This work was financially supported by the Deutsche Forschungsgemeinschaft and by the Fonds der Chemischen Industrie. The HRZ Marburg, HLRZ Darmstadt, and HLRS Stuttgart provided computer time and excellent service.

REFERENCES

1. A Veillard (ed.). The Challenge of Transition Metals and Coordination Chemistry. NATO ASI Series C, Vol. 176. Dordrecht, The Netherlands: Reidel, 1986.
2. ER Davidson. Chem Rev 91:649, 1991.
3. T Ziegler. Chem Rev 91:651–667, 1991.
4. G Frenking, I Antes, M Boehme, S Dapprich, AW Ehlers, V Jonas, A Neuhaus, M Otto, R Stegmann, A Veldkamp, SF Vyboishchikov. In: KB Lipkowitz, DB Boyd (eds.). Reviews in Computational Chemistry. Vol. 8. New York: VCH, 1996, pp 63–144.
5. TR Cundari, MT Benson, ML Lutz, SO Sommerer. In: KB Lipkowitz, DB Boyd (eds.). Reviews in Computational Chemistry. Vol. 8. New York: VCH, 1996, pp 145–202.
6. J Andzelm, E Radzio, D Salahub. J Chem Phys 83:4573–4580, 1985.
7. TV Russo, RL Martin, PJ Hay. J Phys Chem 99:17085–17087, 1995.
8. MJ Frisch, GW Trucks, HB Schlegel, GE Scuseria, MA Robb, JR Cheeseman, VG Zakrzewski, JA Montgomery, RE Stratmann, JC Burant, S Dapprich, JM Milliam,

AD Daniels, KN Kudin, MC Strain, O Farkas, J Tomasi, V Barone, M Cossi, R Cammi, B Mennucci, C Pomelli, C Adamo, S Clifford, J Ochterski, GA Petersson, PY Ayala, Q Cui, K Morokuma, DK Malick, AD Rabuck, K Raghavachari, JB Foresman, J Cioslowski, JV Ortiz, BB Stefanov, G Liu, A Liashenko, P Piskorz, I Komaromi, R Gomberts, RL Martin, DJ Fox, TA Keith, MA Al-Laham, CY Peng, A Nanayakkara, C Gonzalez, M Challacombe, PMW Gill, BG Johnson, W Chen, MW Wong, JL Andres, M Head-Gordon, ES Replogle, JA Pople. Gaussian 98. Pittsburgh: Gaussian, 1998.

9. MW Schmidt, KK Baldridge, JA Boatz, ST Elbert, MS Gordon, JH Jensen, S Koseki, N Marsunaga, KA Nguyen, S Su, TL Windus, M Dupuis, JA Montgomery Jr. J Comput Chem 14:1347–1363, 1993.

10. (a) H Horn, H Weiss, M Häser, M Ehrig, R Ahlrichs. J Comput Chem 12:1058–1064, 1991. (b) M Häser, J Almlöf, MW Feyereisen. Theor Chim Acta 79:115–122, 1991.

11. RD Amos, IL Alberts, JS Andrews, SM Colwell, NC Handy, D Jayatilaka, PJ Knowles, R Kobayashi, GJ Laming, AM Lee, PE Maslen, CW Murray, P Palmieri, JE Rice, ED Simandiras, AJ Stone, MD Su, DJ Tozer. CADPAC: The Cambridge Analytical Derivatives Package. Cambridge, UK.

12. ACES II, an ab initio program system written by JF Stanton, J Gauss, JD Watts, WJ Lauderdale and RJ Bartlett, University of Florida: Gainesville.

13. HJ Werner, PJ Knowles. Users Manual for MOLPRO, Sussex, UK: University of Sussex.

14. K Andersson, MRA Blomberg, MP Fülscher, G Karlsröm, R Lindh, P-Å Malmquist, P Neogrády, J Olsen, BO Roos, AJ Sadlej, M Schütz, L Seijo, L Serrano-Andrés, PRM Siegbahn, P-O Widmark. MOLCAS. Lund University, Lund, Sweden.

15. TITAN, Wavefunction, Inc., 18401 Von Karman Ave., Ste. 370, Irvine, CA 92612.

16. M Klobukowski. Theor Chim Acta 83:239–248, 1992.

17. B Delley. J Chem Phys 92:508–517, 1990. Dmol is available commercially from BIOSYM Technologies, San Diego, CA.

18. JW Andzelm, E Wimmer. J Chem Phys 96:1280–1303, 1992. DGauss is available as part of the UniChem software package from Cray Research, Eagan, MN.

19. (a) A St-Amant, DR Salahub. Chem Phys Lett 169:387–395, 1990. (b) A St-Amant. Ph.D. dissertation, University of Montreal, 1992. (c) ME Casida, C Daul, A Goursot, A Koester, L Pettersson, E Proynov, A St-Armant, DR Salahub, H Duarte, N. Godbout, J Guan, C Jamorski, M Leboeuf, V Malkin, O Malkina, F Sim A Vela. deMon-KS, version 3.4. deMon Software, 1996.

20. (a) EJ Barends, A Bérces, C Bo, PM Boerrigter, L Cavallo, L Deng, RM Dickson, DE Ellis, L Fan, TH Fischer, C Fonseca Guerra, SJA van Gisbergen, JA Groeneveld, OV Gritsenko, FE Harris, P van den Hoek, H Jacobsen, G van Kessel, F Kootstra, E van Lenthe, VP Osinga, PHT Philipsen, D Post, C Pye, W Ravenek, P Ros, PRT Schipper, G Schreckenbach, JG Snijders, M Sola, D Swerhone, G te Velde, P Vernooijs, L Versluis, O Visser, E van Wezenbek, G Wiesenekker, SK Wolff, TK Woo, T Ziegler. ADF 1999. (b) C Fonseca Guerra, JG Snijders, G te Velde, EJ Baerends. Theor Chem Acc 1998, 99:391–403.

21. A Berces, T Ziegler. In: RF Nalewajski (ed.). Topics in Current Chemistry. Vol. 182. Berlin: Springer Verlag, 1996, pp 41–85.

22. W Koch, RH Hertwig. In: PvR Schleyer, NL Allinger, T Clark, PA Kollman, HF Schaefer III, PR Scheiner (eds.) Encyclopedia of Computational Chemistry. Vol. 1. Chichester, UK: Wiley, VCH, 1998, pp 689–700.

23. A Ricca, CW Bauschlicher. Theor Chim Acta 92:123–131, 1995.

24. Representative examples: (a) MN Glukhotsev, RD Bach, CJ Nagel. J Phys Chem A 101:316–323, 1997. (b) I Bytheway, GB Bacskay, NS Hush. J Phys Chem 100: 6023–6031, 1994. (c) EA McCullough Jr, E Aprà, J Nichols. J Phys Chem A 101: 2502–2508, 1997. (d) TV Russo, RL Martin, PJ Hay. J Chem Phys 102:8023–8082, 1995. (e) A Rosa, AW Ehlers, EJ Baerends, JG Snijders, G te Velde. J Phys Chem 100:5690–5696, 1996. (f) MC Holthausen, C Heinemann, HH Cornehl, W Koch, H Schwarz. J Chem Phys 102:4931–4941, 1995. (g) B Delley, M Wrinn, HP Lüthi. J Chem Phys 100:5785–5791, 1994.

25. AD Becke. Phys Rev A 38:3098–3106, 1988.

26. JP Perdew. Phys Rev B 33:8822–8827, 1986.

27. C Lee, W Yang, RG Parr. Phys Rev B 37:785–792, 1988.

28. JP Perdew, Y Wang. Phys Rev B 45:13244–13254, 1992.

29. SH Vosko, L Wilk, M Nuisar. Can J Phys 58:1200–1206, 1980.

30. (a) AD Becke. J Chem Phys 98:1372–1377, 1993. (b) AD Becke. J Chem Phys 98:5648–5652, 1993.

31. PJ Stephens, FJ Devlin, CF Chabalowski, MJ Frisch. J Phys Chem 98:11623–11627, 1994.

32. (a) V Jonas, W Thiel. Organometallics 17:353–360, 1998. (b) V Jonas, W Thiel. J Chem Phys 102:8474–8484, 1995. (c) V Jonas, W Thiel. J Chem Phys 105:3636–3646, 1996. (d) V Jonas, W Thiel. J Phys Chem A 103:1381–1393, 1999.

33. FA Hamprecht, AJ Cohen, DJ Tozer, NC Handy. J Chem Phys 109:6264–6271, 1998.

34. AD Becke. J Chem Phys 109:2092–2098, 1998.

35. AD Becke. J Comput Chem 20:63–69, 1999.

36. (a) CR Landis, TK Firman, DM Root, T Cleveland. J Am Chem Soc 120:1842–1854, 1998. (b) CR Landis, T Cleveland, TK Firman. J Am Chem Soc 120:2641–2649, 1998. (c) TK Firman, CR Landis. J Am Chem Soc 120:12650–12656, 1998. (d) CA Bayse, MB Hall. J Am Chem Soc 121:1348–1358, 1999.

37. (a) V Jonas, G Frenking, MT Reetz. J Comput Chem 13:919–934, 1992. (b) M Couty, MB Hall. J Comput Chem 17:1359–1370, 1996.

38. PJ Hay, WR Wadt. J Chem Phys 82:299–310, 1985.

39. WJ Stevens, M Krauss, H Basch, G Jasien. Can J Chem 70:612–630, 1992.

40. (a) M Dolg, U Wedig, H Stoll, H Preuss. J Chem Phys 86:866–872, 1987. (b) D Andrae, U Häussermann, M Dolg, H Stoll, H Preuss. Theor Chim Acta 77:123–141, 1990.

41. (a) LF Pacios, PA Christiansen. J Chem Phys 82:2664–2671, 1985. (b) MM Hurley, LF Pacios, PA Christiansen, RB Ross, WC Ermler. J Chem Phys 84:6840–6853, 1986. (c) LA LaJohn, PA Christiansen, RB Ross, T Atashroo, WC Ermler. J Chem Phys 87:2812–2824, 1987. (d) RB Ross, JM Powers, T Atashroo, WC Ermler, LA LaJohn, PA Christiansen. J Chem Phys 93:6654–6670, 1990.

42. (a) L Seijo, Z Barandiarán, S Huzinaga. J Chem Phys 91:7011–7017, 1989. (b) Z Barandiarán, L Seijo, S Huzinaga. J Chem Phys 93:5843–5850, 1990.

43. (a) P Pyykkö. Chem Rev 88:563–594, 1988. (b) J Almlöf, O Gropen. Rev Comput Chem 8:203–244, 1996. (c) P Schwerdtfeger, M Seth. In: PvR Schleyer, NL Allinger, T Clark, PA Kollman, HF Schaefer III, PR Scheiner (eds.). Encyclopedia of Computational Chemistry. Vol. 4. Chichester, UK: Wiley-VCH, 1998, pp 2480–2499.
44. (a) LA Barnes, M Rosi, CW Bauschlicher. J Chem Phys 93:609–624, 1990. (b) I Antes, G Frenking. Organometallics 14:4263–4268, 1995.
45. K Balasubramanian. In: PvR Schleyer, NL Allinger, T Clark, PA Kollman, HF Schaefer III, PR Scheiner (eds.). Encyclopedia of Computational Chemistry. Vol. 4. Chichester, UK: Wiley-VCH, 1998, pp 2471–2480.
46. B Hess. In: PvR Schleyer, NL Allinger, T Clark, PA Kollman, HF Schaefer III, PR Scheiner (eds.). Encyclopedia of Computational Chemistry. Vol. 4. Chichester, UK: Wiley-VCH, 1998, pp 2499–2508.
47. C van Wüllen. J Comput Chem 20:51–62, 1999.
48. E van Lenthe, JG Snijders, EJ Baerends. J Chem Phys 105:6505–6516, 1996.
49. S Dapprich, U Pidun, AW Ehlers, G Frenking. Chem Phys Lett 242:521–526, 1995.
50. (a) J Li, G Schreckenbach, T Ziegler. J Phys Chem 98:4838–4841, 1994. (b) J Li, G Schreckenbach, T Ziegler. J Am Chem Soc 117:486–494, 1995.
51. RK Szilagyi, G Frenking. Organometallics 16:4807–4815, 1997.
52. V Jonas, G Frenking, MT Reetz. J Am Chem Soc 116:8741–8753, 1994.
53. AJ Lupinetti, V Jonas, W Thiel, SH Strauss, G Frenking. Chem Eur J 5:2573–2583, 1999.
54. (a) E van Lenthe, A Ehlers, EJ Baerends. J Chem Phys 110:8943–8953, 1999. (b) A Rosa, EJ Baerends, SJA van Gisbergen, E van Lenthe, JA Groeneveld, JG Snijders. J Am Chem Soc 121:10356–10365, 1999.
55. (a) K Wichman. Diplom Thesis, University of Marburg, 1999. (b) K Wichmann, G Frenking. In press.
56. (a) AW Ehlers, S Dapprich, SF Vyboishchikov, G Frenking. Organometallics 15: 105–117, 1996. (b) AW Ehlers. PhD dissertation, University of Marburg, 1994.
57. C van Wüllen. J Comput Chem 18:1985–1992, 1997.
58. AW Ehlers, G Frenking, EJ Baerends. Organometallics 16:4896–4902, 1997.
59. C van Wüllen. J Chem Phys 103:3589–3599, 1995.
60. (a) T Ziegler, JG Snijders, EJ Baerends. J Chem Phys 74:1271–1284, 1981. (b) T Ziegler, V Tschinke, EJ Baerends, JG Snijders, W Ravenek. J Phys Chem 93:3050–3055, 1989.
61. (a) C Chang, M Pelissier, Ph Durand. Phys Scr 34:394–407, 1986. (b) JL Heully, I Lindgren, E Lindroth, S Lundquist, AM Martensson-Pendrill. J Phys B 19:2799–2808, 1986. (c) E van Lenthe, EJ Baerends, JG Snijders. J Chem Phys 99:4597–4610, 1993.
62. RA Fischer, J Weiss. Angew Chem 111:3002–3022, 1999; Angew Chem Int Ed Engl 38:2830–2850, 1999.
63. C Boehme, J Uddin, G Frenking. Coord Chem Rev 197:249–276, 2000.
64. SF Vyboichshikov, G Frenking. Theor Chem Acc 102:300–310, 1999.
65. T Wagener, G Frenking. Inorg Chem 37:1805–1811, 1998.
66. WJ Hehre, L Radom, PvR Schleyer, JA Pople. Ab Initio Molecular Orbital Theory. New York: Wiley, 1986.

67. E Forsellini, U Casellato, R Graziani, L Magon. Acta Cryst B 38:3081–3083, 1982.
68. S Ritter, R Hübener, U Abram. J Chem Soc Chem Commun: 2047–2048, 1995.
69. S Ritter, U Abram. Z Anorg Allg Chem 622:965–973, 1996.
70. CE Laplaza, WM Davis, CC Cummins. Angew Chem 107:2181–2183, 1995; Angew Chem Int Ed Engl 34:2042–2044, 1995.
71. NC Zanetti, RR Schrock, WM Davis. Angew Chem 107:2184–2186, 1995; Angew Chem Int Ed Engl 34:2044–2046, 1995.
72. M Torrent, M Solà, G Frenking. Organometallics 18:2801–2812, 1999.
73. DC Gross, PC Ford. Inorg Chem 21:1704–1706, 1982.
74. LS Sunderlin, RR Squires. J Am Chem Soc 115:337–343, 1993.
75. (a) PO Norrby, HC Kolb, KB Sharpless. Organometallics 13:344–347, 1994. (b) PO Norrby, HC Kolb, KB Sharpless. J Am Chem Soc 116:8470–8478, 1994.
76. A Veldkamp, G Frenking. J Am Chem Soc 116:4937–4946, 1994.
77. U Pidun, C Boehme, G Frenking. Angew Chem 108:3008–3011, 1996; Angew Chem Int Ed Engl 35:2817–2820, 1996.
78. S Dapprich, G Ujaque, F Maseras, A Lledós, DG Musaev, K Morokuma. J Am Chem Soc 118:11660–11661, 1996.
79. M Torrent, L Deng, M Duran, M Sola, T Ziegler. Organometallics 16:13–19, 1997.
80. AJ Del Monte, J Haller, KN Houk, KB Sharpless, DA Singleton, T Strassner, AA Thomas. J Am Chem Soc. 119:9907–9908, 1997.
81. DV Deubel, G Frenking. J Am Chem Soc 121:2021–2031, 1999.
82. (a) A Beste, G Frenking. Z Allg Anorg Chem 626:381–391, 2000. (b) A Beste. Diplom thesis, University of Marburg, 1999.
83. SA Decker, M Klobukowski. J Am Chem Soc 120:9342–9355, 1998.
84. T Wagener. PhD dissertation, University of Marburg, 1998.
85. (a) H Nakatsuji. Chem Phys Lett 167:571–574, 1990. (b) H Nakatsuji, T Inoue, T Nakao. J Phys Chem 96:7953–7958, 1992. (c) M Sugimoto, M Kanayama, H Nakatsuji. J Phys Chem 96:4375–4381, 1992. (d) H Nakatsuji. In: JA Tossell (ed.). Nuclear Magnetic Shieldings and Molecular Structure. Dordrecht, The Netherlands: Kluwer Academic, 1993, pp 263–278.
86. (a) W Kutzelnigg. Isr J Chem 19:193–200, 1980. (b) M Schindler, W Kutzelnigg. J Chem Phys 76:1919–1933, 1982. (c) W Kutzelnigg, U Fleischer, M Schindler. NMR: Basic Princ Prog 23:165–262, 1990.
87. (a) F London. J Phys Radium 8:397–409, 1937. (b) RM Stevens, RM Pitzer, WN Lipscomb. J Chem Phys 38:550–560, 1963. (c) R Ditchfield. J Chem Phys 56:5688–5691, 1972. (d) H Hameka. Mol Phys 1:203–215, 1958.
88. U Fleischer, C van Wüllen, W Kutzelnigg. In: PvR Schleyer, NL Allinger, T Clark, PA Kollman, HF Schaefer III, PR Scheiner (eds.). Encyclopedia of Computational Chemistry. Vol. 3. Chichester, UK: Wiley-VCH, 1998, pp 1827–1835.
89. VG Malkin, OL Malkina, ME Casida, DR Salahub. J Am Chem Soc 116:5898–5908, 1994.
90. G Schreckenbach, T Ziegler. Theor Chem Acc 99:71–82, 1998.
91. M Bühl, M Kaupp, OL Malkina, VG Malkin. J Comput Chem 20:91–105, 1999.
92. M Kaupp, VG Malkin, OL Malkina. In: PvR Schleyer, NL Allinger, T Clark, PA Kollman, HF Schaefer III, PR Scheiner (eds.). Encyclopedia of Computational Chemistry. Vol. 3. Chichester, UK: Wiley-VCH, 1998, pp 1857–1866.

93. (a) M Bühl. Chem Phys Lett 267:251–257, 1997. (b) M Bühl. Organometallics 16: 261–267, 1997. (c) M Bühl, OL Malkina, VG Malkin. Helv Chim Acta 79:742– 754, 1996. (d) M Bühl, FA Hamprecht. J Comput Chem 19:113–122, 1998. (e) M Bühl. Angew Chem 110:153–155, 1998; Angew Chem Int Ed Engl 37:142–144, 1998.

94. (a) M Kaupp. Chem Ber 129:527–533, 1996. (b) M Kaupp. Chem Eur J 2:348– 358, 1996. (c) M Kaupp. Chem Ber 129:535–544, 1996. (d) M Kaupp. J Chem Soc Chem Commun 1141–1142, 1996. (e) M Kaupp. J Am Chem Soc 118:3018– 3024, 1996.

95. (a) N Godbout, R Havlin, R Salzmann, PG Debrunner, E Oldfield. J Phys Chem A 102:2342–2350, 1998. (b) MC McMahon, AC deDios, N Godbout, R Salzmann, DD Laws, H Le, RH Havlin, E Oldfield. J Am Chem Soc 120:4784–4797, 1998. (c) N Godbout, E Oldfield. J Am Chem Soc 109:8065–8069, 1997. (d) R Havlin, M McMahon, R Srinivasan, H Le, E Oldfield. J Phys Chem A 101:8908–8913, 1997.

96. (a) G Schreckenbach, T Ziegler. Int J Quantum Chem 61:899–907, 1997. (b) Y Ruiz-Morales, G Schreckenbach, T Ziegler. J Phys Chem 100:3359–3367, 1996. (c) Y Ruiz-Morales, G Schreckenbach, T Ziegler. Organometallics 15:3920–3923, 1996. (d) AW Ehlers, Y Ruiz-Morales, EJ Barends, T Ziegler. Inorg Chem 36: 5031–5036, 1997.

97. AM Lee, NC Handy, SM Colwell. J Chem Phys 103:10095–10109, 1995.

98. (a) M Kaupp, OL Malkina, VG Malkin. J Chem Phys 106:9201–9212, 1997. (b) M Kaupp, VG Malkin, OL Malkina, DR Salahub. Chem Phys Lett 235:382–388, 1995. (c) M Kaupp, VG Malkin, OL Malkina, DR Salahub. Chem Eur J 2:24–30, 1996. (d) R Salzmann, M Kaupp, M McMahon, E Oldfield. J Am Chem Soc 120: 4771–4783, 1998. (e) GM Bernard, G Wu, RE Wasyleshen. J Phys Chem A 102: 3184–3192, 1998. (f) Y Ruiz-Morales, T Ziegler. J Phys Chem A 102:3970–3976, 1998.

99. (a) PWNM van Leeuwen, K Morokuma, JH van Lenthe (eds.). Theoretical Aspects of Homogeneous Catalysis. Dordrecht, The Netherlands: Kluwer Academic, 1995. (b) DG Truhlar, K Morokuma (eds.). Transition State Modeling for Catalysis. ACS Symposium Series 721. Washington, DC: American Chemical Society, 1999. (c) M Solá, M Torrent, G Frenking. Chem Rev 100:439–493, 2000.

100. AW Ehlers, G Frenking. J Am Chem Soc 116:1514–1520, 1994.

101. LA Barnes, B Liu, R Lindh. J Chem Phys 98:3978–3989, 1993.

102. J Li, G Schreckenbach, T Ziegler. J Phys Chem 98:4838–4841, 1994.

103. KE Lewis, DM Golden, GP Smith. J Am Chem Soc 106:3905–3912, 1984.

104. M Diedenhofen, G. Frenking. Unpublished results.

5

Nondynamic Correlation Effects in Transition Metal Coordination Compounds

Kristine Pierloot
Catholic University of Leuven, Leuven, Belgium

1. INTRODUCTION

Molecules containing transition metals are traditionally considered to be difficult to treat by ab initio methods. Apart from the fact that these systems are often so large that their correlation problem involves many electrons (leading, for example, to problems with size extensivity), it was realized as soon as the first applications began to appear that the occurrence of open shells, together with occasionally strong near-degeneracies, made it impossible to obtain the same kind of accuracy for first-row transition metal systems as could be obtained for small organic molecules using rather simple (e.g., Møller–Plesset perturbation theory) correlation methods. An obvious way to treat open shells and near-degeneracies is to use multireference methods. Scientists involved in the development of such methods have often shown a strong interest in transition metal atoms and molecules as a crucial test for their methods (1), whereas, on the other hand, transition metal chemistry is one of the domains where multireference methods have become most popular as a computational tool (1–9).

However, it is probably also true that a widespread use of multireference methods in this domain and other domains of chemistry has been hampered by the additional intricacies connected to using these methods, i.e., the construction of the appropriate reference wavefunction to be used as a starting point for the treatment of dynamic correlation. There are not, and cannot be, any general rules for constructing such a wavefunction, nor can such rules be implemented in any computer code. The only constant factor is that the reference wavefunction should include all important nondynamic correlation effects. What precisely these effects incorporate is dependent on the specific electronic structure of the molecule to be treated. In other words, the construction of an appropriate reference wavefunction requires at least some a priori knowledge of the answer to be obtained from the calculation. It is therefore often based on trial and error or, in the case of a more experienced user in the field, on "chemical intuition." By making use of the CASSCF (complete active-space self-consistent field) method for constructing the reference wavefunction, the problem can be reduced to selecting a set of active orbitals (10). However, since the number of configurations, and hence the computational effort, increases very rapidly with the size of the active space, a certain skill is still required to define the active space in such a way that all important nondynamic correlation effects are included while at the same time keeping the number of active orbitals to a minimum (with a maximum of 12–14 orbitals, depending on the symmetry of the molecule).

In this chapter, we will try to formulate some general guidelines for treating nondynamic correlation in molecules containing transition metals. The way we will do this is by looking for connections between the appearance of such correlation effects and the specific molecular electronic structure arising from certain metal–ligand combinations, and from there trying to provide some trends. Before starting it should be emphasized, however, that the picture given in the rest of this chapter will by no means be complete. For one thing, we will confine the discussion to systems containing only one transition metal atom or ion, so, for example, the treatment of magnetic interactions between different centers will not be considered. Furthermore, most of the chapter will be devoted to "large" transition metal systems, i.e., molecules containing a transition metal surrounded by at least four ligands. A crucial distinction between the correlation effects appearing in these large systems and the smaller molecules built from only one or two ligands is connected to their ground state electronic structure: the first one or two ligands that bind to a transition metal atom will find the latter in a hybridized state composed of a mixture of the configurations $d^n s^2$, $d^{n+1} s^1$ and d^{n+2}, with a composition that may be strongly dependent on the metal–ligand distance. The description of such systems therefore requires an accurate treatment of differential correlation effects connected to states with such varying configurations. As more ligands surround the metal, the $(n + 1)s$ [and $(n + 1)p$] orbitals are pushed upward in energy (see also Sec. 3) so that the important correlation effects in these larger systems are confined to the nd electrons and their interaction with the ligand environment. For obvious reasons, the correlation problem in smaller transition systems was the first to be recognized and has already been discussed on several occasions (2,5,11–12). We will touch on the problem shortly in Section 2. The rest of the chapter will be devoted to the discussion of nondynamic correlation effects connected to the interaction between the metal nd shell and its environment.

Another introductory remark concerns the distinction between nondynamic and dynamic correlation effects. Starting from the restricted Hartree–Fock (RHF) solution, *dynamic correlation* is defined as the energy lowering due to correlating the motion of the electrons. On the other hand, *nondynamic*, or *static, correlation* is the energy lowering obtained when adding additional flexibility to the (RHF) wavefunction to describe near-degeneracy effects (two or more configurations having almost the same energy). In other words, by dealing with nondynamic correlation, an improved starting point for treating dynamic correlation can be constructed in cases where RHF fails. However, clearly the "failure" of RHF or also the definition of "near"-degeneracy is to some extent dependent on the elaboration of the method used for treating dynamic correlation. For instance, ferrocene, a typical organometallic system, has been found in the past to be hard to treat using rather simple single-reference methods (13,14), e.g. MP2 (Møeller–Plesset second-order perturbation theory) or MCPF (modified coupled-pair functional). It was therefore concluded that nondynamic correlation effects are very important in this molecule. And indeed, accurate results for the bonding may be obtained by using instead second-order perturbation theory based on a CASSCF wavefunction (14), i.e., the so-called CASPT2 method (15). However, it was also shown (16) that a similar accuracy may be reached by using instead the single-reference CCSD(T) approach (coupled-cluster singles-and-doubles with a perturbative correction for triples). In this chapter we will consider nondynamic correlation effects in a broad sense; e.g., all near-degeneracy effects that are too strong to be handled efficiently by second-order perturbation theory, such as those appearing in ferrocene, will be considered as nondynamic correlation effects. This option is based on our experience with the CASSCF/CASPT2 method (1). Indeed, the success of this method is on the one hand critically dependent on whether or not all important correlation effects are included in the CASSCF reference wavefunction, but is on the other hand guaranteed in many large transition metal systems by the ability of the method to combine very extended CASSCF reference wavefunctions (containing up to 1 million or more determinants) with a large number of correlated electrons.

2. THE ATOMIC CASE: THE $3d$ DOUBLE-SHELL EFFECT

One of the most important correlation effects in transition metal systems is the so-called $3d$ double-shell effect. This correlation effect appears in particular in first-row transition metal atoms or ions with a more-than-half-filled $3d$ shell, and is related to the presence of a large number of electrons in a compact $3d$ shell, giving rise to very large $3d$ radial correlation effects. But of course such an effect should be classified as dynamic rather than nondynamic, and one may therefore wonder whether its description really belongs in this chapter. It does, because this is one of the exceptional cases where an accurate description of dynamic correlation effects really benefits from a reoptimization of the orbitals involved, i.e., from a multireference treatment. The effect was first noted in a multiconfigurational Hartree–Fock calculation on the copper atom (17), where it was found that a large fraction of the electron correlation in the $3d^{10}4s$ state could be described by including the electronic configuration $3d^93d'4s$, with $3d'$ a more diffuse shell than $3d$. A more general description of the effect in first-row transition metals was given by Dunning et al. (18,19). The first quantitative calculations, showing that the inclusion of a second d shell in a multireference treatment is indeed a prerequisite to obtaining accurate results, were performed on the relative energies of the $3d^84s^2$, $3d^94s$, and $3d^{10}$ states in the Ni atom, using either the MRCI (20) or CASPT2 (21) approach.

In this section we will illustrate the occurrence of the double-shell effect by a set of CASSCF/CASPT2 calculations on the ground and lowest excited states of the monopositive ions Ti^+, Co^+, and Rh^+. The motivation for this choice is twofold: (1) Since monopositive ions are characterized by low-lying states belonging to either the configurations d^{n+1} or $d^n s^1$, calculations of their spectra allow for an investigation of the double-shell effect on the energy of $d \rightarrow d$ as well as $d \rightarrow s$ excitations. (2) The specific choice of transition metal makes it possible to compare the effect of an increasing number of $3d$ electrons (Ti^+ versus Co^+) as well as an increasing main quantum number (Co^+ versus Rh^+). The double-shell effect is investigated by comparing the results obtained from a CASSCF calculation with an active space including only the Ti, Co $3d$, $4s$ orbitals or Rh $4d$, $5s$ orbitals [denoted as CAS(6)] to a calculation where this active space is extended with a second d shell [denoted as CAS(11)]. Relativistic effects (which are quite important for the $4d \rightarrow 5s$ transitions in Rh^+) were accounted for by performing the calculations using the relativistic core–AIMP (ab initio model potential) of Barandiaran (22). These potentials were used in combination with the corresponding valence basis sets with contraction $[3s3p4d]$ and further enhanced with one f-type function. Given the moderate size of these basis sets, it is certainly not our intention to present quantitative results for the considered excitation energies. It is known from the calculations on the Ni atom (20) that (apart from a multireference treatment) an accurate description of radial correlation effects in the $3d$ shell would require much larger basis sets, including angular momentum functions up to g and higher. The results of our calculations are presented in Table 1. Both Co^+ and Rh^+ are characterized by an ^3F ground state, corresponding to a d^8 configuration, and three $d \rightarrow d$ transitions (to ^1D, ^3P, ^1G)

TABLE 1 Calculated CASSCF and CASPT2 Excitation Energies (eV) of the Lowest Excited States in Ti^+, Co^+, and Rh^+, Using an Active Space Consisting of Either Six Orbitals [$3d$, $4s$ or $4d$, $5s$, Denoted as CAS(6)] or 11 Orbitals [$3d$, $3d'$, $4s$ or $4d$, $4d'$, $5s$, Denoted as CAS(11)]

	CAS (6)		CAS (11)		
State	CASSCF	CASPT2	CASSCF	CASPT2	Experimental
		Ti^+			
^4F ($3d^3$)	0.65	0.22	0.32	0.34	0.09
^2F ($3d^24s$)	0.68	0.66	0.77	0.56	0.56
		Co^+			
^5F ($3d^74s$)	−1.37	0.71	−0.04	0.15	0.43
^3F ($3d^74s$)	−0.49	1.53	0.84	0.94	1.21
^1D ($3d^8$)	1.68	1.24	1.63	1.42	1.36
^3P ($3d^8$)	2.00	1.46	1.85	1.57	1.61
^5P ($3d^74s$)	0.90	2.35	2.06	1.86	2.14
^1G ($3d^8$)	2.69	2.16	2.57	2.35	2.29
		Rh^+			
^1D ($4d^8$)	1.26	1.01	1.25	1.06	0.85
^3P ($4d^8$)	1.38	1.06	1.29	1.08	1.22
^1G ($4d^8$)	2.00	1.84	1.96	1.75	1.68
^5F ($4d^75s$)	1.61	2.14	2.08	2.05	2.17
^3F ($4d^75s$)	2.72	3.16	3.17	3.14	3.33
^5P ($4d^75s$)	3.26	3.39	3.62	3.38	3.49

Source: Taken from Ref. 58 for Ti^+, Ref. 59 for Co^+, and Ref. 60 for Rh^+. The values were obtained as the weighted average of all J levels corresponding to each LS state.

as well as three $d \rightarrow s$ transitions (to ^5F, ^3F, ^5P) have been included in the calculations. On the other hand, Ti^+ has an ^4F ground state corresponding to $3d^24s^1$; here, we only consider the lowest $s \rightarrow d$ (to ^4F) and $d \rightarrow d$ (to ^2F) transition.

The importance of radial correlation effects within the d shell is already obvious from the CASSCF results obtained with the smallest active space [CAS(6)], where any description of such effects is lacking. Because the importance of these correlation effects strongly increases with the number of d electrons present, neglecting them leads to a preferential stabilization of states corresponding to the $d^n s^1$ rather than the d^{n+1} configuration. Thus, at the CAS(6) level we find both the ^3F, ^5F ($3d^74s^1$) states at a considerably lower energy than the ^3F (d^8) ground state in Co^+, while the ^5P state is also calculated more than 1.2 eV too low. The corresponding errors are considerably smaller, up to 0.6 eV, for Rh^+, indicating that correlation effects are much less important within the (less compact), $4d$ shell. In Ti^+ the splitting between the ^4F states corresponding to

$3d^3$ and $3d^24s^1$ is overestimated by 0.56 eV. The smaller error for this ion as compared to Co^+ is of course related to the reduced number of $3d$ electrons. When looking instead at the $d \to d$ transitions, we find that all excitation energies are overestimated at the lowest, CAS(6), level of calculation, although the absolute errors are much smaller than for the $d \to s$ transitions: up to 0.4 eV for Co^+ and Rh^+, and only 0.12 eV for Ti^+.

Including correlation effects should take care of the large errors obtained at the CAS(6) level. Two approaches are presented in Table 1: in a first CASPT2 calculation based on CAS(6), all correlation effects are treated by second-order perturbation theory; in a second approach, the important radial correlation effects are first dealt with by the larger CAS(11) calculation, and perturbation theory is used only for the remaining correlation effects. Let us look at the results for Co^+ first. Here we find that the first, perturbative, approach quite strongly overestimates the differential correlation effects on the $d \to s$ transitions. Indeed, at the CAS(6)/CASPT2 level all three d^7s^1 states are calculated about 0.3 eV too high in energy with respect to the d^8 ground state. On the other hand, considerable improvements of the $d^7s^1 - d^8$ splitting are obtained even at the CASSCF level when including a second d shell in the active space: all $d \to s$ transitions are still calculated too low, but the error has been reduced to less than 0.5 eV. A further improvement is obtained at the CAS(11)/CASPT step. The final CASPT2 results are, however, still too low by up to 0.3 eV. This final error should be traced back to the incapacity of the limited basis set used to describe to its full extent the large $3d$ correlation effects and their variation with the number of $3d$ electrons present.

The mere fact that the CASPT2 results are so strongly dependent on the size of the CASSCF active space clearly points to the necessity of including a second d shell in the reference correlation treatment of any transition involving a change of the number of $3d$ electrons in the considered Co^+ ion. One might argue that the overshooting of correlation effects in the CAS(6)-based calculation is merely a result of the perturbative approach in the second step and that a CI or CC treatment based on the same reference wavefunction should give much superior results. This is of course true; however, the experience with the Ni atom (20) has indicated that at the MRCI level too, including (the most important) $3d-3d'$ excitations in the reference wavefunction is a prerequisite for obtaining quantitative accuracy. To our knowledge, no coupled-cluster treatment of $3d \to 4s$ transitions in transition metal atoms or ions has been performed with a basis set that is larger enough to be able to decide whether or not the $3d$ radial correlation effects can be captured by a single-reference treatment using this method. This is therefore still an open question.

Naturally, the double-shell effect should play a less important role for the Co^+ $3d \to 3d$ transitions. Indeed, the difference between the CASSCF results (Table 1) obtained with or without the $3d'$ shell is much more limited for the d^8

excited states: a lowering by 0.15 eV or less is found. The CASPT2 results are also less affected, although the effect is not negligible: up to 0.2 eV. It brings the final CASPT2 results very close to the experimental values, with deviations of less than 0.1 eV for all three $d \rightarrow d$ transitions. Apparently, the description of $3d \rightarrow 3d$ transitions is less basis-set demanding than transitions involving an alteration of the $3d$-occupation number. This is a fortunate conclusion, considering that the lowest excited states in many transition metal coordination compounds correspond to $3d \rightarrow 3d$ excitations.

With only three valence electrons, the results obtained for Ti^+ are less sensitive to the presence of $3d'$ in the active space. Adding $3d-3d'$ excitations leads to a stabilization of the ^4F (d^3) state with respect to the ^4F (d^2s^1) ground state by 0.3 eV at the CASSCF level. The corresponding effect at the CASPT2 level is 0.12 eV in the opposite direction. A moderate effect is also found for the $^4F-$ ^2F excitation within the $3d$ shell. As for Co^+, the final CASPT2 result for the $d \rightarrow d$ transition is excellent, while the $s \rightarrow d$ transition has a persisting error of 0.26 eV, ascribable to basis-set deficiencies.

An important conclusion to be drawn from the data in Table 1 is that the double-shell effect seems to vanish when moving down to the second-row (and presumably also third-row) transition series. Indeed, for Rh^+ the difference between the CASPT2 results obtained with the small and large active space is never larger than 0.1 eV. Of course, this does not mean that radial $4d$ correlation effects do not affect the relative energies of configurations with a varying number of $4d$ electrons. The effect is still manifested at the CASSCF level, giving results for the $d^7d^1-d^8$ splitting that are increased by up to 0.5 eV in CAS(11) as compared to CAS(6). However, the fact that the CASPT2 relative energies are virtually indifferent to the altered reference wavefunctions indicates that the differential $4d$ correlation effect can in a satisfactory manner be described by second-order perturbation theory. The $4d \rightarrow 4d$ transitions care even less about the presence of the $4d'$ shell than the $4d \rightarrow 5s$ transitions: at both the CASSCF and CASPT2 levels the difference between the CAS(11) and CAS(6) results is less than 0.1 eV. Also note that the CASPT2 results for all three $4d \rightarrow 5s$ excitations are significantly closer to the experimental values for Rh^+ than was the case for Co^+, thus indicating that the basis-set requirements for describing varying d occupations also tend to loosen up when moving down in the transition series. On the other hand, an exceptionally large deviation from experiment is found for the two lowest excited states ^1D, ^3P, $+0.21$ eV and -0.14 eV, respectively. A possible explanation comes from the lack of spin-orbit coupling in the calculations, combined with the averaging procedure used to obtain the experimental values in Table 1. The latter procedure may no longer be justified in heavy metals like Rh^+, where spin-orbit coupling becomes important and may lead to a mixing of the rather close-lying ^1D and ^3P states (C. Ribbing, personal communication, 1998).

Finally, some words concerning the relevance of the correlation problem in TM atomic or ionic spectroscopy for the large coordination compounds we will look at in the next sections. In such compounds the ground state is most often built from the d^{n+2} configuration, and $d \to s$ excitations often become high-lying and unimportant. Instead, many coordination compounds are characterized by their low-lying $d \to d$ (the so-called ligand-field) transitions (23), for which the *double-shell* effect is of less importance (although not absent; see earlier). However, two cases can clearly be distinguished where one cannot avoid having to deal with the second d shell: 1) In electronic spectroscopy, when describing charge-transfer (CT) excitations: such excitations indeed again involve an increase (ligand-to-metal LMCT) or decrease (metal-to-ligand MLCT) of the d-occupation number, so that correlation effects similar to those for the $d \leftrightarrow s$ transitions may be expected; 2) When describing the total bonding energy (e.g., $ML_6 \to M + 6L$) or the consecutive dissociation of the different ligands (e.g., $ML_6 \to ML_5 \to ML_4 \to ML_3 \ldots$), one will certainly at some point be confronted again with a low-lying $4s$ shell and the $3d$ double-shell effect.

3. COVALENT VERSUS IONIC METAL–LIGAND BONDS

In order to understand the occurrence of nondynamic correlation effects on the bonding in transition metal coordination compounds it is useful to begin with a short introduction to the basic ideas behind the molecular orbital approach, as applied to such systems. We do this by considering a qualitative molecular orbital diagram of a hypothetical octahedral ML_6 complex, as shown in Fig. 1. Here, L is assumed to be a ligand that can form one σ and two π bonds with the metal (e.g., an atom or ion with an np valence shell). At the right-hand side of Fig. 1, symmetry-adapted combinations of the ligand orbitals are constructed that transform according to one of the irreducible representations of O_h. These are a_{1g}, e_g, and t_{1u} for the σ orbitals and t_{1g}, t_{2g}, t_{1u}, t_{2u} for the π orbitals. On the other hand, within O_h, the metal $4s$ orbital belongs to the a_{1g} representation, $4p$ belongs to t_{1u}, and the five $3d$ orbitals form two groups: two $d(e_g)$ orbitals (d_{z^2} and $d_{x^2-y^2}$) belonging to the e_g representation, and three $d(t_{2g})$ orbitals (d_{xy}, d_{xz}, and d_{yz}) that belong to t_{2g}. The interaction of the metal atomic orbitals and the ligands is assumed to proceed in two steps. In the first step, the metal is placed at the center of the six L, without allowing any (covalent) interaction. Due to electronic repulsion with the ligands the metal orbitals are destabilized and split. $4s$ and $4p$ undergo the largest repulsion and are strongly pushed up in energy with respect to the $3d$ orbitals, which themselves are split into e_g (directed toward the ligands) at higher energy and t_{2g} (not directed toward the ligands) at lower energy. This

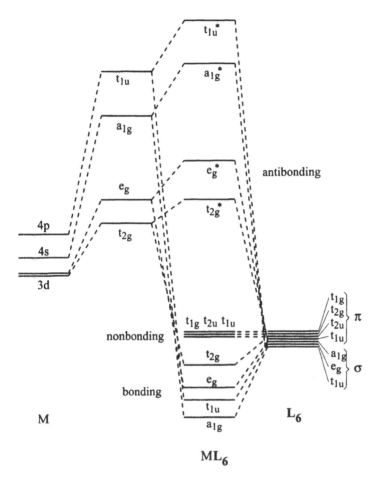

FIGURE 1 Qualitative MO energy-level scheme for regular octahedral complexes ML_6 of a transition metal M with ligands L that have one σ and two π active orbitals each.

first step describes the situation as incorporated in the classical crystal field model.

In a second step, overlap between the metal and ligand orbitals is taken into consideration and the molecular orbitals are constructed, taking into account that only group-symmetrical orbitals belonging to the same O_h symmetry representation may interact. Since the ligand t_{1g} and t_{2u} orbitals have no counterpart on the metal, these orbitals remain nonbonding. The same is (to a first approximation) true for the t_{1u} ligand π orbitals, due to the virtual absence of π interactions with the $4p(t_{1u})$ orbitals on the metal. Within the other representations, bonding (ψ_A) and antibonding (ψ_B) combinations of the metal orbitals (ϕ_M) and the corresponding group-symmetrical ligand orbitals (ϕ_L) are constructed as follows:

$$\psi_B = C_L\phi_L + C_M\,\phi_M \tag{1}$$

$$\psi_A = C'_L\phi_L + C'_M\phi_M \tag{2}$$

The coefficients may be obtained from a solution of the appropriate secular equation:

$$\begin{vmatrix} H_{LL} - E & H_{ML} - EG_{ML} \\ H_{ML} - EG_{ML} & H_{MM} - E \end{vmatrix} = 0 \tag{3}$$

with G_{ML} the so-called group overlap integral (24).

Due to the large zeroth-order energy difference between the metal $3d$ and $4s$, $4p$ orbitals, the antibonding combinations a_{1g}^* and t_{1u}^* fall outside the valence region of our ML_6 molecule, and will not play an important role in the consideration of near-degeneracy effects in the next sections. At this point we therefore focus on the molecular orbitals involving the metal $3d$ orbitals, i.e., the bonding and antibonding combinations of e_g and t_{2g} symmetry.

For the latter two representations, an approximate solution of Eq. (3) may be obtained by assuming that the difference in zeroth-order energy between the metal orbitals and the ligand valence orbitals, $H_{MM} - H_{LL}$, is large while the interaction between them is relatively small, or, $G_{ML} \ll 0$ and $H_{ML} \ll (H_{MM} - H_{LL})$.

Then one obtains the following approximate solutions (24):

$$E_b \cong H_{LL} - \frac{(H_{ML} - H_{LL}G_{ML})^2}{H_{MM} - H_{LL}} \tag{4}$$

$$E_a \cong H_{MM} - \frac{(H_{ML} - H_{MM}G_{ML})^2}{H_{MM} - H_{LL}} \tag{5}$$

while for the coefficients C_M, C_L, C'_M, C'_L one finds that:

$$\frac{C_M}{C_L} \approx \frac{G_{ML}}{H_{MM} - H_{LL}} \ll 1 \tag{6}$$

$$\frac{C'_L}{C'_M} \approx \frac{G_{ML}}{H_{MM} - H_{LL}} \ll 1 \tag{7}$$

Or the bonding orbital is of predominantly ligand origin, the antibonding orbital of predominantly metal $3d$ origin. Furthermore, the ratio C_M/C_L (or C'_L/C'_M) gives a qualitative measure of the extent of covalency of the M–L interaction. In the extreme case that $C_M = C'_L = 0$ ($G_{ML} = 0$), both molecular orbitals entirely retain their atomic character and the bonding is purely ionic. On the other hand, an increasing value of C_M/C_L is indicative of an increasing degree of covalency. Also notice that the group overlap G_{ML} will in general be larger for the σ orbitals belonging to the e_g representation than for the π orbitals in t_{2g}, thus giving rise to more covalent σ than π bonds.

In order to determine the ground state electronic structure of the ML_6 compound, the available valence electrons have to be distributed over the molecular orbitals. Thereby, the bonding and nonbonding orbitals get first priority, leaving only the remaining electrons to be assigned to the antibonding orbitals. Let us consider CrF_6^{3-} as an example. Here, 39 electrons have to be distributed over the molecular orbitals in Figure 1. Of these 39, 36 fit into the 18 bonding and nonbonding fluorine orbitals, leaving three electrons for the antibonding t_{2g}^* orbitals with Cr3d character. Even if this MO occupation scheme is independent of the way one imagines the constituent parts on both sides of the complex, i.e., either as neutral atoms Cr + 6F (+3 extra electrons) or as ions Cr^{3+} + $6F^-$, the latter picture is obviously more consistent with the actual situation presented by the proposed MO diagram. Furthermore, the ionic starting point forms the basis for concepts like "formal metal oxidation state" and "formal metal 3d occupation number," used throughout the literature. Thus, CrF_6^{3-} is always described as a 3d^3 system, with a formal oxidation state (III) or charge (+3) on chromium. The same formal charge and 3d occupation number is also found, for example, in $Cr(H_2O)_6^{3+}$. Whether or not this formal picture indeed corresponds to the actual situation in the complex, depends of course on the extent of covalency of the metal–ligand bonds. As we will show later, the ionic picture is in the case of both CrF_6^{3-} and $Cr(H_2O)_6^{3+}$ quite close to reality. However, in other TM complexes the M-L interaction may acquire a much more covalent character, and in those cases the ionic picture becomes nothing but a formal starting point. In the next section we will also show that such covalent M-L interactions are responsible for the occurrence of strong nondynamic correlation effects in the latter TM complexes.

4. M–L COVALENCY AND CORRELATION EFFECTS

The ideal starting point of any multireference correlation treatment is a CASSCF calculation including (at least) all valence orbitals and electrons. In the octahedral ML_6 complex described in Figure 1 this would mean including 23 orbitals: 18 nonbonding or bonding molecular orbitals with predominant ligand character and five antibonding orbitals with predominant metal 3d character. An active space of this size is, however, not even close to what can be handled by today's hardware and software. Therefore, restrictions are in order. Such restrictions can be accomplished in two ways: 1) Restrict the number of orbitals included in the active space, 2) Restrict the number of configurations included in the CI space.

The latter may, for example, be accomplished by performing instead a RASSCF (restricted active space SCF) calculation, in which the active space is divided into three subspaces, RAS1, RAS2 and RAS3. All possible occupations of RAS2 are then still allowed, but excitations out of RAS1 and into RAS3 are restricted to a maximum number of electrons. When designed economically, such calculations can handle a considerably larger total active space than a regular CASSCF calculation. However, to our knowledge no method is currently available that is capable of treating dynamic correlation based on an RASSCF reference treatment. Such RASSCF calculations are, however, often useful by themselves and may help the user to decide which of the active orbitals give rise to important near-degeneracies and should therefore be selected as active in a subsequent CASSCF reference treatment including a more limited number of such orbitals (if such a selection can be made at all). In the present section, we will present the results of a series of RASSCF calculations on some representative octahedral and tetra-hedral complexes of first-row TM. These calculations [denoted as RASSCF (all valence)] were designed as follows: The five (antibonding) orbitals with predomi-nant metal $3d$ character are included in the RAS2 space, while all ligand valence orbitals (bonding and nonbonding) are included in RAS1. Up to quadruple excita-tions from RAS1 into RAS2 are included in the calculations. This should suffice to provide information as to which of the valence orbitals indeed give rise to strong near-degeneracies; from there we can try to design a more economical CASSCF active space.

However, before looking at the results of these calculations, we believe that a few important points can already be made based on the octahedral MO scheme in Figure 1 and the considerations of the previous section. Thus, as a first case, suppose that we are dealing with the extreme situation of a truly ionic transition metal system. In such a case all molecular orbitals are either entirely ligand or entirely metal based. The ligand valence orbitals are fully occupied and at considerably lower energy than the metal $3d$ orbitals, with which they do not interact. Therefore, one may in this case expect important correlation effects to occur only within the $3d$ valence shell. This means that a reference CASSCF calculation on such a system should include only the metal $3d$, and possibly a second $3d'$ shell (see Sec. 2).

A second case is what we will call the case of weak covalency. This case occurs when there is a distinct overlap between the metal $3d$ and valence orbitals ($G_{ML} \neq 0$), which are, however, still well separated [$(H_{MM} - H_{LL}) \gg 0$]. According to Eqs. (6) and (7) one then finds a significant contribution of ligand character in the antibonding molecular orbitals of e_g^* and t_{2g}^* symmetry, and a corresponding $3d$ contribution in the bonding e_g and t_{2g} orbitals. In other words, the metal $3d$ electrons are delocalized in the ligand valence shell and vice versa. A CASSCF calculation on such a system should include the bonding e_g, t_{2g} and antibonding e_g^*, t_{2g}^* combinations of the metal $3d$ and ligand valence orbitals in the representations. The other, bonding and nonbonding, valence orbitals in Figure 1 can be expected to be of minor importance, since they are still well separated from the open-shell antibonding orbitals.

However, suppose that, in a third case, we also give up the restriction that $(H_{MM} - H_{LL}) \gg 0$ and allow the ligand valence orbitals to approach the metal $3d$ shell. Apart from observing a further strengthening of the covalent interactions within the molecular orbitals of e_g and t_{2g} symmetry, we may now also get confronted with important contributions in the ground state wavefunction coming from excitations out of the bonding and nonbonding orbitals of symmetry a_{1g}, t_{1u}, t_{2u}, and t_{1g} into the open-shell e_g^*, t_{2g}^* orbitals. Such contributions can obviously be accounted for only by including the entire ligand valence shell into the multiconfigurational treatment of nondynamic correlation. Since a CASSCF calculation with such a large active space is out of the question, systems like this are out of the reach of CASPT2 and other presently available multiconfigurational correlation methods.

The foregoing considerations are further illustrated by the results obtained from the previously described set of RASSCF calculations. A first set of calculations concerns the series CrF_6^{x-}, with x varying between 4 and 0. CrF_6^{4-} is a formal $3d^4$ complex with a $t_{2g}^{*3} e_g^{*1}$ quintet ground state and a formal charge of $(+2)$ on chromium. Each consecutive withdrawal of an electron from the antibonding molecular orbitals (obviously e_g^* is depopulated first) brings about a reduction in the formal metal $3d$ occupation and a simultaneous increase in the formal oxidation state on the metal, until in CrF_6 we find a formal charge of $(+6)$ on chromium, with no $3d$ electrons left. With this ionic picture in mind, one can see how this CrF_6^{x-} series reflects a decrease in the difference between the zeroth-order energies H_{MM} and H_{LL} at both sides of Figure 1. Indeed, by identifying these zeroth-order energies with ionization potentials (Koopman's theorem) it becomes clear that the higher the formal charge on chromium, the more the energy of the $3d$ valence shell is pushed down toward the F^- valence shells.

The results obtained form a series of RASSCF calculations on the CrF_6^{x-} complexes are included in Tables 2 and 3. Table 2 shows the composition of the valence natural orbitals and the corresponding occupation numbers; Table 3 lists the number of configurations included and the correlation energy obtained from these RASSCF calculations. In Table 2 the first, t_{2u}, and t_{1g} nonbonding shells are 100% fluorine based in all complexes, while in a_{1g} and both t_{1u} shells we find a small and almost constant contribution of Cr s and p character, respectively. However, more important is the composition of the molecular orbitals of e_g and t_{2g} symmetry. Here we clearly observe a growing admixture within the series of chromium $3d$ character in the bonding e_g, t_{2g} shells and of fluorine $2p$ character in the antibonding e_g^*, t_{2g}^* shells. According to the definition of covalency from the previous section, this growing admixture with an increasing formal charge on chromium is the reflection of a concomitant increase in covalency of

TABLE 2 Composition and Occupation Numbers of the Natural Orbitals Resulting from an RASSCF (All-Valence)[a] Calculation on a Series of Octahedral Chromium Complexes

	CrF_6^{4-}					CrF_6^{3-}					CrF_6^{2-}				
	Occupation no.	Composition (%)				Occupation no.	Composition (%)				Occupation no.	Composition (%)			
		Cr			F		Cr			F		Cr			F
MO		s	p	d			s	p	d			s	p	d	
t_{2u}	6.00	—	—	—	100	6.00	—	—	—	100	6.00	—	—	—	100
t_{1g}	6.00	—	—	—	100	6.00	—	—	—	100	6.00	—	—	—	100
a_{1g}	2.00	0	—	—	100	2.00	2	—	—	100	2.00	3	—	—	97
t_{1u}	6.00	—	0	—	100	6.00	—	2	—	100	5.99	—	0	—	100
t_{1u}	6.00	—	0	—	100	6.00	—	2	—	98	5.98	—	3	—	97
t_{2g}	6.00	—	—	0	99	6.00	—	—	2	98	5.98	—	—	4	96
e_g	4.00	—	—	4	96	3.99	—	—	11	89	3.94	—	—	15	85
t_{2g}^*	3.00	—	—	9	91	3.00	—	—	98	2	2.01	—	—	91	9
e_g^*	1.00	—	—	97	2	0.01	—	—	60	40	0.08	—	—	83	17

MO	CrF_6^- Occupation no.	Composition (%) Cr s	Cr p	Cr d	F	CrF_6 Occupation no.	Composition (%) Cr s	Cr p	Cr d	F	$CrCl_6^{2-}$ Occupation no.	Composition (%) Cr s	Cr p	Cr d	Cl
t_{2u}	5.98	—	—	—	100	5.89	—	—	—	100	6.00	—	—	—	100
t_{1g}	5.97	—	—	—	100	5.91	—	—	—	100	6.00	—	—	—	100
a_{1g}	1.99	3	—	—	97	1.98	3	—	—	97	2.00	15	—	—	85
t_{1u}	5.95	—	2	—	98	5.91	—	3	—	97	6.00	—	4	—	96
t_{1u}	5.94	—	3	—	97	5.85	—	2	—	98	5.98	—	5	—	95
t_{2g}	5.92	—	—	16	84	5.84	—	—	26	74	5.93	—	—	10	90
e_g	3.90	—	—	34	66	3.88	—	—	38	62	3.82	—	—	34	66
t_{2g}^*	1.18	—	—	78	22	0.46	—	—	69	21	2.06	—	—	93	7
e_g^*	0.18	—	—	66	34	0.28	31	—	63	37	0.22	—	—	70	30

a For the description of the RASSCF (all-valence) calculation, see text.

TABLE 3 Number of Included Configuration State Functions (CSF) and Calculated Correlation Energy (a.u.) Obtained from Two RASSCF Calculations on a Set of Octahedral and Tetrahedral Complexes

Complex	Metal formal charge	RASSCF (all valence)		RASSCF $(e_g, t_{2g}, e_g^*, t_{2g}^*)^a$		Difference	
		Number of CSF	Correlation energy	Number of CSF	Correlation energy	Number of CSF	Correlation energy
CrF_6^{4-}	+2	32 864	.001473	560	.001283	32 304	.000190
CrF_6^{3-}	+3	127 492	.012747	1 881	.008668	125 611	.004079
CrF_6^{2-}	+4	251 015	.070053	3 489	.052583	247 526	.017470
$CrCl_6^{2-}$	+4	251 015	.100166	3 489	.108689	247 526	.008593
CrF_6^-	+5	265 586	.201012	3 460	.127509	262 126	.073503
CrF_6	+6	133 981	.401547	1 699	.218086	132 282	.183461
VCl_4^-	+3	101 273	.030320	3 428	.027108	97 845	.003212
CrF_4	+4	101 273	.103309	3 428	.073406	97 845	.029903
MnO_4^{3-}	+5	101 273	.256321	3 428	.198441	97 845	.057880
VO_4^{3-}	+5	53 173	.235151	1 699	.145562	51 474	.089590
CrO_4^{2-}	+6	53 173	.396291	1 699	.246475	51 474	.148916
MnO_4^-	+7	53 173	.563461	1 699	.344010	51 474	.219451

^{a} RASSCF (e, t_2, e^*, t_2^*) in the tetrahedral complexes.

the Cr–F bonds. Also note that the admixture is always larger for e_g than for t_{2g}, consistent with the larger metal–ligand σ- than π-overlap.

As expected, the decreasing gap between metal and ligand orbital energies with an increasing formal charge on the metal also gives rise to increasing correlation effects. This can be seen from the occupation numbers in Table 2 and from the correlation energy obtained from the RASSCF calculation, shown at the left-hand side in Table 3. The correlation energy is almost insignificant for $CrF_6{}^{4-}$, increases strongly with the formal charge on the metal, and becomes quite important (>0.4 a.u.) for the neutral CrF_6 molecule. The occupation numbers reveal the same trend. For $CrF_6{}^{4-}$, all occupation numbers in Table 2 are equal to the ROHF numbers. This molecule could therefore, without significant loss of accuracy, be described starting from a single reference configuration. On the other hand, for CrF_6 all orbitals are either considerably populated or depopulated. Adopting the strict rule that all orbitals with an occupation different by more than 0.01 from its ROHF occupation (or 0.02 for a doubly degenerate and 0.03 for a triply degenerate shell) should be included in a multiconfigurational reference CI, we must conclude that for CrF_6 such a calculation would require the entire valence space.

However, it is important to note that when going from $CrF_6{}^{4-}$ to CrF_6, important correlation effects first become apparent in the t_{2g}, e_g shells, and only afterwards in the other bonding and nonbonding orbitals. Thus, as the orbital occupation numbers in Table 2 show, both $CrF_6{}^{3-}$ and $CrF_6{}^{2-}$ would still be satisfied with a 10-orbital active space, including only the bonding (e_g, t_{2g}) and antibonding (e_g^*, t_{2g}^*) combinations of Cr $3d$ and F $2p$, while excitations out of the other orbitals become important in $CrF_6{}^{-}$. This is also shown by a comparison (Table 3) of the correlation energy obtained from the full-valence RASSCF calculation with a similar, smaller calculation, containing only e_g and t_{2g} in the RAS1 space. As one can see, the first three molecules in the $CrF_6{}^{x-}$ series are indeed almost equally well described by the small active space: the (much larger) number of omitted configurations is responsible for a total contribution of less than 0.02 a.u. to the correlation energy. For the other two molecules, $CrF_6{}^{-}$ and CrF_6, this contribution is considerably larger and cannot be overlooked. The present considerations are also corroborated by previous, more quantitative treatments of the considered molecules. The ligand field spectra of the complexes $CrF_6{}^{x-}$ ($x = 2$–4) were successfully calculated using either an MRCI (25,26) or CASPT2 treatment (27) based on a limited reference active space. However, a CASPT2 calculation of the relative stability of an octahedral and trigonal prismatic structure for CrF_6 (28), based on an active space of only 10 orbitals (e_g, e_g^*, t_{2g}, t_{2g}^*), turned out to produce results that deviate considerably from similar studies performed using either the coupled-cluster method (29,30) or density functional theory (31). The failure of the CASPT2 treatment for this problem must undoubtedly be traced back to the inadequacy of the employed 10-orbital active space.

Looking back at the considerations made at the beginning of this section, we note that CrF_6^{4-} can be classified as an almost purely ionic complex (case 1), CrF_6^{4-} and CrF_6^{3-} both belong to the weakly covalent case (case 2), while CrF_6^{-} and CrF_6 were classified as a third, strongly covalent, case. As already noted, the increasing extent of covalency within this series is due to a decreasing difference in energy (ionization potential) between the metal and ligand valence orbitals. A second factor determining the extent of covalency of the M–L bonds is the overlap between these orbitals [G_{ML} in Eqs. (6) and (7)]. In order to further investigate this second factor, we have added one more example to our series of octahedral test molecules, i.e., $CrCl_6^{2-}$. Cl^- should indeed give a more covalent M–L bond, not because of its valence energy (F and Cl having a similar electron affinity), but because of a stronger M–L overlap. This is confirmed by the composition of the orbitals of e_g and t_{2_g} symmetry in $CrCl_6^{2-}$ as compared to CrF_6^{2-}. And also in this case, stronger covalency leads to stronger correlation effects: the correlation energy obtained from the full-valence RASSCF calculation is indeed larger for $CrCl_6^{2-}$ than for CrF_6^{2-}. It is, however, gratifying to see that the correlation effects in $CrCl_6^{2-}$ are still limited to the orbitals of e_g, t_{2g} symmetry. Indeed, the occupation numbers of the other orbitals remain very close to their ROHF values, while the difference in correlation energy between the small and large RASSCF calculation in Table 3 is even slightly smaller for $CrCl_6^{2-}$ than for CrF_6^{2-}. This indicates that the dimension of the orbital space involved in nondynamic correlation effects in TM systems is not determined by the covalency of the M–L interactions as such, but rather by the difference in valence orbital energies between the metal and ligand ions (starting from an ionic picture).

In order to show that the foregoing findings are not limited to six-coordinate octahedral complexes, we have also performed a similar set of test calculations on a series of tetrahedral molecules. The molecular orbital scheme for a tetrahedral ML_4 complex is shown in Figure 2. The metal $4s$ orbital is now found in representation a_1, $4p$ transforms as t_2, while the tetrahedral ligand environment splits the $3d$ orbitals into e and t_2. On the other hand, the group-symmetrical combinations of ligand orbitals are found in a_1, t_2 for σ and e, t_1, t_2 for π. As such, in ML_4 only the t_1 shell remains nonbonding and purely ligand based. The metal $3d$ orbitals belonging to representation e (d_z^2, d_{x2-y2}) may only be involved in π interactions with the ligands, while the t_2 orbitals (d_{xy}, d_{xz}, d_{yz}) can form a mixture of σ and π bonds. This means that the latter shell is more strongly destabilized, by both a stronger repulsion and a stronger overlap with the ligands.

In Tables 3 and 4, RASSCF results have been included for three formal $3d^2$ molecules—VCl_4^-, CrF_4, and MnO_4^{3-}, containing metals in a formal oxidation state $(+3)$, $(+4)$, and $(+5)$, respectively—and for three formal $3d^0$ complexes, i.e., the isoelectronic series VO_4^{3-} (vanadate), CrO_4^{2-} (chromate), MnO_4^- (permanganate), with formal charges on the metal of $(+5)$, $(+6)$, and $(+7)$, respectively. The calculations were performed in a similar way as for the octahedral

TABLE 4 Composition and Occupation Numbers of the Natural Orbitals Resulting from an RASSCF (All-Valence)[a] Calculation on a Series of Tetrahedral Complexes

VCl_4^-

MO	Occupation no.	Composition (%) V s	p	d	Cl
t_1	6.00	–	–	–	100
a_1	2.00	9	–	–	91
t_2	6.00	–	4	–	96
e	3.99	–	–	6	94
t_2	5.95	–	2	14	84
e^*	2.00	–	–	96	4
t_2^*	0.06	–	2	72	26

CrF_4

MO	Occupation no.	Composition (%) Cr s	p	d	F
t_1	5.99	–	–	–	100
a_1	1.99	3	–	–	98
t_2	5.98	–	2	0	98
e	3.98	–	–	9	91
t_2	5.90	–	2	22	76
e^*	2.02	–	–	91	9
t_2^*	0.14	–	0	72	28

MnO_4^{3-}

MO	Occupation no.	Composition (%) Mn s	p	d	O
t_1	5.95	–	–	–	100
a_1	1.96	0	–	–	100
t_2	5.95	–	2	2	96
e	3.94	–	–	17	83
t_2	5.71	–	1	36	63
e^*	2.07	–	–	84	16
t_2^*	0.42	–	0	62	38

VO_4^{3-}

MO	Occupation no.	Composition (%) V s	p	d	O
t_1	5.91	–	–	–	100
a_1	1.97	0	–	–	100
t_2	5.93	–	1	2	97
e	3.94	–	–	23	77
t_2	5.86	–	1	28	71
e^*	0.15	–	–	66	34
t_2^*	0.24	–	0	64	36

CrO_4^{2-}

MO	Occupation no.	Composition (%) Cr s	p	d	O
t_1	5.85	–	–	–	100
a_1	1.95	0	–	–	97
t_2	5.88	–	3	3	94
e	3.88	–	–	25	75
t_2	5.79	–	1	35	64
e^*	0.28	–	–	64	36
t_2^*	0.37	–	0	60	40

MnO_4^-

MO	Occupation no.	Composition (%) Mn s	p	d	O
t_1	5.77	–	–	–	100
a_1	1.93	1	–	–	99
t_2	5.83	–	5	7	88
e	3.85	–	–	44	56
t_2	5.76	–	1	39	60
e^*	0.40	–	–	57	43
t_2^*	0.46	–	0	54	46

[a] For a description of the RASSCF (all-valence) calculation, see text.

complexes: the $e*$, t_2^* orbitals with predominant metal $3d$ character are put in RAS2, while up to quadruple excitations are allowed out of the RAS1 space, including all ligand-based valence orbitals in a full-valence RASSCF calculation and only one set of (bonding) e, t_2 orbitals in a smaller RASSCF calculation. Note that the full-valence RASSCF calculation now includes a total of 17 active orbitals, which is still too large to be handled by a regular CASSCF calculation. In a tetrahedral complex, covalent M–L interactions can be formed in the molecular orbitals of symmetry t_2 and e. A difference with the octahedral situation is that the ligand valence orbitals within T_d symmetry give rise to *two* group-symmetrical t_2 shells. This means that, at least in principle, the metal $3d$ orbitals may become delocalized in 13 valence molecular orbitals (two t_2,e, t_2^*, $e*$) as opposed to only 10 (t_{2g}, e_g, t_{2g}^*, e_g^*) in an octahedron. However, the results in Table 4 indicate that in practice only one of the bonding t_2 shells, i.e., the one with the lowest occupation number, contains a significant amount of metal $3d$ character. The other t_2 shell remains almost pure L in all cases, with a maximum $3d$ contribution of 7% in MnO_4^-. As for the other shells of e, t_2 symmetry, Table 4 again indicates an increasing covalent interaction with an increasing formal charge on the metal. A limiting case is permanganate, with a formal charge of (+7) on Mn. Here, both the e, $e*$ and t_2, t_2^* couples contain an almost equal mixture of metal $3d$ and L $2p$ character. As concerns the appearance of nondynamic correlation effects, the results in Tables 3 and 4 are also consistent with the octahedral situation. Both in VCl_4^- [formal charge (+3) on V] and CrF_4 [formal charge (+4) on Cr] such effects are limited to the 10 orbitals (e, $e*$, t_2, t_2^*) containing a mixture of ligand and metal $3d$ character: The other natural orbitals, t_1, a_1, t_2, remain close to doubly occupied (Table 4), and excitations out of these orbitals do not contribute much to the correlation energy (Table 3). On the other hand, the latter orbitals become increasingly more important in the complexes MnO_4^{3-} and the d^0 series VO_4^{3-} CrO_4^{2-}, MnO_4^-, where we find a growing formal charge on the metal. The present results are consistent with an earlier study on the permanganate MnO_4^- ion (32), painting a detailed picture of the bonding and correlation effects in this system. We would also like to refer to a recent RASSCF study on the bonding and spectroscopy of the tetraoxoferrate(VI) FeO_4^{2-} ion (9), where the presence of strong near-degeneracies between the Fe($3d$) and O($2p$) levels was also recognized and studied in detail.

On the whole an important, though negative, conclusion is to be drawn from the preceding considerations, i.e., that transition metal complexes containing metals in high formal oxidation states (+5 or higher) may demonstrate severe nondynamic correlation effects involving a large number of orbitals, and are therefore utterly hard to treat by multireference methods (and of course more so by single-reference methods). The origin of these correlation effects should be brought back to the near-degeneracy of the metal $3d$ and ligand valence orbitals, which is ultimately due to the high ionization energy (or electron affinity) connected with the high oxidation state of the metal. Obviously, these near-degeneracies should also be manifested by the presence of low-lying ligand-to-metal charge-transfer (LMCT) states in the experimental optical spectra of the complexes under consideration. Indeed, MnO_4^- is intensely purple due to the presence of low-lying LMCT transitions starting at 18 000 cm^{-1} (23,33) (first allowed excitation to $^1T_2$). The CT states are shifted upward by around 5 000 cm^{-1} in the yellow CrO_4^{2-} ion and even more so in the colorless VO_4^{3-} ion (23). It is interesting to note that the lowest CT states in these d^0 systems indeed correspond to excitations from the nonbonding t_1 shell (23,34), consistent with the qualitative ordering of the orbitals in Figure 2. Looking at the chromium-fluoride compounds, we find LMCT bands (35) at 32 700 cm^{-1} in CrF_4 and from 30 000 to 40 000 cm^{-1} in CrF_6^{2-}, while in CrF_6 the onset of the charge-transfer band is found even at ca. 20 000 cm^{-1}, with the first prominent band appearing at 26 700 cm^{-1} (35). On the other hand, going to lower formal charges on the metal, e.g., in CrF_6^{3-} and CrF_6^{4-}, the lowest CT bands are shifted strongly upward in energy, e.g., above 45 000–50 000 cm^{-1} (36).

Fortunately, many transition metal complexes contain metals in rather low [(+4) or lower] formal oxidation states. The results presented in this section have also indicated that for such complexes all important nondynamic correlation effects may be efficiently dealt with in a reference CASSCF calculation with an active space of at most ten orbitals, i.e., the antibonding molecular orbitals of predominantly metal $3d$ character and their bonding ligand counterparts. In the next section we will look in more detail at these "weakly covalent" systems and further investigate the connection between the extent of covalency of the metal–ligand interactions on the one hand and the importance of nondynamic correlation effects on the other hand.

5. THE CASE OF WEAK COVALENCY

Transition metal complexes containing metals in oxidation states ranging between (+2) and (+4), combined with "classical" (i.e., noncarbon) neutral or negatively charged ligands, constitute a class of coordination compounds that are often designated *Werner* complexes (referring to Alfred Werner, who, at the beginning of the twentieth century, developed the modern picture of coordination complexes) (37). This is the class of complexes that can be described with considerable success by the semiempirical *ligand field theory* and related methods (e.g., the currently still commonly used *angular overlap model* [24]). The success of these methods is related to the fact that the metal–ligand interactions in this group of complexes range from ionic to weakly covalent, a range that is within the limits of the approximations assumed by such models (leading, for example, to energy expressions as given by Eqs. 4 and 5).

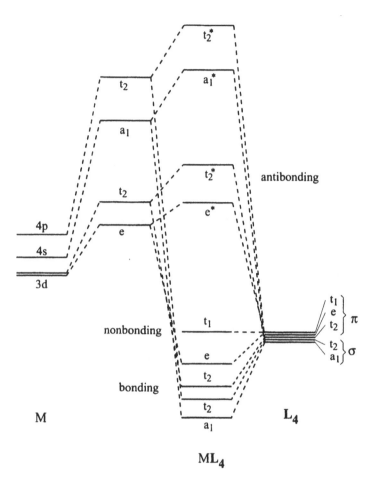

Figure 2 Qualitative MO energy-level scheme for regular tetrahedral complexes ML_4 of a transition metal M with ligands L that have one σ and two π active orbitals each.

In this section we will take a short walk through the domain of the Werner complexes and look for electronic structure–correlation relationships. We will do this by considering two plots, shown in Figure 3. This figure is based on a series of CASSCF calculations on octahedral ML_6 transition metal complexes with different ligands, such as NH_3, H_2O, and the halides F^-, Cl^-, Br^-, I^-. Figure 3a includes results obtained for the $^4A_{2g}((t_{2g}^*)^3)$ ground state of some formal d^3 complexes, with M = (V, Nb, Ta)$^{2+}$, (Cr, Mo, W)$^{3+}$, or (Mn, Tc, Re)$^{4+}$; the results

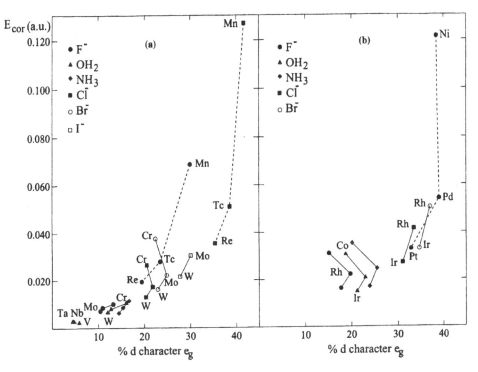

FIGURE 3 Plot of correlation energy (E_{cor}, obtained from a CASSCF calculation; see text) versus the M d contribution (in terms of percentage) in the bonding e_g orbital in a series of octahedral ML_6 complexes, with L = H_2O, NH_3, F^-, Cl^-, Br^-, and I^-, and M a transition metal with a formal d^3 (a) or d^6 (b) occupation number. Solid lines connect metals with a formal charge (+3); dashed lines connect metals with a formal charge (+4). For simplicity, the formal charges on the metals have been omitted from the plots.

presented in Figure 3b are for formal d^6 complexes, with M = (Co, Rh, Ir)$^{3+}$, or (Ni, Pd, Pt)$^{4+}$ in the $^1A_{1g}((t_{2g}^*)^6)$ state (which is not the ground state in all cases; e.g., CoF_6^{3-} has a high-spin $^5T_{2g}((t_{2g}^*)^4(e_g^*)^2)$ ground state [38]). The CASSCF calculations were performed as follows: for the d^3 systems 13 electrons were correlated in an active space consisting of 10 orbitals: e_g^*, e_g^*, t_{2g}, t_{2g}^*. On the other hand, in the d^6 systems, the t_{2g}^* shell is fully occupied, so $t_{2g} \rightarrow t_{2g}^*$ excitations cannot contribute to the wavefunction. The t_{2g} shell was therefore left out of the active space, leaving 7 active orbitals, including 10 electrons. The (x, y) plots in

Figure 3 combine two sets of results obtained from these CASSCF calculations. The x-axis represents the metal $3d$ contribution (in terms of percentage) in the bonding e_g orbitals, as a measure of the extent of covalency of the $M-L(\sigma)$ interactions, while the y-axis represents the correlation energy (in a.u.) obtained from the calculations, i.e., the energy lowering with respect to a single-configurational SCF (in case of d^6) or CASSCF (in case of d^3) treatment. The lines in the plots connect points obtained for metals of the first- to third-row TM series, in the same oxidation state, and coordinated to the same ligands.

A first glimpse at both plots confirms the trend already discussed in the previous section, i.e., that, generally speaking, static correlation energy is an increasing function of the extent of covalency of the M–L bonds. When looking for trends in correlation effects we should therefore start by looking for factors affecting the M–L covalency. Considering both plots in detail we note the following.

1. For the same metal, the M–L covalency and connected correlation effects increase in the following order of ligands:

$$F^- < OH_2 < NH_3 < Cl^- < Br^- < I^- \qquad (8)$$

The increasing tendency to form covalent bonds with an increasing ligand polarizability is not unexpected. Actually, the series presented here is closely related to the *nephelauxetic series*, originating from *ligand field theory* (23,24,39). The *nephelauxetic effect* was originally defined by C. K. Jørgensen (39) as the reduction with respect to the free ion of the interelectronic repulsion in the ligand field states of a transition metal coordination compound. This reduction, expressed as the ratio of the Racah parameter B in the complex and in the free ion, $\beta = B_{complex}/B_{ion}$, depends on the character of the surrounding ligands. From spectroscopic data the following *nephelauxetic series* was obtained, ordering the ligands with respect to decreasing β values (for the same TM in the same oxidation state):

$$F^- > OH_2 > (NH_2)_2CO > NH_3 > H_2NCH_2CH_2NH_2$$
$$\approx (COO)_2^{2-} \approx (CO_3)^{2-} > NCS^- > Cl^- > CN^- \qquad (9)$$
$$> Br^- > N_3^- > I^- > S^{2-}$$
$$\approx (C_2H_2O)PS_2^{2-} > \text{diarsine}$$

The fact that the two preceding series are equivalent is by no means surprising. Indeed, the nephelauxetic effect is related directly to covalency: the reduction of B by complex formation is caused by delocalization of the d-electron cloud on the ligands, which is in turn caused by the formation of covalent bonds. Even if not complete, the nephelauxetic series can come in handy when having to construct the reference space of a multireference calculation, since it helps to decide which ligands are likely to give covalent bonds and hence give rise to important nondynamic correlation effects.

2. All curves systematically appear at higher (x, y) in Figure 3b than in Figure 3a, indicating that more covalent M−L bonds and concomitant correlation effects occur for the d^6 than for the d^3 complexes. Again, this is only a small representation of a more general trend within a row of TM ions: the increasing polarizing power of the ions (in the same oxidation state) from left to right in the same row (corresponding to a decreasing $(H_{MM} - H_{LL})$ in Eqs. 4–7) gives rise to a growing tendency to form covalent M−L bonds and hence also to increasingly more important nondynamic correlation effects. As such, the most strongly covalent M−L bonds are to be expected for TM ions at the right-hand side of their series, combined with soft and easily polarizable ligands. A typical example is the Cu(II)−cysteine combination in the so-called blue copper proteins (40–43). The intense blue color of these proteins is due to a strongly covalent Cu(II)−thiolate bond giving rise to the intense "blue" cysteine → Cu LMCT band in the visible region. As was shown recently (44), substituting Cu(II) by Co(II) in these proteins goes together with a considerable weakening of the covalency of the M−cysteine interaction, consistent with the trends predicted in this chapter.

3. The plots also clearly show (see also Sec. 4) the steeply increasing covalency and concomitant correlation effects with an increasing formal charge on the metal. Thus we find that the MX_6^{2-} lines are strongly shifted in the $(+x, +y)$ direction as compared to the MX_6^{3-} lines for X = F, Cl in the d^3 plot and for X = F in the d^6 plot, while the d^3 $M(H_2O)_6^{2+}$ complexes are found at the bottom left side of Figure 3a, below at and at the left of the $M(H_2O_6)^{3+}$ complexes.

4. Finally we can also compare TM ions from the same column but between different rows of the periodic table, i.e., the data connected by the lines in the plots. Considering first the trends in covalency, we see that in all calculated complexes the third-row metals give a more ionic M−L bond than the second-row metals. However, the observed shifts in M−L covalency between the first- and second-row TM are not unequivocal: in the d^6 systems and also in the d^3 MCl_6^{3-} complexes second-row metals give the most covalent bonds, whereas in the other d^3 systems the strongest covalency is found for the first-row metals. One thing is clear, however: In all considered series, nondynamic correlation effects become less important when moving down in the periodic table. This is also the case for those complexes where the first-row TM do not give the most covalent M−L bonds. The general rule "increasing M−L covalency → increasing correlation effects" is obviously not always valid when considering transition metals belonging to different rows of the periodic system.

Before finishing this section, we would like to remind the reader of another trend between different rows of the TM, i.e., the strongly reduced $4d$ as compared to $3d$ double-shell effect (see Sec. 2 and Table 1). The latter trend, together with the decreasing importance of nondynamic correlation effects connected to the M–L interaction, may explain a rule that is well illustrated in the literature (45–47), i.e., that second- and third-row TM systems are much easier to treat by single-reference methods than first-row TM systems.

6. ORGANOMETALLIC COMPLEXES

So far we have limited our study to cases where the ligands surrounding the metal do not possess their own π-bonding system. In such systems, all covalent interaction types involve a transfer of electrons from the ligands to the metal (σ- and π-donation), thereby reducing the metal formal charge. In the previous sections we have seen that nondynamic correlation effects connected to such covalent interactions involve electron excitations out of fully occupied ligand orbitals into the empty or partly filled metal d shell. When considering organometallic complexes, a second type of covalent interaction has to be added to this description. Indeed, many organic ligands are characterized by low-lying virtual π^* orbitals. When coordinated to a central metal atom or ion, these low-lying empty orbitals may form group-symmetrical combinations with the right symmetry to overlap with the filled metal d orbitals. This second type of covalent interaction is called π-backbonding, because it involves a transfer of electrons from the metal to the ligands, thus resulting in an increase of the formal charge on the metal. Obviously, when considering nondynamic correlation effects connected to M–L bonding in organometallics, π-backbonding cannot be overlooked.

A qualitative picture of the M–L bonding scheme in organometallic complexes is given in Figure 4, showing a molecular orbital diagram for octahedral $Cr(CO)_6$. For the sake of simplicity, only the interaction between CO and the Cr $3d$ orbitals is included, while interactions with $4s$, $4p$ were omitted. The description of the σ-interaction between Cr and the six CO is the same as given previously (Sec. 3 and Fig. 1). However, for the description of the π-interactions a different pattern must be used. Indeed, the most important π-interaction in this case is not built from the Cr $3d$ and the CO π orbital, but instead from Cr $3d$ and CO π^*. The interaction again gives rise to a set of bonding and antibonding molecular orbitals within the octahedral t_{2g} representation; however, since the CO π^* orbitals are located higher in energy than the Cr $3d$ orbitals ($H_{MM} < H_{LL}$ in Eq. 3) the bonding molecular orbitals will be of predominant Cr $3d$ character, and the antibonding orbitals of predominant CO π^* character. When distributing the electrons over the molecular orbitals, all levels up to t_{2g} are filled, while e_g^* remains empty, thus giving Cr a formal d^6 occupation number and charge = 0. Also note the presence of the t_{1u}^*, t_{2u}^*, t_{1g}^*, CO π^* shells below t_{2g}^* in Figure 4. These shells are nonbonding (although within t_{1_u} Cr $4p$ and CO π^* may interact; see Table 5). However, populated out of t_{2g} they may give rise to low-lying excited states with MLCT character, and they are in fact responsible for the appear-

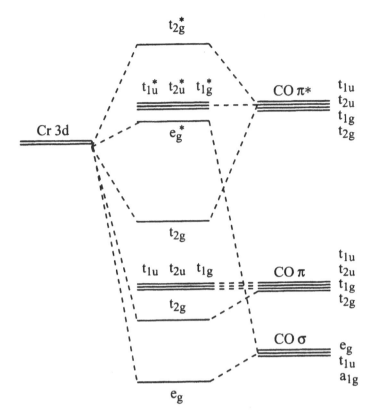

Figure 4 Qualitative MO energy-level scheme for Cr(CO)₆.

ance of intense absorption bands in the UV region of the spectrum of $Cr(CO)_6$ and other organometallic systems (see also the next section).

When trying to predict the most important correlation effects on the bonding in a molecule like $Cr(CO)_6$, the most obvious line of thinking (given the results and considerations of the previous sections) is to start by considering the orbitals that involve Cr $3d$ character, i.e., ten bonding and antibonding molecular orbitals formed as bonding and antibonding combinations of CO σ and Cr $3d$ within e_g and CO π^* and Cr $3d$ within t_{2g}. However, there is a catch here: with a d^6 central Cr, the $3d$ double-shell effect will come into play (see Sec. 2). Therefore, the virtual t_{2g} shell included in this ten-orbital active space might have either Cr $3d'$ character or CO π^* character, depending on the relative importance of the radial $3d$ correlation energy versus the correlation energy connected to the covalent Cr $3d$–to–CO π^*-backbonding. Does this mean that in order to describe

TABLE 5 Composition and Occupation Numbers of the Natural Orbitals Resulting from an RASSCF[a] Calculation on a Series of Organometallic Complexes

Cr(CO)$_6$

MO	Occupation no.	Composition (%) Cr p	d	CO
e_g	3.93	–	24	76
t_{2g}	5.71	–	71	29
e_g^*	0.05	–	81	19
t_{2g}^*	0.22	–	63	37
t_{1u}^*	0.05	34	–	66
t_{2u}^*	0.03	–	–	100
t_{1g}^*	0.01	–	–	100
t'_{2g}	0.00	–	91	9

Mo(CO)$_6$

MO	Occupation no.	Composition (%) Mo p	d	CO
e_g	3.95	–	24	76
t_{2g}	5.80	–	70	0
e_g^*	0.03	–	76	4
t_{2g}^*	0.11	–	59	1
t_{1u}^*	0.06	54	–	6
t_{2u}^*	0.04	–	–	00
t_{1g}^*	0.01	–	–	00
t'_{2g}	0.00	–	82	8

W(CO)$_6$

MO	Occupation no.	Composition (%) W p	d	CO
e_g	0.96	–	24	76
t_{2g}	0.80	–	67	33
e_g^*	0.02	–	74	26
t_{2g}^*	0.09	–	60	40
t_{1u}^*	0.07	51	–	49
t_{2u}^*	0.05	–	–	100
t_{1g}^*	0.01	–	–	100
t'_{2g}	0.00	–	91	9

Ni(CO)$_4$

MO	Occupation no.	Composition (%) Ni p	d	CO
e	3.89	–	91	9
t_2	5.80	6	85	9
e^*	0.11	–	59	41
t_2^*	0.19	13	47	40
t_1^*	0.01	–	–	100
t'_2	0.01	1	90	9
e'	0.01	–	89	11

Cr(NO)$_4$

MO	Occupation no.	Composition (%) Cr p	d	NO
e	3.70	–	56	44
t_2	5.46	5	35	60
e^*	0.22	–	51	49
t_2^*	0.24	0	62	38
t_1^*	0.36	–	–	100
t'_2	0.01	23	53	24
e'	0.01	–	–	82

[a] For a description of the RASSCF calculation. See text.

both correlation effects at the CASSCF level one might have to add two virtual t_{2g} shells to the active space? Fortunately, the answer to this question is negative.

We will illustrate this further by considering a set of RASSCF calculations on a few representative complexes, the results of which are shown in Table 5. Included are the series of d^6 octahedral $M(CO)_6$ ($M = Cr, Mo, W$) complexes as well as two tetrahedral complexes, $Ni(CO)_4$ and $Cr(NO)_4$, both of which are formal d^{10} systems (with a formal charge of (-4) on Cr in the latter complex, see further). In both cases, 10 electrons were included in the RASSCF treatment. In the d^6 octahedrons, the 10 electrons are residing either in the e_g (CO σ) orbitals or in the t_{2g} (M d) orbitals, whereas in the d^{10} tetrahedrons they are from the e, t_2 (M d) shells. The five doubly occupied orbitals were put into RAS1, from which up to quadruple excitations were allowed into RAS3, consisting of the CO or NO π^* orbitals (t_{2g}^*, t_{1g}^*, t_{1u}^*, t_{2u}^* in O_h; t_2^*, e^*, t_1^* in T_d) as well as a d' orbital for each doubly occupied M d orbital (t_{2g}' in O_h; t_2', e' in T_d).

Table 5 shows the composition and occupation numbers of the natural orbitals resulting from the RASSCF calculations. The results for $Ni(CO)_4$ nicely illustrate the way a RASSCF (or CASSCF) calculation deals with the competition between the $3d'$ and CO π^* orbitals for a spot in the active space: they are both mixed into one orbital! Indeed, we find a large population only in the (t_2^*, e^*) shells, both of which contain an almost equal mixture of Ni d and CO character. The second set of (t_2', e') orbitals are almost pure Ni d, but have a small occupation number and are therefore unimportant for correlation. The latter is also true for the nonbonding t_1^*(CO π^*) shell. The problem of nondynamic correlation in $Ni(CO)_4$ is therefore limited to 10 electrons in 10 orbitals, at least for the ground state (48) (see the next section and Refs. 49 and 50 for a discussion of the excited states in $Ni(CO)_4$).

Also in the $M(CO)_6$ series, the most important correlation effects involve the t_{2g}, t_{2g}^* π system, with t_{2g}^* containing a mixture of M d and CO π^* character, while the almost pure M d t_{2g}' shell is left empty. Apart from that, CO $\sigma \rightarrow$ M d excitations within the e_g representation are also important. The composition of the orbitals of symmetry e_g and t_{2g} indicates a slightly increasing M–L bond covalency when moving down from Cr to W, while their populations clearly point to a decreasing importance of nondynamic correlation. The latter is consistent with the trends between different TM rows observed in the previous section. To a first approximation, the ground state of $Cr(CO)_6$ can adequately be described by a multireference treatment based on a (10-in-10) active space; CASPT2 calculations based on this active space have indeed proven to give quite accurate results for the structure and total binding energy (48). However, Table 5 also indicates a significant population of the t_{1u}^*, t_{2u}^* nonbonding orbitals in the $M(CO)_6$ complexes (note that t_{1u}^* also contains a considerable amount of metal p character). Furthermore, the importance of excitations into these orbitals grows when moving from Cr to W, where they in fact become more important than the excitations within the e_g, e_g^* σ system. CASPT2 calculations on $Mo(CO)_6$ and $W(CO)_6$ have not yet been reported. These results in Table 5 suggest that it might be a good idea to base such calculations on an alternative active space than the one used for $Cr(CO)_6$.

Excitations into the nonbonding $t_1^*(NO\ \pi^*)$ orbitals become extremely important, and can therefore certainly not be neglected in $Cr(NO)_4$. The extraordinary large t_1^* population can be understood by starting again (see also Sec. 3 and 4) from an ionic picture of the Cr–NO bond, i.e., a picture based on closed-shell NO^+ ligands. Considering that the ground state electronic configuration of $Cr(NO)_4$ is indeed the same as for the isoelectronic $Ni(CO)_4$ molecule, picturing the former molecule as $Cr^{4-}(d^{10})$ + $4NO^+$ in fact seems a reasonable starting point. However, it is clear that such a picture gives an unrealistically high negative formal charge on Cr, such that the difference in zeroth-order energy between the Cr^{4-} $3d$ and NO^+ π^* orbitals will be small and therefore the actual extent of Cr–NO covalency will be large (cf. Eqs. 6–7). This is confirmed by the compositions in Table 5, showing indeed a much larger M–L mixing in the orbitals of t_2 and e symmetry in $Cr(NO)_4$ than in $Ni(CO)_4$. Furthermore, the near-degeneracy between the metal and NO valence orbitals also explains the important contributions of Cr $3d \rightarrow t_1^*(NO\ \pi^*)$ excitations in the ground state wavefunction. Obviously, the nondynamic correlation problem in $Cr(NO)_4$ is similar to the problem met in Section 4 for molecules with metals in high positive oxidation states, e.g., CrF_6 and MnO_4; although the direction of the involved charge flow is reversed. To our knowledge, no experimental or theoretical information is available concerning the position of the electronically excited states in $Cr(NO)_4$. Based on the present considerations, however, we believe we can safely predict the occurrence of MLCT bands at considerably lower wavenumbers than found for the isoelectronic $Ni(CO)_4$ (i.e., probably well below 30 000 cm^{-1} [49]).

7. CALCULATION OF ELECTRONIC SPECTRA WITH MULTICONFIGURATIONAL METHODS

In this last section we will take a short sidestep from the main subject of this chapter to look at another aspect of computational chemistry where multiconfigurational methods, in particular CASSCF, come in very handy, i.e., the description of excited states. Indeed, one of the nice features of the CASSCF method is that it can generally be used for excited states as well as for the ground state. All it takes is to optimize a set of orbitals for the excited state in question (which is not always straightforward in cases where different roots are close in energy and may "flip" [10]) or, alternatively, for an average of a set of excited states. It is important to realize, however, that the calculation of excited states imposes additional demands on the active space, other than to include all near-degeneracy effects. The rule is simple and self-evident: all orbitals that are either populated or depopulated in any of the considered excited states should be present in the active space. The actual practice is less straightforward, considering that the orbitals involved in excitations are not necessarily also those orbitals involved in near-degeneracies and therefore already included in the ground state active space. This means that for the calculation of electronic spectra, additional orbitals have to be included on top of the ones already discussed in the previous sections. Needless to say, this may easily lead to unmanageably large active spaces if one wants to consider a large number of excited configurations. As we will illustrate, when calculating electronic spectra of TM coordination compounds, compromising on the active space is therefore a rule rather than an exception.

Without going into details or presenting any results, let us look at the possibility of calculating excited states for some of the complexes already discussed. The most straightforward cases are the *Werner* complexes of Section 5, with their typical ligand field spectra. Ligand field transitions occur between the molecular orbitals with predominant d character, e.g., the t_{2g}^*, e_g^* orbitals in Figure 1 or the e^*, t_2^* orbitals in Figure 2. Obviously, the description of nondynamic correlation effects of the previous sections is generally valid for any state belonging to a d^n configuration. Excitations within the d shell therefore do not add any additional demands on the active space, other than the ones already considered in the previous sections. In general, ligand field spectra can be handled with an active space of at most ten orbitals, i.e., the bonding and antibonding combinations of the metal d orbitals and their ligand counterparts. This also holds for ligand field excitations in organometallic systems, except that here (within O_h) the ligand field excitations are between the bonding t_{2g} and the antibonding e_g^* orbitals (see Fig. 4). This is confirmed by the quality of the results obtained in previous CASSCF/CASPT2 studies of the ligand field spectra of the hexacyanides of first-row TM (51) and of $Cr(CO)_6$ (49), based on this active space of 10 orbitals.

However, things get more complicated if one also wants to include charge-transfer states in the calculations. The origin of the problem lies in the fact that the ligand-based orbitals already included in the basic ten-orbital active space are usually not the HOMO or LUMO orbitals. This is clearly illustrated in all three MO diagrams presented (Figs. 1, 2, and 4). The highest doubly occupied ligand orbitals (involved in LMCT; see Figs. 1 and 2) or lowest virtual orbitals (with L π^* character, involved in MLCT; see Fig. 4) are instead the nonbonding orbitals. These orbitals should therefore be included in the active space of a calculation aiming at describing the lowest charge-transfer states. Since adding all of them at once unavoidably leads to too large an active space, the only solution is to find a compromise (based, for example, on RASSCF calculations) by using different active spaces for different excited states. For instance, in $Cr(CO)_6$, the active space needed for a full description of the Cr $3d \rightarrow CO$ π^* spectrum would have to include, on top of the basic 10-orbital active space, the CO π^* shells of symmetry t_{1u}, t_{2u}, t_{1g}, leading to too large a number of 19 active orbitals. However, within O_h symmetry MLCT states of either *gerade* or *ungerade* symmetry are strictly separated, since they involve excitations (from the *gerade* 3d orbitals) into different orbitals: t_{1g}^*, t_{2g}^* for the *gerade* states, t_{1u}^*, t_{2u}^* for the *ungerade* states. Therefore, the spectrum of *gerade* states can be described by an active space of 13 orbitals, i.e., the basic ten plus t_{1g}^*. A similar procedure would still give too many (16) active orbitals for the *ungerade* states. Therefore, the t_{1u}^* and t_{2u}^* have to be added in turn or, alternatively, may be included together at the expense of giving in on the basic ten orbitals (by omitting the e_g, e_g^* couple). Both alternatives were tested in a CASPT2 study of the electronic spectrum of $Cr(CO)_6$ (49), where they turned out to produce similar (and quite accurate) results for the excitation energies (but not for the calculated oscillator strengths; see Ref. 49).

A second complicating factor affecting the calculation of MLCT states (at least for first-row TM) is the double-shell effect. In Section 2 we saw that including a second d shell in the active space is a prerequisite for obtaining accurate results for transitions involving a change in the number of 3d electrons, e.g., for charge-transfer states. On the other hand, in Section 6 we saw that in the ground state active space of organometallic complexes, metal $3d'$ character and ligand π^* character are combined within one orbital. However, what will happen when exciting an electron into one of these orbitals? Obviously the orbital under consideration will lose all d' character and turn into a pure π^* orbital. Therefore, a strictly balanced CASSCF treatment of both the ground and excited state would still require two sets of virtual orbitals, i.e., the $3d'$ shell to describe the double-shell effect, and the ligand π^* orbitals to describe the actual excitations. In $Ni(CO)_4$, for example, this would lead to an active space of 18 orbitals (five 3d, five $3d'$, eight CO π^*). Fortunately, here also a compromise could be found (49). Indeed, as it turned out, the loss of $3d'$ character in just the one orbital receiving the electron does not have a dramatic effect on the CASPT2 result for the excited state (although again the calculated oscillator strengths do suffer; see also Refs. 49 and 50, so the calculation of the $Ni(CO)_4$ spectrum could still be performed with 13 active orbitals. However, a prerequisite for this type of calculation to be successful is that a set of CASSCF orbitals is optimized separately for each of the excited states. If instead one average set of CASSCF orbitals is used, all $3d'$ character gets lost (since all virtual orbitals are populated in one of the excited states and therefore turn into π^*), and the corresponding CASPT2 results are afflicted with large errors (50). We would therefore like to express a clear warning against using average CASSCF orbitals for the calculation of MLCT states in organometallics.

The preceding examples have indicated that the calculation of charge-transfer states in transition metal coordination compounds is certainly far from straightforward. However, a positive message is that, provided appropriate choices are made to keep the size of the CASSCF active space within limits, the subsequent CASPT2 are still accurate enough (with errors of at most 0.5 eV and usually smaller than 0.3 eV) to provide an assignment and interpretation of experimentally observed electronic transitions in TM systems.

8. CONCLUSION

We have tried to describe the most important nondynamic correlation effects in transition metal coordination compounds and to provide some guidelines for the construction of the appropriate multiconfigurational wavefunction as a starting point for multireference ab initio calculations. We have made the connection between nondynamic correlation effects and the covalency of the M–L bond, and have shown that in complexes with weakly covalent M–L bonds these correlation effects can be included in an active space containing the molecular orbitals with predominant metal d character and their ligand counterparts, either bonding or antibonding. A first rule therefore is to make sure that "all metal d character is included in the active space." We have also shown that the presence of low-lying excited states may give rise to strong near-degeneracies in complexes containing metals in very high (positive or negative) formal oxidation states. Hence a second rule is to "be alert when charge-transfer states appear at low wavenumbers in the experimental electronic spectrum, and to include the orbitals involved in the active space." We have illustrated the double-shell effect in the atomic case and have indicated how it affects the calculation of MLCT states. Also, the results presented have indicated that correlation effects tend to become considerably less important for second- and third-row than for first-row TM.

All examples shown in this chapter were for high-symmetric, either octahedral or tetrahedral, complexes. One may therefore wonder if the present considerations still remain valid in cases without symmetry. For instance, will the metal d contributions still be confined to a limited set of molecular orbitals (i.e., the "basic" ten) in cases where such limitations are not enforced by symmetry? That this is indeed the case was already illustrated by the tetrahedral examples (Table 4), where the M $3d$ orbitals can in principle be delocalized over two bonding t_2 shells, but in practice significantly contribute only to one of these shells.

And finally, what happens if the size of the calculated complexes increases? Only cases with small ligands were presented in this chapter, but it is clear that the present considerations remain valid if the size of these ligands grows, since important correlation effects are confined to the region between the metal and the coordinating atoms. This of course implies that intraligand nondynamic correlation effects are unimportant. However, a cause for greater worry are molecules or clusters containing more than one metal atom or ion. One cannot get around the fact that, at least in principle, all demands on the active space increase proportionally to the number of metal centers. This means that systems with three or more metals are virtually out of reach of the CASSCF method. Systems with two metals are still possible if they contain only ionic M–L bonds so that the active space can be limited to the metal d shells. Studies along these lines were performed for the magnetic interactions in $M_2Cl_9^{3-}$ (M = Ti, Cr) (52,53) or in copper and nickel oxides (54–56). For a case with more covalent bonds, i.e., the Cu_2O_2 center in the proteins hemocyanin or tyrosinase (57), a CASSCF/CASPT2 study could be performed for the ground state, by excluding from the active space all Cu $3d$ orbitals not involved in the Cu–O interactions. Such limitations on the active space of course inherently preclude the study of the excited states of this system.

ACKNOWLEDGMENTS

This investigation has been supported by grants from the Flemish Science Foundation (FWO), the Concerted Research Action of the Flemish Government, and by the European Commission through the TMR program (grant ERBFMRXCT960079).

REFERENCES

1. BO Roos, K Andersson, MP Fülscher, PÅ Malmqvist, L Serrano-Andrés, K Pierloot, M Merchan. In: I Prigogine, SA Rice, eds. Advances in Chemical Physics: New Methods in Computational Quantum Mechanics. Vol. XCIII. New York: Wiley, 1996, pp 219–331.
2. SR Langhoff, CW Bauschlicher. Ann Rev Phys Chem 39:181–212, 1988.
3. A Veillard. Chem Rev 91:743–766, 1991.
4. C Daniel, A Veillard. In: A Dedieu, ed. Transition Metal Hydrides. New York: VCH, 1992, pp 235–262.
5. CW Bauschlicher, SR Langhoff. In: A Dedieu, ed. Transition Metal Hydrides. New York: VCH, 1992, pp 103–126.
6. C Sousa, WA De Jong, R Broer, WC Nieuwpoort. J Chem Phys 106:7162–7169, 1997.
7. D Guillaumont, C Daniel. Coord Chem Rev 177:181–199, 1998.
8. C De Graaf, WA De Jong, R Broer, WC Nieuwpoort. Chem Phys 237:59–65, 1998.
9. A Al Abdalla, L Seijo, Z Barandiaran. J Chem Phys 109:6396–6405, 1998.
10. BO Roos. In: BO Roos, PO Wildmark, eds. European Summerschool in Quantum Chemistry. Book I. Lund University, Lund, Sweden, 1999, pp 307–388.
11. PEM Siegbahn. In: BO Roos, PO Wildmark, eds. European Summerschool in Quantum Chemistry. Book I. Lund University, Lund, Sweden, 1999, pp 255–305.
12. H Partridge, SR Langhoff, CW Bauschlicher. In: SR Langhoff, ed. Quantum Mechanical Electronic Structure Calculations with Chemical Accuracy. Dordrecht, The Netherlands: Kluwer Academic, 1995, pp 209–260.
13. C Park, J Almlöf. J Chem Phys 95:1829–1833, 1991.
14. K Pierloot, BJ Persson, BO Roos. J Phys Chem 99:3465–3472, 1995.
15. K Andersson, PÅ Malmqvist, BO Roos. J Chem Phys 96:1218–1226, 1992.

16. H Koch, P Jørgensen, T Helgaker. J Chem Phys 104:9528–9530, 1996.
17. C Froese-Fischer. J Phys B 10:1241–1251, 1977.
18. TH Dunning Jr, BH Botch, JF Harrison. J Chem Phys 72:3419–3420, 1980.
19. BH Botch, TH Dunning Jr, JF Harrison. J Chem Phys 75:3466–3471, 1981.
20. CW Bauschlicher, P Siegbahn, LGM Pettersson. Theor Chim Acta 74:479–491, 1988.
21. K Andersson, BO Roos. Chem Phys Letters 191:507–514, 1992.
22. Z Barandiaran, L Seijo, S Huzinaga. J Chem Phys 93:5843–5850, 1990.
23. ABP Lever. Inorganic Electronic Spectroscopy. Amsterdam: Elsevier, 1984.
24. IB Bersuker. Electronic Structure and Properties of Transition Metal Compounds. New York: Wiley, 1996.
25. K Pierloot, LG Vanquickenborne. J Chem Phys 93:4154–4163, 1990.
26. K Pierloot, E Van Praet, LG Vanquickenborne. J Chem Phys 96:4163–4170, 1992.
27. K Pierloot, E Van Praet, LG Vanquickenborne. J Chem Phys 102:1164–1172, 1995.
28. K Pierloot, BO Roos. Inorg Chem 31:5353–5354, 1992.
29. A Neuhaus, G Frenking, C Huber, J Gauss. Inorg Chem 31:5355–5356, 1992.
30. CJ Marsden, D Moncrieff, GE Quelch. J Phys Chem 98:2038–2043, 1994.
31. LG Vanquickenborne, A Vinckier, K Pierloot. Inorg Chem 35:1305–1309, 1996.
32. MA Buijse, EJ Baerends J Chem Phys 93:4129–4141, 1990.
33. L Holt, CJ Ballhausen. Theor Chim Acta 7:313–320, 1967.
34. SJA van Gisbergen, JA Groeneveld, A Rosa, JG Snijders, EJ Baerends. J Phys Chem A 103:6835–6844, 1999.
35. EG Hope, PJ Jones, W Levason, JS Ogden, M Tajik, JW Turff. J Chem Soc Dalton Trans 1443–1449, 1985.
36. GC Allen, KD Warren. Structure Bonding 9:49–138, 1972.
37. RH Crabtree. The Organometallic Chemistry of the Transition Metals. New York: Wiley, 1994.
38. LG Vanquickenborne, K Pierloot, E Duyvejonck. Chem Phys Lett 224:207–212, 1994.
39. CK Jørgensen. Modern Aspects of Ligand Field Theory. Amsterdam: North Holland, 1971.
40. K Pierloot, JOA De Kerpel, U Ryde, BO Roos. J Am Chem Soc 119:218–226, 1997.
41. MHM Olsson, U Ryde, BO Roos, K Pierloot. J Biol Inorg Chem 3:109–125, 1998.
42. K Pierloot, JOA De Kerpel, U Ryde, MHM Olsson, BO Roos. J Am Chem Soc 120: 13156–13166, 1998.
43. U Ryde, MHM Olsson, BO Roos, K Pierloot, JOA De Kerpel. Encycl Comp Chem 3:2255–2270, 1998.
44. JOA De Kerpel, K Pierloot, U Ryde, BO Roos. J Phys Chem B 103:8375–8382, 1999.
45. PEM Siegbahn. In: I Prigogine, SA Rice, eds. Advances in Chemical Physics: New Methods in Computational Quantum Mechanics. Vol. XCIII. New York: Wiley, 1996, pp 333–387.
46. G Frenking, I Antes, M Böhme, S Dapprich, AW Ehlers, V Jonas, A Neuhaus, M Otto, R Stegmann, A Veldkamp, SF Vyboishchikov. In: KB Lipkowitz, DB Boyd, eds. Reviews in Computational Chemistry. Vol. 8. New York: VCH, 1996, pp 63–144.

47. AW Ehlers, S Dapprich, SF Vyboishchikov, G Frenking. Organometallics 15:105–117, 1996.
48. BJ Persson, BO Roos, K Pierloot. J Chem Phys 101:6810–6821, 1994.
49. K Pierloot, E Tsokos, LG Vanquickenborne. J Phys Chem 100:16545–16550, 1996.
50. BO Roos, K Andersson, MP Fülscher, L Serrano-Andrés, K Pierloot, M Merchán, V Molina. J Mol Struct (Theochem) 388:257–276, 1996.
51. K Pierloot, E Van Praet, LG Vanquickenborne, BO Roos. J Phys Chem 97:12220–12228, 1993.
52. A Ceulemans, GA Heylen, LF Chibotaru, TL Maes, K Pierloot, C Ribbing, LG Vanquickenborne. Inorg Chim Acta 251:15–27, 1996.
53. A Ceulemans, LF Chibotaru, GA Heylen, K Pierloot, LG Vanquickenborne. Chem Rev 100:787–806, 2000.
54. C De Graaf, F Illas, R Broer, WC Nieuwpoort. J Chem Phys 106:3287–3291, 1997.
55. C De Graaf, R Broer, WC Nieuwpoort. Chem Phys Lett 271:372–376, 1997.
56. C De Graaf, IDR Moreira, F Illas, RL Martin. Phys Rev B Cond Mat 60:3457–3464, 1999.
57. M Flock, K Pierloot. J Phys Chem A 103:95–102, 1999.
58. J Sugar, J Corliss. J Phys Chem Ref Data 8:1–62, 1979.
59. J Sugar, J Corliss. J Phys Chem Ref Data 10:1097–1174, 1981.
60. CE Moore. Atomic Energy Levels. Vol. III. Washington, DC: Circular of the National Bureau of Standards 467, 1958.

6

Quantitative Consideration of Steric Effects Through Hybrid Quantum Mechanics/ Molecular Mechanics Methods

Feliu Maseras

Universitat Autònoma de Barcelona, Barcelona, Catalonia, Spain

1. INTRODUCTION

Steric effects have been largely absent from the spectacular progress experienced by computational organometallic chemistry in the last decades. There is good reason for this. The methodological and computational struggle to properly describe the properties at transition metal centers leaves little space for the introduction of the bulky ligands responsible for steric effects. This situation is, however, currently changing, in part because of the entry of hybrid quantum mechanics/ molecular mechanics (QM/MM) methods into this field of chemistry.

The power of QM/MM methods is based directly on that of pure QM and pure MM methods. There are standard QM methods that describe reliably all chemical features of small molecular systems. There are standard MM methods that describe reliably some chemical features of large molecular systems. Both types of methods, pure QM [either Hartree–Fock (HF)–based or density functional theory (DFT)] and pure MM, are steadily expanding the range of systems to which they can be applied, and up-to-date notice of this can be found in other chapters of this same volume. The QM/MM approach is simpler: The chemical system is divided in two regions, and the more convenient method is applied to each region. In this way, the heavy computational cost of QM methods can be concentrated only in the regions where it is strictly required while keeping an overall correct MM description for the rest of the system. The potential application of QM/MM methods to organometallic chemistry is enormous, because the electronic complexity of the systems is usually concentrated in a small region, namely, the metal and its immediate environment.

This chapter is intended for an audience of computational organometallic chemists interested in the practical use of hybrid QM/MM methods. Because of this the description of the methodological details will be kept to a minimum, condensed in the second section. Similarly, the chapter is not intended to be a review of published applications, which can be found in another recent review (1). Instead, this chapter illustrates a series of practical aspects of QM/MM calculations that make them different from other, traditional QM or MM approaches. These features are shown mostly through the presentation of selected aspects of different examples of calculations, some of them carried out specifically for this text, some of them taken from previous publications, mostly by the author.

Most of the applications presented use the integrated molecular orbital/ molecular mechanics (IMOMM) method (2), which therefore will be briefly described in the next section. Afterwards, two sections will discuss particular aspects of the calculation setup, and another section will present specific ways to analyze QM/MM results. A final section will offer concluding remarks.

2. INTEGRATED MOLECULAR ORBITAL/MOLECULAR MECHANICS METHOD WITHIN THE CONTEXT OF QUANTUM MECHANICS/MOLECULAR MECHANICS METHODS

Hybrid QM/MM methods already have a certain history of their own in computational chemistry (3–8). Early work in this field has been on the introduction of solvation effects, with special focus on biochemical systems. The general application of this approach to transition metal systems, where there are often chemical bonds across the frontier between QM and MM regions, has been more recent. The IMOMM method, proposed in 1995 (2), has been remarkably successful, as shown by the large number of applications, concentrated mostly in transition metal chemistry (9–28), although not exclusively (29–31). The method has also been the starting point of other QM/MM methodological developments (32–36). One must cite in this regard the IMOMO and ONIOM methods. The IMOMO method (32) is the extension of the method to the use of two different-quality QM descriptions. The ONIOM method (33,34) is essentially a generalization that encompasses both the IMOMM and IMOMO methods, with the significant addition of the possibility of using more than two layers. The reason why the label IMOMM (instead of the more general ONIOM) is used throughout this chapter is fundamentally practical: the program used in most of the calculations presented was that of the original IMOMM implementation.

Good general discussions on hybrid QM/MM methods can be found in presentations of methodological novelties (8,34,37) and also in recent reviews of these methods (1,38–41). Because of this, the discussion here will be very brief.

The main difference between the current implementations of IMOMM, IM-OMO, and ONIOM and the majority of other available QM/MM methods is related to the handling of the interaction between the QM and the MM regions. In principle, in any hybrid QM/MM method the total energy of the whole system can in all generality be expressed as:

$$E_{tot}(QM, MM) = E_{QM}(QM) + E_{MM}(MM) + E_{interaction}(QM/MM)$$

where the labels in the subscript refer to the type of calculation and the labels in parentheses correspond to the region under study. Both the QM and MM methods can in principle compute the interaction energy between the QM and MM regions, and the previous expression becomes:

$$E_{tot}(QM, MM) = E_{QM}(QM) + E_{MM}(MM)$$
$$+ E_{QM}(QM/MM) + E_{MM}(QM/MM)$$

The energy expression in a general hybrid QM/MM method thus has four components. Two of them correspond simply to the pure QM and MM calculations of the corresponding regions. And the other two correspond to the evaluation of the interaction between both regions, in principle at both computational levels. Different computational schemes are defined by the choice of the method to compute the $E_{QM}(QM/MM)$ and $E_{MM}(QM/MM)$ terms.

One of the defining characteristics of current implementations of IMOMM and derived methods is the neglect of the $E_{QM}(QM/MM)$ term. The neglect of this term obviously introduces an error in the reproduction of experimental reality. But it simplifies the calculation enormously, from a technical point of view, and it leads to an easier interpretation of the results.

This $E_{QM}(QM/MM)$ term is usually critical in solvation problems, because one of the points of interest is precisely how the quantum mechanical properties of the solute are modified by the presence of the solvent. In the case of a transition metal complex, this term would account mainly for the electronic effects of the ligand substituents on the metal center. A common way to introduce the $E_{QM}(QM/MM)$ term is to put electrostatic charges in the positions occupied by the MM atoms, introducing in practice a term in the monoelectronic Hamiltonian (8). One problem with this kind of approach is the choice of the electrostatic charges, which is by no means trivial, since its validity is usually confined to the consistency of a force field. A more serious problem appears when there are chemical bonds between the QM and MM regions, where this approach breaks down in the proximity of the interface and needs to be reformulated. A more elaborate scheme to answer this problem has been proposed through the introduction of localized orbitals, but so far it has been applied mostly to organic systems (42,43). As mentioned earlier, IMOMM neglects this E_{QM}(QM/MM).

The E_{MM}(QM/MM) term, which is considered explicitly by IMOMM and derived methods, accounts for the direct effect of the atoms in the QM region on the MM energy of the system. This term is usually critical in transition metal complexes, because it accounts for the geometrical constraints introduced by the presence of the metal center in the arrangement of the ligands. In other words, it is related mostly to the steric effect of the ligand substituents. The E_{MM}(QM/MM) term is usually introduced through the parameterization of the QM atoms with the same force field used in the MM region. This parameterization is much simpler than it would be in the case of a pure MM calculation, because the MM part of the QM/MM calculation neglects the interactions within the QM region.

The presence of chemical bonds between the QM and MM regions poses a problem to the use of hybrid QM/MM methods, with different approaches taking different solutions. The particular method applied by IMOMM and derived methods, the introduction of additional link atoms to saturate the dangling bonds, will be discussed in more detail in Section 3.4 of this chapter.

It must also be mentioned that IMOMM and derived methods involve a full multistep optimization (1). This means that the geometry of both the QM regions and the MM regions is modified to minimize the total energy, with the result that the final geometry corresponds neither to the optimal QM arrangement nor to the optimal MM arrangement. Simpler one-step methods, where the MM geometry is optimized on a frozen QM geometry, also have their value but will not be discussed here.

The IMOMM method can in principle be applied to any combination of QM and MM methods. Both the QM and the MM level of the calculation are indicated in this text through a compact terminology of the type IMOMM(QM level:MM level).

3. DEFINITION OF THE COMPUTATIONAL LEVEL

The chemist interested in performing a hybrid QM/MM calculation must make a number of choices that may affect substantially the result of the computation. These choices, ranging from the obvious to the subtle, will be discussed in this section.

3.1. Quantum Mechanics/Molecular Mechanics Partition

A fundamental decision to be taken in the planning of a QM/MM calculation is that of the QM/MM partition: Which atoms are going to be included in the QM region and which atoms are going to be included in the MM region? The guiding idea for this choice must be to use a QM region as small as possible yet containing all the fundamental interactions that cannot be described by the MM method. Chemical knowledge is the main guiding line for this choice, although calibration tests will be necessary in doubtful cases. These tests should ideally be carried out through comparison of the hybrid QM/MM results with those of pure QM calculations for the whole system. This would provide a more reliable criterion than comparison with experiment, because there are a number of reasons unrelated to the QM/MM partition why QM/MM results may differ from experimental data, like inaccuracies in the QM description or the presence of solvent or packing effects in the experimental data. Comparison with experiment is nevertheless still a useful criterion, especially when the agreement is satisfactory.

An interesting example showing the importance of the QM/MM partition can be found in a study containing IMOMM(MP2:MM3) calculations on potentially agostic $Ir(H)_2(PR_3)_3^+$ complexes (20). An agostic interaction is the intramolecular interaction that takes place within one complex between the metal center and a C–H bond of one the ligands (44). These $Ir(H)_2(PR_3)_3^+$ complexes have an empty coordination site at the metal and CH bonds in the ligands, but present agostic interactions only for certain combinations of R substituents in the phosphines.

In particular, we will discuss here the results on $Ir(H)_2(PCy_2Ph)_3^+$. In this complex, X-ray data (20) indicate the existence of an agostic distortion, with one of the Ir–P–C angles being 100.9°, significantly different from the standard tetrahedral angle of ca. 109°. In the IMOMM(MP2:MM3) calculation of this species, two different partitioning schemes between the QM and MM domains were applied (Fig. 1). In model I, all atoms not directly bound to the metal were

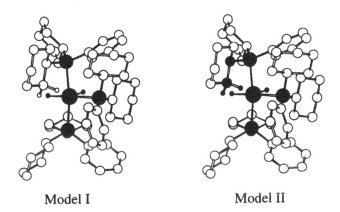

Model I   Model II

FIGURE 1 Different QM/MM partitions used in the IMOMM calculations of Ir $(H)_2(PCy_2Ph)_3^+$. Atoms in the QM part are shown in black.

calculated at the MM level, the QM part therefore being $Ir(H)_2(PH_3)_3{}^+$. In model II, the QM part of the phosphine included the agostically distorted chain, the QM part thus being $Ir(H)_2(P(Et)H_2)_3{}^+$. The essential difference between both models is thus in the description of the potentially agostic C–H bond.

The geometry of $Ir(H)_2(PCy_2Ph)_3{}^+$ was fully optimized at the IM-OMM(MP2:MM3) level with both QM/MM partitions I and II. The IMOMM(MP2/I:MM3) calculation gave results in qualitative agreement with X-ray data, with the largest discrepancy being in the value of 105.6° for the agostic Ir–P–C angle. This was larger than the experimental X-ray value of 100.9° but already smaller than the computed average Ir–P–C value at this phosphorus center of 109.9°. This result is interesting because it proves that the bulk of the ligands alone is able to push one of the C–H bonds of the cyclohexyl group to the proximity of the metal, but it is still somehow removed by 4.5° from the experimental value. Use of the more elaborate model IMOMM(MP2/II:MM3), with the C–H bond in the QM part, led to results much closer to the X-ray values. The Ir–P–C bond angle improved to 99.8°, only 1.1° from the experimental value. This change proves that the interaction between the Ir center and the C–H bond is not properly reproduced by the MM3 force field, and must therefore be included in the QM region. The problem is obviously that the force field is not parameterized to describe agostic interactions.

The definition of the QM and MM regions is therefore critical for the validity of the QM/MM calculation. The smaller the QM region, the more affordable the computational cost, but care must be taken not to leave any critical electronic interactions out of the QM region. The importance of the choice of the QM/MM partition must, however, not hide the fact that both the QM and the MM descriptions must describe with sufficient accuracy interactions within the respective regions.

3.2. Quantum Mechanics Level

There is a wide variety of QM levels that can be applied to organometallic compounds, ranging from semiempirical methods like PM3(tm) to high-level methods such as multireference-configuration interaction, passing through all the derivations of HF and DFT methods. Information about these methods can be found in other chapters of this book. The choice of the most appropriate QM method for each chemical problem is by no means trivial. One of the main features of the use of hybrid QM/MM methods is the use of a small QM region, allowing the use of QM levels that would be unaffordable with pure QM calculations for all the system. The QM level chosen must nevertheless be sufficiently accurate to describe all significant interactions within the QM region, or else the whole QM/MM calculation will fail. In what follows, we will discuss one example showing how this can be critical.

This example concerns calculations on heme species. Understanding the features of this type of complexes is important because heme groups are at the active center of a number of biochemically very relevant proteins and enzymes, like hemoglobin, myoglobin, cytochromes, peroxidases, and catalases (45). However, its theoretical study has been hindered so far by the large size of the system, containing a porphyrin ring (4 nitrogen atoms plus 20 carbon atoms) attached to an iron center. Because of this, it would be appealing to apply a QM/MM method to reduce the computational effort. This requires introducing the QM/MM partition within the porphyrin ring, which will obviously worsen the modeling of the ring aromaticity. The question of the magnitude of the effect of the introduction of the QM/MM partition on the description of the electronic properties of iron has no straightforward answer, and required performing a series of systematic test calculations (22). The conclusion of that study was that the partition presented in Fig. 2, cutting the QM part of the heme group to [Fe(NH(CH)₃NH)₂], leads to satisfactory results. This study nevertheless had an interesting spinoff concerning the validity of the RHF description that merits discussion here.

The particular complex discussed here is [Fe(P)(Im)(O₂)] (P = porphyrin, Im = imidazole) (22). This species is a model for the active center of oxygen transport proteins hemoglobin and myoglobin, and both in the biological and bioinorganic systems is able to bind oxygen. In spite of that, IMOMM(RHF: MM3) calculations on this system did not yield any stable minimum with the oxygen bound to the iron, the optimal Fe–O distance being above 3.5 Å. A weakly bound state (with bond distances around 2.2 Å) could be obtained by

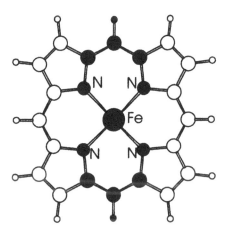

FIGURE 2   QM/MM partition used in IMOMM calculations of systems involving the heme group. Atoms in the QM part are shown in black.

using a minimal basis set on oxygen, but this value is far from the experimental 1.746 Å found in related bioinorganic models and can almost certainly be attributed to a computational artifact associated with basis set superposition error. Therefore, in view of these results, one can see that the calculations fail to reproduce, even at a qualitative level, experimental reality. The QM/MM partition was proposed to be responsible for this failure.

However, this hypothesis was proved wrong by additional calculations. When IMOMM(Becke3LYP:MM3) calculations were carried out with the same QM/MM partition, the Fe–O distance became 1.759 Å, only 0.01 Å away from the X-ray value. Therefore, the inaccuracy of the IMOMM(RHF:MM3) calculation had nothing to do with the fact that part of the system was described with an MM method, but was due to the failure of the RHF method in the description of the iron–oxygen interaction in this particular system.

3.3. Molecular Mechanics Level

As with the QM level, the MM level utilized can also affect decisively the outcome of the calculation. However, the choice of an appropriate MM level has some complications that were absent in the case of the choice of the QM level described earlier. The first of these complications is that there is no clear hierarchy of MM methods. In contrast with QM methods, all MM methods have similar computational costs, and the differences between them are concentrated mostly in the type of system for the which they are parameterized.

An additional problem, of a technical nature, is that while QM methods are usually available as options of a single program, MM force fields are usually available only in independent programs. On one hand, this poses a serious limitation in the comparison of their performances, although some efforts have been reported (46). On the other hand, this requires the programmer interested in building a QM/MM code to make a specific interface for each of the force fields, and to have access to the source code of each of them. An unfortunate outcome of this situation is that most applications of QM/MM methods to organometallic chemistry have been carried out with a single force field, namely, MM3 (47). This force field, devised especially for organic systems, seems appropriate for the description of steric effects, and in fact the comparison with experiment of the results obtained in its application in QM/MM methods is mostly correct. However, it would be highly desirable to carry out systematic comparisons with the performance of other force fields, and so far this has seldom been possible. The most significant results in this direction are probably the two independent sets of QM/MM calculations on the particular problem of olefin polymerization via homogeneous catalysis. One set of calculations (25,26) was carried out with the MM3 force field; the other set (27,28) was carried out with the AMBER force field. The results were in remarkable agreement, indicating that, at least in this particular case, the choice of one of the two force fields was not critical.

A second possible approach to the topic of the quality of the MM description is the tuning of some parameter of the force field to the particular problem under study. Although this has seldom been done in applications of the IMOMM method to transition metal chemistry, because it can obscure interpretation of the result, there is at least one particular example where it has been helpful (12).

The problem arose from the fact that IMOMM(Becke3LYP:MM3) calculations of a series of compounds containing chlorine ligands yielded abnormally poor results. Among these, we are going to discuss here the case of the Ir(H)$_2$Cl(P-(^{t}Bu$_2$Ph))$_2$ complex (Fig. 3). This complex has an experimental asymmetry (48) between the two H–Ir–Cl bond angles (156° and 131°), which was not reproduced by the IMOMM(Becke3LYP:MM3) calculation with the standard MM3 parameters (bond angles of 147° and 146°). This problem was solved from the observation that the van der Waals radius of chlorine used in the MM3 force field, which is fundamental in defining its steric activity, is defined for organic systems, where chlorine must have less anionic character than in inorganic compounds.

The relationship between the van der Waals radii of organic and inorganic chlorine was explored through high-level calculations on the CH$_3$–Cl $\cdots$ He and Na–Cl$\cdots$He model systems, and it was found that the van der Waals radius for inorganic chlorine is larger by 0.4Å. This increase was introduced in the corresponding MM3 radius, and the resulting IMOMM(Becke3LYP:MM3) calculations provided results in much better agreement with experiment. In particular, for the iridium complex described earlier the computed H–Ir–Cl bond angles became clearly different, with values of 162.6° and 122.0°.

3.4. Bond Distances of Connecting Atoms

One of the subjects that has taken a good deal of space in methodological discussions of the design of QM/MM methods is the way to deal with the connection between the QM and MM regions when there are chemical bonds across the boundary. In a number of methods, including IMOMM and derived methods, additional link atoms (usually hydrogen) are introduced in the QM calculation

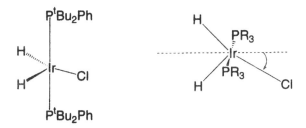

FIGURE 3 Two views of the geometry of the Ir(H)$_2$Cl(P^{t}Bu$_2$Ph)$_2$ complex.

because of the impossibility of having dangling bonds. The position of these atoms can be very critical if electronic effects of the MM region are going to be introduced in the calculation through the introduction of electrostatic charges or related parameters at their positions. Fortunately, things are not so dramatic when methods like IMOMM are used.

The approach chosen by IMOMM and related methods consists of associating the position of these additional link atoms to those of the atoms they are replacing. The two bonds are forced to lie in the same direction, and this solves the problem of two of the three degrees of freedom associated with the coordinates of each link atom. There remains the problem concerning the bond distance between the atoms. It must have different values for the QM atom–link atom and for the QM atom–MM atom distances.

Different answers to this problem have been proposed by IMOMM and derived methods (34). The simplest consists of freezing this value in both the QM and MM calculations, at corresponding different values, while other, more elaborate approaches define either a constant factor between the two values or a constant difference between them. In the simplest case, which is the one used in the examples presented throughout this chapter, there is the problem of choosing the particular values for both frozen distances. This choice, which will affect the numerical outcome of the calculation, is to a certain extent arbitrary. It would be particularly troublesome if it had a large effect on the outcome of the calculations.

In order to analyze this topic, a set of IMOMM(RHF:MM3) calculations has been carried out on the relative stabilities of the cis and trans isomers of $Pt(P^tBu_3)_2(H)_2$. This is a square planar complex that has been used before for validation tests of the IMOMM method (23). The QM/MM partition applied here uses $Pt(PH_3)_2(H)_2$ for the QM part, as in the previous tests, and the basis set is also the same (23). In this IMOMM calculation there are six P–H distances to be frozen in the QM part, with the associated six P–C distances to be frozen in the MM part. The values used previously for these parameters were 1.420 and 1.843 Å, respectively, taken from the equilibrium values used by the MM3 force field for the corresponding atom types. Additional calculations presented here consist of making a rather large displacement of 0.05 Å in each direction for each of the two parameters. Thus, the P–H distance has been assigned values of 1.370, 1.420, and 1.470 Å; and the P–C distance has been set to 1.793, 1.843, and 1.893 Å. The energy differences for the cis/trans pair in each of the resulting nine cases are collected in Table 1. The differences in relative stabilities of the two isomers do not deviate by more than 1.5 kcal/mol from those obtained with the standard values, even taking into account these abnormally large displacements from the equilibrium distances.

The conclusion therefore is that the choice of the particular values for the frozen distances related to the connecting atoms has a very minor effect on the outcome of the calculation, as far as the values applied are reasonable. A standard

TABLE 1 Dependence of the IMOMM(RHF:MM3) Computed Energy Difference (kcal/mol) Between cis and trans Isomers of $Pt(P^tBu_3)_2(H)_2$ with Respect to the Frozen Values of Distances in the QM/MM Boundary Region

	P–H = 1.370 Å	P–H = 1.420 Å	P–H = 1.470 Å
P–C = 1.793 Å	26.4	26.6	26.8
P–C = 1.843 Å	25.2	25.4	25.6
P–C = 1.893 Å	24.1	24.3	24.5

The different values (in Å) for the P–C distance in the MM calculation are presented in the rows, and the different values (in Å) for the P–H distance in the QM calculation are presented in the columns.

equilibrium value taken from a force field can in principle be an acceptable option. Care must be taken, however, always to use the same value throughout the whole set of calculations that must be compared.

4. PERFORMANCE CONSIDERATIONS

Apart from the initial setup of the calculation discussed in the previous section, there are a number of additional technical features that, without being specific to the application of QM/MM methods to transition metal chemistry, find particular importance in the performance of this type of calculation. They are briefly reviewed in this section

4.1. Use of Microiterations

In a typical pure QM or pure MM geometry optimization, all geometrical degrees of freedom are treated equally. At a given step of the optimization cycle, the energy is computed at the corresponding geometry, and so is the gradient, the first derivative with respect to each geometrical parameter, and, eventually, the hessian or second derivative. Geometry convergence is checked; and if it is not achieved, a new step is defined by the displacement of each of the variables.

The use of microiterations supposes a substantial breaking of this algorithm. The flux diagram of the alternative optimization scheme is shown in Figure 4. The geometrical degrees of freedom are divided into two sets, which can be labeled as M macrovariables and m microvariables. While the typical algorithm just described would consider the eventual convergence of the $M + m$ total variables at each geometry optimization step and generate a new value for each of them, the microiteration scheme works in a different way. At a given step in the optimization cycle the energy, gradient, and, eventually, hessian are computed; and the stationary-point condition (zero gradient) is checked for all variables, in

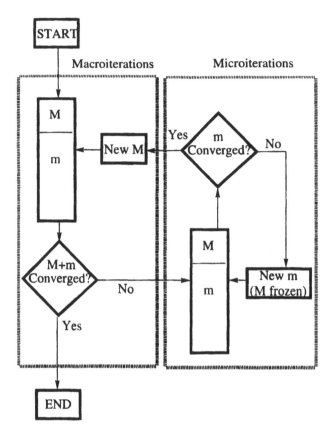

Figure 4   Flux diagram of an optimization scheme making use of microiterations.

much the same way as in a standard optimizer. The microiterations appear in the process of generation of the new geometry. Instead of modifying all the $M + m$ variables at once, only the m microvariables are modified (microiteration) in a frozen environment of the M macrovariables. The micro-optimization continues in a conventional way until the m microvariables are converged. Only after this process is finished are new values for the M macrovariables generated (macroiteration). The convergence of the total gradient is checked again, and the microiterations then have to be repeated for the new value of the macrovariables, the whole process continuing until the macrovariables are converged.

In principle, the optimizing scheme just described is independent of the QM/MM methodology and could also be applied to pure QM or pure MM calculations. However, its main effect would be to increase the total number of iterations and, consequently, the required computer time, thus making little sense. Things are very different in QM/MM methods like IMOMM and related methods. The QM energy is not affected by the position of atoms in the MM region, and therefore there is no need to recompute the QM energy during the micro-optimization process. As a result, a thousand microiterations will usually cost less than one single macroiteration. The use of a small number of macrovariables will reduce the number of macrosteps as compared to those required in a conventional optimization, with a consequent saving in the number of QM calculations and amount of computer effort required.

The efficiency of microiterations is well exemplified by test IMOMM calculations on the cis isomers of $Pt(P(^tBu)_3)_2(H)_2$, a system already used earlier and taken from an IMOMM calibration study (23). In this molecule, there are a total of 83 atoms. Therefore, the total number of geometrical degrees of freedom in the system is 243 (83 × 3 coordinates per atom, minus the 6 degrees of freedom corresponding to translation and rotation). If one further subtracts the 6 frozen distances between the connecting atoms, there are a total of 237 variables to be optimized in the calculation. If one uses a standard optimization scheme, with no microiterations, after 99 steps and 144 minutes of computer time the calculation is still far from convergence, with the maximum gradient 10 times above the threshold.

When one applies microiterations to this same problem, with the QM/MM partition described earlier, there are 21 macrovariables and 216 microvariables. The optimization is finished after 20 macrosteps, and it consumes only 41 minutes of computer time on the same workstation. The number of microiterations is of course much larger, with some macrosteps taking over 100 microsteps. This is, however, not a problem, because the microiterations, requiring only MM calculations, take much less computer time.

Although microiterations are a powerful tool that can save substantial computational effort in QM/MM calculations, it can also, in certain cases, introduce specific problems to the calculation. The user must take care to prevent them.

The first problem concerns the optimization of transition states. The use of microiterations implies that the microvariables are going to be fully optimized to a minimum energy in each step. Therefore, components of these variables in the transition vector will be neglected. This is not a problem if the atoms involved in the transition vector are in the QM region, and care should be taken so this is accomplished. At any rate, the problem affects only the computation of the second derivative, and the identification of geometries as stationary points (zero gradient) will still be rigorously accurate regardless of the use of microiterations.

A second problem with the use of microiterations appears when this method leads to different local minima for the same distribution of the macrovariables. This type of situation will usually make the calculation fail. The geometry optimizer assumes that each geometry (defined here exclusively by the macrovariables) has a unique energy and a unique gradient. If this is not the case at some point, the optimization procedure usually collapses in endless loops. There is, however, one easy way out of this problem: the addition of selected geometrical variables from the MM region into the macrovariables set, because there is certainly no methodological requirement to restrict the set of macrovariables to those in the QM and link regions.

4.2. Other Technical Aspects

The QM/MM programs that commonly have been applied to organometallic systems make use of geometry optimizers taken from pure QM programs. These optimizers are reliable in terms of locating the local minimum (or transition state) closer to the starting geometry, but offer no warranty in terms of whether this is the absolute minimum. This problem can be critical if there are several possible conformations available to the chemical system. Although this is true for any computational method, the problem is more likely to appear when one deals with very large interlocked ligands. And these are precisely the kind of systems that will be more commonly calculated with hybrid QM/MM methods.

There are at least two possible solutions to this problem. One is to carry out a conformational search. A number of possible minima are examined through a certain algorithm, and only the lowest of them is taken. This procedure is available in a number of MM programs, and in principle it should be required only for the MM region. Therefore, one could carry this search with the QM region frozen at a reasonable computational cost. The second option is to start from a known crystal structure. If the program starts with a geometry in the correct conformation, it will stay there. In any case, one must be aware that the option of defining a geometry with a graphical program from scratch can lead to an incorrect conformation and can seriously hinder the validity of the computation of relative energies.

A final comment on the practical use of QM/MM methods concerns the need to apply sophisticated features in the optimizer. In general, the QM/MM geometry optimization will be more challenging to the program than that of the corresponding QM optimization of the model system, for a variety of reasons. For example, one should expect a larger coupling between the geometrical variables and a less accurate initial estimation of the hessian. As a result, the user will probably have to exploit more uncommon optimization features than when interested in a pure QM calculation. In the particular case of the implementation of IMOMM in the mmabin program, using the gaussian92/dft code for the geometry optimizer, the features often required by QM/MM calculations included special care in the definition of the Z-matrix, numerical calculations of hessian elements for selected geometrical parameters, and application of additional features like restricting both the smallest and the largest acceptable absolute value of the eigenvectors of the hessian.

5. ANALYSIS OF RESULTS

Hybrid QM/MM methods have been developed essentially as a tool for the calculation of reliable geometries and energies at a reasonable computational cost. However, the results of their application can also be analyzed in a way that can provide further insight into the properties of chemical systems.

5.1. Identification of Steric Effects

Like several other concepts in chemistry, the definition of steric effects is clear from an intuitive point of view but difficult to put into a rigorous mathematical definition. The traditional division between electronic and steric effects loses precision when one realizes that steric effects are also included in the wavefunction and that they are also ultimately caused by the presence of electrons and nuclei. A possible definition of electronic effects as those going through bonds and steric effects as those going through space is appealing. But this has also problems in a number of cases. Should one label as steric electrostatic interactions between charged parts of different ligands in an organometallic complex? Are the agostic intramolecular $M \cdots H–C$ interactions electronic or steric?

This problem in the definition of electronic and steric effects has no clear solution, because it is essentially a problem of semantics, and this is not the subject of computational chemistry. However, hybrid QM/MM approaches like IMOMM and derived methods, neglecting the $E_{QM}(QM/MM)$ term (Sec. 2), can be useful in providing an arbitrary definition, as good as many others, that can be quantified in mathematical terms. Our proposal consists of defining steric effects as the perturbations introduced by the MM region in the properties of the QM region of the system. The rationale behind this definition is that these interactions consist exclusively of geometrical strains, and that corresponds precisely to the usual understanding of steric effects. This definition is not absolute, because it depends on the particular force field applied in each case, but can lead to a consistent quantitative separation of qualitatively different effects on chemical systems.

A good example of the possibilities of hybrid QM/MM methods in the separation of electronic and steric effects was provided by a study on the relative stability of the cis and trans isomers of $Ru(CO)_2(PR_3)_3$ complexes (11). The two isomers considered, presented in Figure 5, have a trigonal bipyramidal geometry, and they are labeled following the arrangement of the two carbonyl ligands. The two forms have been observed experimentally (11,49) by changing the nature of the phosphine ligand PR_3. In particular, when $PR_3 = PEt_3$, the cis isomer is the

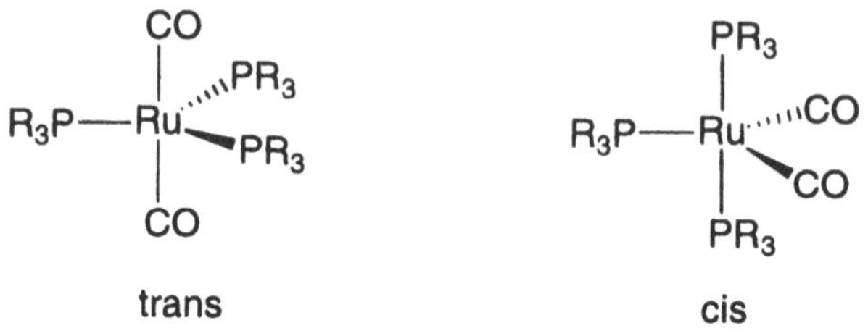

FIGURE 5   The two observed isomers in Ru(CO)$_2$(PR$_3$)$_3$ complexes.

species present in the crystal. When PR$_3$ = P(^{i}Pr)$_2$Me, two independent molecules are present in the crystal, one of them trans and the other cis. In this second case, IR studies in solution seem to indicate a larger proportion of the trans isomer. IMOMM(MP2:MM3) calculations were carried out on these complexes using Ru(CO)$_2$(PH$_3$)$_3$ for the QM part. Both isomers, cis and trans, were found to be local minima for each complex, with geometries in agreement with available X-ray data.

The most relevant part of this study was, however, the comparison of the relative energies. For Ru(CO)$_2$(PEt$_3$)$_3$, the cis isomer, the only one existing in the crystal, was computed to be more stable than the trans isomer by 3.0 kcal/mol. The relationship between the two isomers was reversed for complex Ru(CO)$_2$(P(^{i}Pr)$_2$Me)$_3$, with the trans species being more stable by 2.8 kcal/mol. Therefore, the QM/MM calculation reproduced the experimental observation accurately. This result would likely also have been obtained through the performance of much more expensive pure QM calculations. But the use of a QM/MM method had the additional advantage of allowing the clarification of the origin of the difference between both complexes, by a simple decomposition of the total QM/MM energy in its two terms, QM and MM. The QM contribution, representing the electronic effects, always favors the cis isomer, by 3.1 kcal/mol in the case of Ru(CO)$_2$(PEt$_3$)$_3$ and by 1.7 kcal/mol in the case of Ru(CO)$_2$(P(^{i}Pr)$_2$Me)$_3$. The MM contribution, representing the steric effects, marks the difference between both complexes. This part always favors the trans isomer, but it does so by quite different magnitudes: 0.1 kcal/mol when the phosphine is PEt$_3$, and 4.5 kcal/mol when the phosphine is P(^{i}Pr)$_2$Me. The conclusion from these results is straightforward. There is always a small electronic preference for the placement of the π-acidic carbonyl ligands in equatorial positions (cis isomer). There is a steric preference toward the placement of the bulkier phosphine ligands in the equatorial positions (trans isomer), with the weight of this preference depending on the nature of the phosphine ligand. Only in the case of bulkier phosphines are the steric effects strong enough to overcome the electronic preference for the cis isomer.

It can certainly be argued that the conclusions for $Ru(CO)_2(PR_3)_2$ complexes could have been deduced from the experimental results alone without need of calculation. Apart from the fact that the performance of calculations always strengthens the validity of qualitative arguments, it is worth mentioning that this example proves precisely the validity of the assignment of the MM part of the calculation to the steric effects.

A second example of identification of steric effects by QM/MM calculations concerns a case when their presence was not obvious a priori. This is the case of a joint experimental and theoretical study (19) on $[Ir(biph)X(QR_3)_2)]$ (biph = biphenyl-1,2-diyl; X = Cl, I; Q = P, As). These five-coordinate complexes have a distorted trigonal bipyramidal geometry, as shown in Figure 6. The phosphine (or arsine) ligands occupy the axial sites, and the chelating ligand biph and the halide are in the equatorial sites. One of the most intriguing features of these compounds is the deviation that the halide presents from the symmetrical arrangement between the two coordinating carbons of biph, a deviation that is characterized by values of ϕ (Fig. 6) different from zero. Previously reported calculations at the extended Hückel level seemed to indicate an electronic origin for the deviation. However, pure Becke3LYP calculations on the [Ir(biph) Cl(PH$_3$)$_2$] model system produced a symmetrical structure ($\phi = 0$), making unlikely an electronic origin for the distortion. The steric origin of the distortion was proved by IMOMM(Becke3LYP:MM3) calculations on the real system $[Ir(biph)Cl(PPh_3)_2]$, which yielded a distortion angle ϕ of $11.4°$ (as compared with the experimental value of $10.1°$).

Again, this result would probably also have been obtained through the performance of very expensive pure QM calculations on the full system. A pure QM calculation on the real system would nevertheless only prove the decisive role of the phenyl substituents of phosphine in the distortion but could not decide on whether their effect was electronic or steric. The fact that this distortion appears in the IMOMM calculation, where the QM effects of phenyls are neglected, in itself constitutes a direct proof that the origin of the distortion is purely steric.

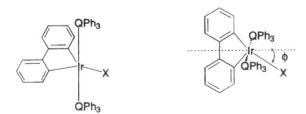

Figure 6 Two views of the geometry of [Ir(biph)X(QPh$_3$)$_2$] complexes, with indication of the definition of the distortion angle ϕ.

Another analysis showed that the symmetrical arrangement of chlorine corresponds to a sterically hindered position and that its distortion had a very low electronic cost. In this particular case, the steric origin of the distortion indicated by the hybrid QM/MM calculation was further confirmed by new experiments. In one of them, PPh_3 was replaced by $AsPh_3$, and in the other Cl was replaced by I. The logic behind these tests was that the modification of the size of the ligands should affect the magnitude of the steric effects and, therefore, the value of ϕ, and this was indeed observed in the crystal structures. In this way, this work constituted a nice example where the importance of steric effects was unearthed by hybrid QM/MM calculations and afterwards confirmed by new experiments.

5.2. Quantification of Steric Effects

Not only do QM/MM calculations lead to the identification of steric effects, they can also lead to their quantification and to its assignment to specific regions of the chemical systems.

An example of the characterization of intraligand steric effects in transition metal complexes was provided by the IMOMM(MP2:MM3) study of $[ReH_5(PR_3)_2(SiR_3)_2]$ systems (10). Two different complexes of this type, $[ReH_5(PPh(^iPr)_2)_2(SiHPh)_2]$ and $[ReH_5(PCyp_3)_2(SiH_2Ph)_2]$ (Cyp = cyclopentyl), were analyzed, with the QM part being constituted by the $[ReH_5(PH_3)_2(SiH_3)_2]$ model system. The two complexes, which had been experimentally characterized by X-ray and, in one case, neutron diffraction (50), are nine-coordinate and have a capped square antiprism structure, with qualitatively similar structures. However, there are some quantitative differences between the crystal structures, especially concerning the bond angles. The largest difference between the experimental structures is in the Si–Re–Si angle, which was well reproduced by the IMOMM calculations.

More to the point, the same study (10) exemplifies the possibilities of a quantitative analysis of steric effects through hybrid QM/MM calculations. This was carried out through a simple computational scheme consisting of several steps: 1) separation of the geometrical variables of the model system into two sets (A, consisting of the geometrical variables to be analyzed, and B, consisting of the other geometrical variables); 2) full IMOMM geometry optimization on the real system; 3) full QM optimization on the model system; 4) restricted IMOMM optimization on the real system, with the geometrical variables of set A frozen at the values of the MO optimization in the model system; 5) comparison of the results of steps 2 (full IMOMM optimization) and 4 (restricted IMOMM optimization). The basic idea in this scheme is that the restricted IMOMM optimization yields the geometry the system would take in the absence of steric effects. The initial separation of geometrical variables in step 1 is necessary to put aside geometrical variables that are deemed unimportant by chemical common sense but that could heavily influence the analysis. A typical example would be ligand rotation around M–P bonds.

This scheme was applied to both the $[ReH_5(PPh(^iPr)_2)_2(SiHPh)_2]$ and $[ReH_5(PCyp_3)_2(SiH_2Ph)_2]$ complexes (10). The comparison between the total energies of the restricted and full IMOMM optimized geometries showed energy differences of 4.9 kcal/mol for the first complex and 7.3 kcal/mol in the case of the second. This total energy was decomposed into its QM and MM contributions. In the case of $[ReH_5(PPh(^iPr)_2)_2(SiHPh)_2]$, the energy difference of 4.9 kcal/mol was obtained from an MM stabilization of 5.3 kcal/mol and an MO destabilization of 0.4 kcal/mol. The corresponding numbers for $[ReH_5(PCyp_3)_2(SiH_2Ph)_2]$ were 10.8 and 3.5 kcal/mol, respectively. Steric effects are therefore clearly more important in the second complex, a result that was not obvious before the calculations.

The calculations can nevertheless go further into the analysis of the steric effects. This is possible because of the mathematical structure of the MM energy, which is just a summation of terms. In the case of complex $[ReH_5(PCyp_3)_2(SiH_2Ph)_2]$, presented in Figure 7, the term-by-term decomposition of the 10.8 kcal/mol of MM difference between the restricted and the full optimization shows that there are two clearly dominant contributions, one of 7.9 kcal/mol associated with the "van der Waals" (VdW) nonbonding interaction, and another of 2.9 kcal/mol from the "bending" interaction. This is not surprising, because the traditional view of steric repulsions associates them precisely to nonbonding interactions through space, which are represented in the MM3 force field by the "van der Waals" term. This VdW term can be further decomposed into each of

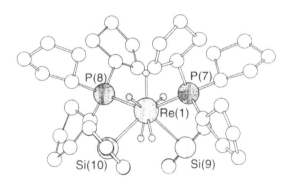

FIGURE 7 IMOMM(MP2:MM3) optimized structure of complex $[ReH_5(PCyp_3)_2(SiH_2Ph)_2]$. Hydrogen atoms not directly attached to the metal are omitted.

the single interatomic repulsions that contribute to it. Because of the difficulty of dealing with 6693 such interactions present in this particular complex, a filter was applied to choose only those differences larger than 0.05 kcal/mol. This reduced the number of interactions to 58, which were further grouped by the ligand to which each of the two atoms belonged. The result provided the contribution of each pair of ligands to the VdW repulsion in this complex. The leading terms happened to be those associated with the P(7)–Si(9) and P(8)–Si(10) ligand pairs (Fig. 7), which are related by symmetry, with values of 2.28 and 2.30 kcal/mol, respectively. The next largest repulsions in size correspond to the P(7)–P(7) and P(8)–P(8) pairs, associated with intraligand reorganization, with significantly smaller values of 0.61 kcal/mol each. A similar analysis of the "bending" term showed that the differences were also concentrated within the P(7) and P(8) ligands. This is therefore an example of how QM/MM methods can allow the quantification of intraligand steric effects and can allow their separation into specific ligand-to-ligand contributions in a given organometallic complex.

A second application of QM/MM methods to the quantification of steric effects is provided by the IMOMM(Becke3LYP:MM3) study on the origin of enantioselectivity in the dihydroxylation of styrene by the $(DHQD)_2PYDZ \cdot OsO_4$ complex (21). The size and complexity of the catalyst, which can be seen in Figure 8, preclude the possibility of accurate pure QM calculations on the problem, because the selectivity is decided precisely by the bulky substituents in the NR_3 cinchona group. The selectivity is defined by the initial approach of the substrate to the catalyst to form an osmate ester intermediate, and consequently a number of possible paths were analyzed. In particular, there are 12 such paths, defined by the three possible regions (A, B, and C) of approach of the substrate

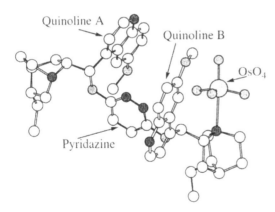

FIGURE 8   Schematic presentation of the $(DHQD)_2PYDZ \cdot OsO_4$ catalyst, with indication of the most significant areas of the cinchona ligand.

to the catalyst and the four possible orientations (I, II, III, and IV) of the phenyl ring of the substrate within each region, as shown in Figure 9. Each of these 12 possible paths was theoretically characterized through the location of the corresponding transition state, with its associated energy. The lowest-energy saddle point, therefore the most likely transition state for the reaction, was B-I, followed at close distance by B-III and B-IV, 0.1 and 2.7 kcal/mol above. The next saddle point in energy, A-IV, came 4.7 kcal/mol above B-I. The preference for paths B-I and B-III has an immediate consequence on the enantioselectivity of the reaction, because both lead to the R product. The minor S product must be obtained from a path through the B-IV saddle point. This is in excellent agreement with experimental data (51), which give a high enantiomeric excess of R product.

This agreement would in principle also be obtained by costly pure QM calculations. But the IMOMM method allows a further analysis of which regions of the catalyst affect the outcome of the reaction the most. In order to do that, the interaction energy between catalyst and styrene in the different saddle points B-I, B-III, and B-IV must be decomposed and compared. The first part of this decomposition consists of separating the interaction energy between substrate and catalyst into binding and distortion contributions. To do this, the process of formation of each saddle point from the separate reactants is divided into two imaginary steps: 1) a first step where the substrate and the catalyst are kept at infinite distance but distorted from their respective equilibrium geometries to the one they have in the saddle point (distortion energy), 2) and a second step where they are put together to yield the saddle point structure (binding energy). This division shows that the similar total interaction energies are reached through the addition of terms of different magnitude. For instance, a total interaction energy

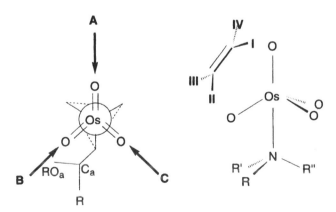

FIGURE 9 Definition of the criteria for labeling the 12 possible reaction paths in the reaction of H_2C=CHPh with $(DHQD)_2PYDZ \cdot OsO_4$.

of -3.3 kcal/mol for B-I is reached by adding a distortion of 15.4 kcal/mol and a binding of -18.7 kcal/mol; the interaction energy of -3.2 kcal/mol in B-III is reached from the addition of 13.0 kcal/mol (distortion) and -16.2 kcal/mol (binding). The fact that the total interaction energies are negative, with the transition state having an energy below the reactants, is explained through the presence of an intermediate in the reaction profile (13).

The analysis can be further refined to see which are the specific parts of the catalyst contributing to this binding energy. This can be done because the IMOMM partition in this particular system leaves most of the binding energy between catalyst and substrate in the MM part. The MM binding energies oscillate widely between -6.6 and -11.43 kcal/mol, while the QM change is comparatively much smaller, with changes between -6.9 and -7.6 kcal/mol. The analysis of the MM contribution to binding energies is straightforward, because it is composed mostly of "van der Waals" contributions, much in the same way as in the previous example. This VdW contribution comes again from a summation of atom-pair interactions. The number of atom pairs is very large, but a grouping of them in the regions depicted in Fig. 8 is instructive. In the case of B-I, 53% of the interaction happens with quinoline A, 21% with quinoline B, 12% with PYDZ, and 15% with the rest of the system. The respective values for B-III are 42%, 12%, 22%, 24%; B-IV gives 57%, 15%, 11%, 17%. The large role of the two quinoline and the pyridazine substituents in the selectivity of the reaction is clear from these results, which leave at most 24% of the interaction to the rest of atoms in the catalyst. The main role furthermore is played by one single group, quinoline A. These results indicate the most sensitive point of the catalyst for alterations in its selectivity.

A final comment on this example of catalyzed olefin dihydroxylation concerns the fact that "steric" effects appear to be stabilizing. This is no surprise if one realizes that they correspond to the parallel (or perpendicular) placement of aromatic rings, which is expected to yield a stabilizing interaction. The validity of the labeling of these interactions as steric effects is arguable, but it goes back to the discussion on the nature of steric effects at the beginning of this section. At any rate, these interactions are properly reproduced by the MM calculations and, therefore, correspond to steric effects according to our criteria defined earlier.

Another aspect concerning both examples presented in this subsection is the importance that van der Waals interactions appear to have in this topic. The dominance of this term, which can be surprising, very likely is affected by the choice of the MM3 force field. Other force fields grant a lesser importance to van der Waals terms and give more weight to electrostatic contributions, for instance. If such other force fields had been applied, the decomposition would likely be substantially different, and other terms should be more important in defining the difference. In any case, the total difference would have to be similar, as far as the different force fields would properly describe the same chemical reality. So this result is merely used in the sense that the more significant MM contributions correspond to what MM3 calls van der Waals interactions, without entering in the real chemical meaning of such terms.

The two examples described in this subsection show how the use of hybrid QM/MM methods gives access to analysis tools that are absent in pure QM calculations, even if those are more accurate. They also show how the particular chemical problem under study can require slightly different handling of the QM/MM results.

6. CONCLUSION

This chapter has presented a summary of specific aspects of the application of QM/MM methods to organometallic chemistry that should be of help to the researcher interested in their use. Emphasis has been put on practical examples of each feature. Although most of the examples correspond to calculations with one particular method, IMOMM, the general features should have application to the new developments that will surely appear in the near future.

Hybrid QM/MM methods should certainly be expected to experience a fast expansion, in large part because any improvement in QM or MM techniques will have a direct effect on them. The natural progress in both methodology and computer power, allowing calculations on larger and larger QM systems, will probably bring a time when systems that now require QM/MM methods will be studied with more accurate full QM methods. There will, however, always be a place for QM/MM methods, as a perpetual attempt to bridge the still enormous chasm between what can be computed and what actually exists.

ACKNOWLEDGMENTS

Financial support is acknowledged from the Spanish DGES through Project No. PB95-0639-CO2-01 and from the Catalan CIRIT through grant No. 1997SGR-00411. Thanks are due to Prof. Morokuma, Emory, for introducing me to this subject, and for continued discussions on methodological improvements. Prof. Lledós, Bellaterra, and Prof. Eisenstein, Montpellier, are also thanked for the steady flow of interesting chemical problems through the years, the solution to which has been the bulk of our application of QM/MM methods. Thanks finally to users of the mmabin program, Toshiaki Matsubara, Dima Khoroshun, Thom Vreven, Stephan Irle, Eric Clot, Helène Gérard, John McGrady, Gregori Ujaque, Guada Barea, Jaume Tomàs, Jean-Didier Maréchal, Lourdes Cucurull-Sánchez, Jordi Carbó, Nicole Dölker, Isabelle Demachy, Núria López, Simona Fantacci, Jordi Vázquez, Rainer Remenyi, Olivier Maresca, Fahmi Himo, and Antonio Morreale, because discussion with them has shaped the ideas that constitute this chapter.

REFERENCES

1. F Maseras. Topics Organomet Chem 4:165, 1999.
2. F Maseras, K Morokuma. J Comput Chem 16:1170, 1995.
3. A Warshel, M Levitt. J Mol Biol 103:227, 1976.
4. UC Singh, PA Kollman. J Comput Chem 7:718, 1986.
5. MH Field, PA Bash, M Karplus. J Comput Chem 11:700, 1990.
6. J Gao. Acc Chem Res 29:298, 1996.
7. J Gao. Rev Comput Chem 7:119, 1996.
8. D Bakowies, W Thiel. J Phys Chem 100:10580, 1996.
9. G Ujaque, F Maseras, A Lledós. Theor Chim Acta 94:67, 1996.
10. G Barea, F Maseras, Y Jean, A Lledós. Inorg Chem 35:6401, 1996.
11. M Ogasawara, F Maseras, N Gallego-Planas, K Kawamura, K Ito, K Toyota, WE Streib, S Komiya, O Eisenstein, KG Caulton. Organometallics 16:1979, 1997.
12. G Ujaque, F Maseras, O Eisenstein. Theor Chem Acc 96:146, 1997.
13. G Ujaque, F Maseras, A Lledós. J Org Chem 62:7892, 1997.
14. F Maseras, O Eisenstein. New J Chem 22:5, 1998.
15. G Ujaque, AC Cooper, F Maseras, O Eisenstein, KG Caulton. J Am Chem Soc 120:361, 1998.
16. F Maseras. New J Chem 22:327, 1998.
17. G Barea, A Lledós, F Maseras, Y Jean. Inorg Chem 37:3321, 1998.
18. J Jaffart, R Mathieu, M Etienne, JE McGrady, O Eisenstein, F Maseras. Chem Commun: 2011, 1998.
19. G Ujaque, F Maseras, O Eisenstein, L Liable-Sands, AL Rheingold, W Yao, RH Crabtree. New J Chem 22:1493, 1998.
20. AC Cooper, E Clot, JC Huffman, WE Streib, F Maseras, O Eisenstein, KG Caulton. J Am Chem Soc 121:97, 1999.
21. G Ujaque, F Maseras, A Lledós. J Am Chem Soc 121:1317, 1999.
22. JD Maréchal, G Barea, F Maseras, A Lledós, L Mouawad, D Pérahia. J Comput Chem 21:282, 2000.
23. T Matsubara, F Maseras, N Koga, K Morokuma J Phys Chem 100:2573, 1996.
24. Y Wakatsuki, N Koga, H Werner, K Morokuma. J Am Chem Soc 119:360, 1997.
25. RDJ Froese, DG Musaev, K Morokuma. J Am Chem Soc 120:1581, 1998.
26. DG Musaev, RDJ Froese, K Morokuma. Organometallics 17:1850, 1998.
27. L Deng, TK Woo, L Cavallo, PM Margl, T Ziegler J Am Chem Soc 119:6177, 1997.
28. TK Woo, PM Margl, PE Blöchl, T Ziegler J Phys Chem B 101:7878, 1997.
29. N Lopez, G Pacchioni, F Maseras, F Illas. Chem Phys Letters 294:611, 1998.
30. RDJ Froese, JM Coxon, SC West, K Morokuma. J Org Chem 62:6991, 1997.
31. RDJ Froese, K Morokuma. Chem Phys Letters 305:419, 1999.
32. S Humbel, S Sieber, K Morokuma. J Chem Phys 105:1959, 1996.
33. M Svensson, S Humbel, RDJ Froese, T Matsubara, S Sieber, K Morokuma. J Phys Chem 100:19357, 1996.
34. S Dapprich, I Komáromi, KS Byun, K Morokuma, MJ Frisch. J Mol Struct (Theochem) 461:21, 1999.
35. TK Woo, L Cavallo, T Ziegler. Theor Chem Acc 100:307, 1998.
36. JR Shoemaker, LW Burggraf, MS Gordon. J Phys Chem A 103:3245, 1999.

37. U Eichler, CM Kölmel, J Sauer. J Comput Chem 18:463, 1996.
38. RDJ Froese, K Morokuma. In: PvR Schleyer, ed. Encyclopedia of Computational Chemistry. Vol 4. New York: Wiley, 1998, p 1257.
39. J Gao. In: PvR Schleyer, ed. Encyclopedia of Computational Chemistry. Vol 4. New York: Wiley, 1998, p 1257.
40. KM Merz Jr, RV Stanton. In: PvR Schleyer, ed. Encyclopedia of Computational Chemistry. Vol 4. New York: Wiley, 1998, p 2330.
41. J Tomasi, CS Pomelli. In: PvR Schleyer, ed. Encyclopedia of Computational Chemistry. Vol 4. New York: Wiley, 1998, p 2343.
42. I Tuñon, MTC Martins-Costa, C Millot, MF Ruiz-López, J-L Rivail. J Comput Chem 17:19, 1996.
43. M Strnad, MTC Martins-Costa, C Millot, I Tuñon, MF Ruiz-López, J-L Rivail. J Chem Phys 106:3643, 1997.
44. M Brookhart, MLH Green, L Wong. Prog Inorg Chem 36:1, 1988.
45. L Stryer. Biochemistry. New York: Freeman, 1995.
46. MD Beachy, D Chasman, RB Murphy, TA Halgren, RA Friesner. J Am Chem Soc 119:5908, 1997.
47. NL Allinger, YH Yuh, J-H Lii. J Am Chem Soc 111:8551, 1989.
48. A Albinati, VI Bakhmutov, KG Caulton, E Clot, J Eckert, O Eisenstein, DG Gusev, VV Grushin, BE Hauger, WT Klooster, TF Koetzle, RK McMullan, TJ O'Loughlin, M Pélissier, JS Ricci, MP Sigalas, AB Vymenits. J Am Chem Soc 115:7300, 1993.
49. M Ogasawara, F Maseras, N Gallego-Planas, WE Streib, O Eisenstein, KG Caulton. Inorg Chem 35:7468, 1996.
50. JAK Howard, PA Keller, T Vogt, AL Taylor, ND Dix, JL Spencer. Acta Crystallogr B 438:4338, 1992.
51. EJ Corey, MC Noe. J Am Chem Soc 118:319, 1996.

7

HIV Integrase Inhibitor Interactions with Active-Site Metal Ions: Fact or Fiction?

Abby L. Parrill, Gigi B. Ray,
Mohsen Abu-Khudeir, Amy Hirsh,
and Angela Jolly
The University of Memphis, Memphis, Tennessee

1. INTRODUCTION

The human immunodeficiency virus (HIV) is a retrovirus that is responsible for acquired immunodeficiency syndrome (AIDS) (1). Human immunodeficiency virus requires the activity of three enzymes during its life cycle (2). Pharmaceutical agents targeting two of these enzymes, reverse transcriptase and protease, are in clinical use. These pharmaceutical agents are not ideal, however, due to the rise of resistant viral strains after treatment is initiated (3,4). The third enzyme, integrase (IN), is the least explored enzyme target for HIV treatment (5), with one agent undergoing clinical trials, although evidence indicates that IN is not its primary target in vivo (6). Integrase is responsible for two essential (7,8) catalytic activities that result in incorporation of viral DNA into host DNA. These activities are shown schematically in Figure 1. The first is 3′ processing, in which the 3′ ends of viral DNA are recessed by removal of two bases. The second is strand transfer, in which the viral DNA is joined to the 5′ phosphate of a staggered cut in the host DNA. This activity places the IN enzyme into the polynucleotidyl-

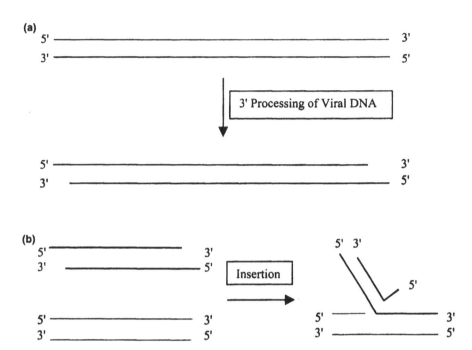

FIGURE 1   Schematic of reactions catalyzed by the integrase enzyme. (a) Processing of the 3′ ends of viral DNA to generate staggered ends. (b) Insertion of staggered viral DNA, shown as thick lines, into double-stranded host DNA, shown as thin lines.

transferase family, which also includes RNase H, the Mu transpose, and Ruv C (9,10). The polynucleotidyltransferases characteristically contain an organometallic active site, with a divalent metal cation playing a critical role in the catalysis of the DNA or RNA strand transfer reactions.

Pharmaceutical development of IN-targeting agents lags behind the development of agents targeting protease and reverse transcriptase, due to incomplete structural data on IN. Complexes of the reverse transcriptase and the protease with their inhibitors as characterized by X-ray crystallography (2) have been available for many years, thus providing the basis for rational design of additional therapeutic agents. The IN structure, however, has been characterized less thoroughly than other HIV enzymes, with crystallographic characterization of the catalytic core domain in the absence of inhibitors (9,11,12) and NMR characterization of the N-terminal and C-terminal domains (13–15). Only very recently has the first crystal structure of the HIV IN catalytic core in the presence of an inhibitor become available (16). Structural information for the intact enzyme and for regions connecting the three domains is not available.

Strand transfer activity of the IN enzyme is consistent with the polynucleotidyltransferase family requirement for a divalent metal ion cofactor. Manganese or magnesium can serve as this cofactor (1,10). Studies on the impact of metal ions on the binding of domain-specific antibodies have shown that the metal ion induces a conformational change (17). This work indicates that the conformational change involves a reorganization of the catalytic core and C-terminal domains based on two separate experiments. Interactions with antibodies specific for the apoenzyme are lost upon metal ion binding, and an increased resistance to proteolysis is observed. These changes were independent of the presence of the N-terminal domain.

The active-site metal ion plays an accepted role in the catalytic function of IN, although its role in inhibition is poorly understood. The role of the metal ion in inhibition has been a topic of speculation since the first inhibitors began to appear in the literature in 1993. Early work identified some DNA-binding and DNA-intercalating molecules as IN inhibitors (18). These early inhibitors include doxorubicin (Fig. 2, **1**), caffeic acid phenethyl ester (CAPE, Fig. 2, **2**), and quercetin (Fig. 2, **3**). This work established, however, that there was no clear relationship between DNA-binding or DNA-intercalating activities and inhibition. It was noted that active inhibitors contained polyhydroxylated aromatic regions that

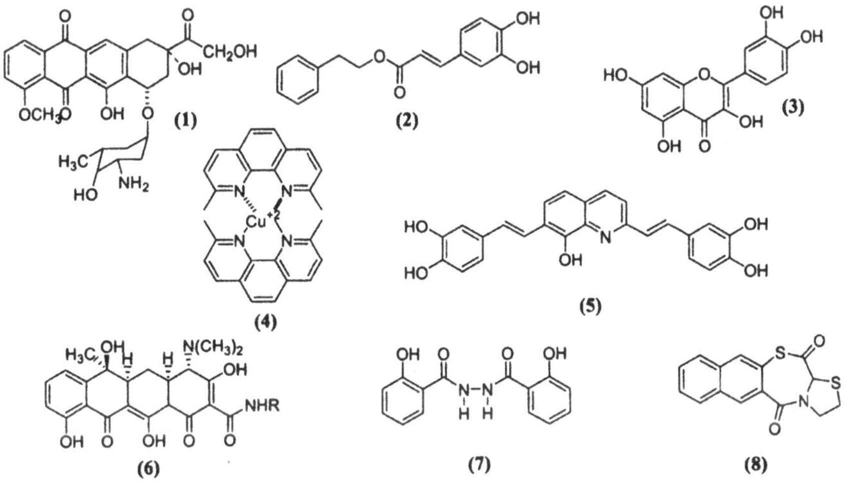

FIGURE 2   Structures of HIV IN inhibitor classes.

might be involved in metal ion chelation at the zinc finger region in the amino terminus of IN (18). Subsequent work with flavones and CAPE analogs identified other active inhibitors (19,20). The ability of these molecules to inhibit the disintegration reaction catalyzed by integrase lacking the amino-terminal domain clarified that inhibitors bind in the catalytic core domain. Speculation on the role of metal ions in inhibition moved to the organometallic active site. The role of the metal ion in inhibition was questioned by the discovery that Cu^{2+}-phenanthroline complexes (Fig. 2, 4) were also able to inhibit IN (21). The inhibition observed was selective for copper over magnesium, indicating that the inhibition by the phenanthroline complex did not involve the divalent cation in the organometallic active site. Independent observation that most polynucleotidyltransferases contain two organometallic active sites separated by 4.4 Å, as well as the commonly found diaryl functionality in known inhibitors, led to explorations of linker type and length between two metal chelating functionalities (Fig. 2, 5) (22–24). Computations on the strength of the cation–π interaction of benzene and catechol with the divalent cations of manganese and magnesium were undertaken in 1997 (25). These calculations showed that the interaction energy is much larger than for previously studied cation–π interactions with monovalent cations and provided no reason to rule out such interactions in the organometallic active site of IN. Experimental work probing metal ion selectivity showed that different metal cations can be used to separate the stages involved in the activity of IN (26). Integrase activity can be divided into an assembly stage, during which the enzyme complexes with DNA to form a preintegration complex, and a catalytic stage, during which the 3′ processing and strand transfer reactions occur. It was shown that although both magnesium and manganese can serve as cofactors for both stages, calcium promotes only assembly and cobalt promotes only catalysis. Further studies used sequential addition of calcium and cobalt to allow addition of inhibitors between the stages as well as prior to both. These studies showed that selected inhibitors, many containing polyhydroxylated aromatic systems, were effective only if added prior to the assembly stage and showed no effect on catalysis in a preassembled complex. These results neither support nor contradict the hypothesis that the inhibitors interact at the active-site metal ion. Additional research identified tetracyclines (Fig. 2, 6) as a new class of IN inhibitors (27). Not only was a new class of inhibitors identified, their inhibition was also evaluated in the presence of both manganese and magnesium ions. No difference in potency was observed, although the earlier computational work had identified a significant difference in gas-phase interaction energies between aromatic systems and these two metal ions (25). These data seem to add support to a lack of a direct interaction between the organometallic active site and enzyme inhibitors. Our understanding of the metal ion role in the inhibition of IN was further muddied in 1998. Neamati and coworkers published the discovery of another class of

IN inhibitors, the salicylhydrazines (Fig. 2, 7) (28). These inhibitors, however, showed selective inhibition in the presence of manganese, but not magnesium. This result indicates that the metal ion can certainly influence the binding of some inhibitors, although not specifically proving that the metal ion must interact directly with the inhibitors. No comparative structural information is currently available for the IN protein containing different divalent metal ions in the active site. Thus the metal ion may influence the shape of a distant inhibitor binding site (an allosteric site). A final result to add to this history is the recent publication of another structural class of inhibitors that show no preference between manganese and magnesium, the thiazolothiazepines (Fig. 2, 8) (29).

The prior history of evidence both supporting and contradicting a direct role for metal ion interaction with IN inhibitors demonstrates the need to perform additional research on this issue. This chapter presents our multipronged computational approach to elucidate the exact involvement of the organometallic active site in the binding of inhibitors to the catalytic core of the IN enzyme. Quantum mechanical calculations are applied to examine differences in metal ion interactions with inhibitors and differences in metal ion interactions with the IN active site. Molecular mechanics calculations are used to explore the best geometric (steric) and electrostatic fit of the IN inhibitors to the IN catalytic core. Current results from each of these three avenues of investigation are described in the following sections.

2. METAL ION COMPLEXES WITH INHIBITORS

This section describes our studies on the interactions of Mg^{2+} and Mn^{2+} with two classes of inhibitors, the salicylhydrazines (Fig. 2, 7) (28) and the thiazolothiazepines (Fig. 2, 8) (29). Inhibitors used in these studies as well as in the docking studies described later are shown in Table 1 along with their biological activities. The salicylhydrazines are of interest because they show an inhibitory effect only when assayed with Mn^{2+} as the active-site metal ion. The thiazolothiazepines, on the other hand, are equally active with either Mg^{2+} or Mn^{2+} present in the active site. Clearly the metal ion has an impact on inhibition by some inhibitors. The modeling studies described in this section seek to evaluate two different mechanisms through which this impact could arise from the direct metal ion interaction with the inhibitors that has been postulated in the literature since 1993. First, the impact could be due to geometric differences that arise from alternate sites of chelation to the inhibitor for different metals. Second, a direct interaction with the metal ion might be reflected in a correlation between the strength of the interaction and the biological activity of inhibitors in a structural class. It is entirely possible that the metal ion impact is a complex mix of these two factors, although that possibility is not tested in this work.

TABLE 1 Structures and Activities of Salicylhydrazines Used in Metal Ion Chelation Site Studies and Docking Studies

	Activity (IC_{50}, µm)		
Structure	3′ processing[a]	Strand transfer[b]	Qualitative
SH30 (Ref. 28)	0.9	0.6	Active
SH1 (Ref. 28)	2.1	0.7	Moderate
SH2 (Ref. 28)	>100	>100	Inactive
T1 (Ref. 29)	110	146	Weak
T19 (Ref. 29)	40	47	Moderate-weak

[a] Concentration at which 50% of normal 3′ processing activity of the enzyme is observed.
[b] Concentration at which 50% of normal strand transfer activity of the enzyme is observed.

The inhibitor classes under investigation both contain multiple sites at which metal ions could interact. Thus initial studies focused on identifying the energetically favored site for Mg^{2+} or Mn^{2+} to interact. These studies probe the possibility that direct metal ion interactions impact inhibition through the first mechanism outlined earlier. Specifically, salicylhydrazines differentially inhibit the enzyme based on the metal ion present, but the thiazolothiazepines show no such dependence. Therefore, results showing a difference in geometry for metal ion chelates of salicylhydrazines but no geometry difference for the metal ion chelates of the thiazolothiazepines would provide support for this mechanism. Metal ion complexes of two salicylhydrazines and two thiazolothiazepines were built and optimized using the semiempirical PM3(tm) method implemented in the Spartan program (30). Figure 3 shows the optimized complexes of Mg^{2+} with the salicylhydrazines. Complexes containing Mn^{2+} had similar structures, with the exception that SH30 had a more stable π complex with Mn^{2+} from the phenol ring rather than the diazolone ring. Table 2 includes the energies of the complexes of both metal ions with two salicylhydrazines and two thiazolothiazepines.

These results demonstrate that Mg^{2+} and Mn^{2+} show energetic preferences for different chelation sites on both the salicylhydrazine and thiazolothiazepine structures. The Mg^{2+} ion forms a more stable σ complex in both salicylhydrazine structures. These complexes are, respectively, 83 and 31 kcal/mol more stable than the complexes formed when Mg^{2+} forms a π complex with the phenol or dirazolone ring. The Mn^{2+} ion, however, shows an 11–94-kcal/mol preference to form π complexes with the salicylhydrazines. Similar geometries are preferred for the thiazolothiazepine structures. These structures show an average 45-kcal/mol preference for Mg^{2+} to form σ rather than π complexes. The opposite prefer-

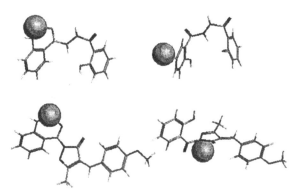

Figure 3 Optimized structures of salicylhydrazine complexes with Mg^{2+}. Top left, SH1 σ complex. Top right, SH1 π complex. Bottom left, SH30 σ complex. Bottom right, SH30 π complex.

TABLE 2   PM3(tm) Heats of Formation for Metal Ion Complexes of
Integrase Inhibitors

	ΔH_f (kcal/mol) Mg^{2+}			ΔH_f (kcal/mol) Mn^{2+}		
Inhibitor	π	σ	$\Delta\Delta H_f$[a]	π	σ	$\Delta\Delta H_f$[a]
SH 1	370	287	83	290	301	−11
SH 30	343	312	31	250	344	−94
T 1	449	400	49	389	413	−24
T 19	452	411	41	389	413	−24

[a] $\Delta\Delta H_f = \Delta H_f(\pi) - \Delta H_f(\sigma)$.

ences are seen for Mn^{2+}. There is a 24-kcal/mol preference for π over σ complex formation.

These results fail to provide support for a mechanism of direct metal ion interaction in which geometric differences for complexes of different metals lead to differences in inhibition of the enzyme.

Further studies were performed using the single preferred chelation site for each metal interacting with the salicylhydrazine series of inhibitors. These studies evaluated the metal ion binding energy using the PM3(tm) semiempirical method as well as the 3-21G* ab initio basis set in the Spartan program. The metal ion binding energy was determined by subtracting the energies of the isolated ion and the isolated inhibitor from the energy of the ion:inhibitor complex. The semiempirical calculations provided energies in the form of heats of formation, whereas the ab initio calculations provided electronic energies. However, the trends in the differences are expected to be similar by both methods. The results of the calculations on 10 salicylhydrazines are shown graphically in Figure 4. The correlation coefficients (R^2) for Mn^{2+} binding versus biological activity determined in the presence of Mn^{2+} are 0.211 and 0.416 by the semi-empirical and ab initio methods, respectively. The correlation coefficients for Mg^{2+} binding versus biological activity determined in the presence of Mn^{2+} are essentially the same, 0.284 and 0.405 by the semiempirical and ab initio methods, respectively. These data clearly show that the correlations for both metal ions are the same at a given level of theory even though biological activity is only observed experimentally in the presence of Mn^{2+}. Thus a direct interaction with the metal ion is *not* reflected in a correlation between the strength of the interaction and the biological activity of these inhibitors.

The current results provide no support for either mechanism by which the metal ion impact on inhibition could arise from direct interactions between the metal ion and the inhibitors. The results reported here used medium-sized basis

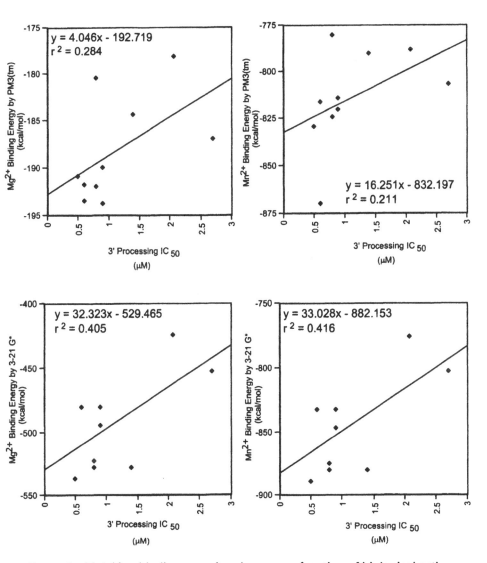

FIGURE 4   Metal ion binding energies shown as a function of biological activity measured in the presence of Mn^{2+}. Top: Semiempirical results. Bottom: Results of 3-21G* calculations.

sets and did not include a full solvation sphere for each metal ion. Additional studies are under way using larger basis sets as well as explicit water molecules completing the inner coordination sphere. The next section describes studies in progress designed to assess the impact of metal ion differences on the integrase active site.

3. INFLUENCE OF METAL IONS ON THE INTEGRASE ACTIVE SITE

The interactions of Mg^{2+}, Mn^{2+}, Ca^{2+}, and Co^{2+} with the active site of IN were investigated. These metal ions clearly have different impacts on the ability of the enzyme to assemble and catalyze the strand transfer reaction as well as showing selectivity toward some structural classes of inhibitors (26,28,29). These effects may be due to minor geometric differences in the inner coordination sphere that propagate into nearby regions of the enzyme structure, thus affecting allosteric sites at which other viral proteins involved in the preintegration complex need to interact. Because Ca^{2+} is only able to promote assembly, and Co^{2+} is only supportive of catalysis, these two ions are expected to induce the structural ex- tremes, with Mg^{2+} and Mn^{2+} having more similar structural impact. The crystallo- graphic complex of the IN catalytic core with Mg^{2+} shows that two carboxylates from aspartate residues 64 and 116 chelate in an η^1 fashion with the metal ion (31). Four water molecules occupy the remaining sites of the octahedral inner coordination sphere. The crystallographic geometry with hydrogen atoms added by the MOE program (32) was used to initiate quantum mechanical optimization of the active site with each metal ion. Optimizations were performed at the Hartree–Fock level of theory using the SBK basis set (33–35) and effective core potentials (36) as implemented in the GAMESS program (37). Figure 5 shows the truncated model of the active site that was included in these computations.

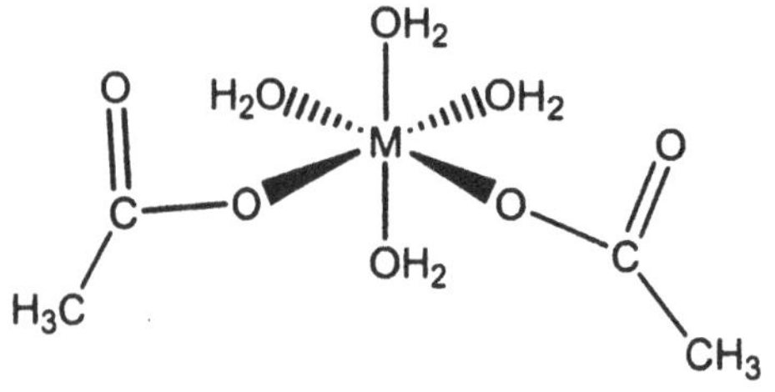

FIGURE 5   Truncated active-site region modeled, with M = Mn^{2+}, Mg^{2+}, Ca^{2+}, and Co^{2+}.

Differences in the calculated geometries of the active-site model with different metal ions were modest, in agreement with crystallographic results for the avian sarcoma virus (ASV) integrase with Mn^{2+}, Mg^{2+}, Ca^{2+}, Zn^{2+}, and Cd^{2+} (38,39). Our results indicate that the distances between the metal ion and the acetate oxygens averaged 2.06, 2.10, 2.15, and 2.26 Å in the Mg^{2+}, Co^{2+}, Mn^{2+}, and Ca^{2+} complexes, respectively. Distances between the metal ion and the water ligands averaged 2.12, 2.17, 2.25, and 2.31 Å in the Mg^{2+}, Co^{2+}, Mn^{2+}, and Ca^{2+} complexes, respectively. Thus distances to the metal ions follow the same trend regardless of the ligand. The distances to the transition metals, cobalt and manganese, are consistent with those determined in an evaluation of the PM3(tm) method, in which the bond lengths between these metals and water ligands in octahedral complexes were predicted within 5% and 3% of the experimental reference value (40). A more significant difference can be seen in the angle formed at the metal ion by the bonds to the two acetate ligands. These angles are 85.0°, 103.5°, 106.1°, and 107.9° in the Ca^{2+}, Co^{2+}, Mg^{2+}, and Mn^{2+} complexes, respectively. This angle in the calcium complex is clearly quite different from that observed in the three other complexes. The previously mentioned crystal studies of the ASV IN, in fact, identified that the calcium complex was not actually octahedral, but had only three water ligands (38,39). This difference may relate to the inability of the integrase enzyme to perform its catalytic function with calcium in the active site. These results, however, do not clearly delineate a reason why cobalt is unable to promote assembly of the viral complex even though it is able to facilitate catalysis. An overlay of the four optimized model active sites is shown in Figure 6.

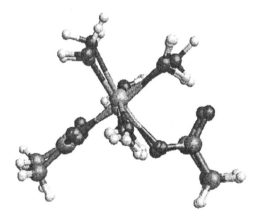

FIGURE 6   An overlay of the model active sites containing Mn^{2+}, Mg^{2+}, Ca^{2+}, and Co^{2+}.

Additional calculations are under way with full aspartate amino acid residues rather than acetate ligands. Finally, calculations need to be done with Ca^{2+} as an incomplete octahedron, as was observed in the ASV IN crystal structure (38,39). The quantum mechanically optimized geometries of the active-site models will be used to determine if the small changes induced by the presence of different metal ions have an effect on the larger HIV IN catalytic core structure. The optimized active-site geometries will be fitted back into the catalytic core and will be held fixed during molecular dynamics simulations that allow the remainder of the IN structure to adapt to the different metal ion environments. It will be of interest to see whether or not molecular dynamics calculations will perpetuate minor geometry differences around the metal ion into the nearby catalytic loop region. These studies may provide insight into a possible allosteric role of the metal ion in IN inhibition by certain inhibitor classes.

4. DOCKING INHIBITORS TO THE INTEGRASE CATALYTIC CORE

An independent series of calculations was performed to explore the steric and electrostatic complementarity of the salicylhydrazine inhibitors for different regions of HIV IN. These calculations used empirical docking and binding affinity evaluations to find favorable regions on the crystal structure of the IN catalytic core (31) for inhibitor binding. Two regions of the IN catalytic core were emphasized in these docking studies, the region around the metal ion and the region above the highly flexible catalytic loop. The active-site metal region and the catalytic loop are labeled in Figure 7. The direct and indirect evidence for a role of the metal ion in both catalysis and inhibition provided motivation to closely examine the inhibitor interactions with the active site. The crystal structure of an HIV IN inhibitor complexed with ASV IN (41) that showed inhibitor binding above the catalytic loop provided motivation to closely examine this second region. Figure 7 shows an overlay of the HIV IN catalytic core crystal structure used in our docking studies on the ASV IN complexed with an HIV IN inhibitor. This figure demonstrates that differences between the two structures are most significant in the region occupied by the inhibitor. Thus the ASV IN crystallographic complex does not definitively identify the inhibitor binding site in the HIV IN structure.

Several methods are available to explore the conformations and configurations of a small molecule in the environment of a larger, rigid biomolecule. These are called *docking* methods. One method used in these studies is the docking module within the MOE program. This docking method precalculates a grid of steric and electrostatic interaction energies for probe atoms within a user-defined volume, the docking box, containing all or only part of the biomolecule. Pre-

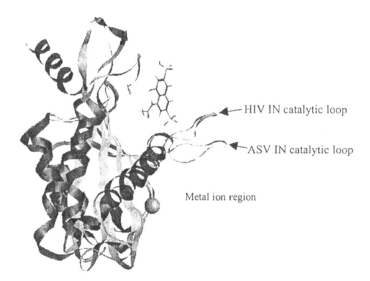

HIV IN catalytic loop

ASV IN catalytic loop

Metal ion region

FIGURE 7   Overlay of the ASV IN:inhibitor crystallographic complex on the HIV IN crystal structure.

computation of these interaction energies allows for rapid evaluation of ligand orientations in the environment of the biomolecule by summing the steric and electrostatic energies at grid points occupied by ligand atoms. Random initial ligand conformations and orientations are subjected to Monte Carlo optimization of the interaction energy. A second docking method was also used in order to investigate if a consensus on the optimal inhibitor binding site would be achieved. The second docking method uses fast Fourier transforms to optimize the orientations of a rigid ligand in the environment of a rigid target biomolecule using geometric complementarity (42). These two docking methods have complementary strengths and weaknesses. The docking method implemented in the MOE program offers the advantages of ligand flexibility and consideration of electrostatic complementarity. The fast Fourier transform method has the advantage of speed and can be used to scan the entire protein surface for favorable binding pockets.

    In order to determine the most likely location for ligand binding in the HIV IN catalytic core, it is necessary not only to fit the ligand into potential binding sites, but also to energetically evaluate the resulting complexes, known as *scoring*. Both docking methods generate multiple geometries for the complex. Fifty geometries were generated by MOE and 100 by the fast Fourier transform in these studies. They each also provide a relative ranking based on empirical interaction energies, either steric and electrostatic in the case of the MOE docking method, or van der Waals interaction energy in the case of the fast Fourier transform method. These rankings both fail to consider the importance of directionality in hydrogen bonding interactions as well as entropic contributions to the binding free energy from desolvation and loss of conformational freedom. Thus the docked configurations generated by the MOE program were additionally evaluated using the SCORE program (43). The SCORE program is an empirical method for calculation of binding affinities and was calibrated using crystallographic complexes from the Protein DataBank with known binding affinities (43). It was therefore not specifically developed to determine the actual binding site for a ligand given several potential complexes, but rather to compare binding affinities among correct geometries of different protein:ligand complexes. Thus this use of this program is outside the range of activities for which it was validated. Nevertheless, the SCORE program includes terms for desolvation, loss of ligand flexibility, metal ion interactions, and for three different strengths of hydrogen bonds. These terms are complementary to those utilized by the other tools, and SCORE provides a valuable additional tool for the determination of whether there is a single binding site that multiple methods agree upon.

Several salicylhydrazines with a range of biological activity were selected for use in docking studies. The structures and biological activities of these inhibitors are shown in Table 1. For each inhibitor studied, both docking methods generated complexes having the inhibitor in both regions of interest. The fast Fourier transform method, which scanned the entire protein surface, did not significantly populate other regions of the protein structure. The MOE method was also used to scan the entire protein surface for the largest salicylhydrazine (structure not shown) and also failed to significantly populate other regions. This indi-

TABLE 3 Relative Rankings for IN Complexes with Salicylhydrazines

Inhibitor and location	MOE rank	SCORE rank	Most structurally similar FFT rank
SH30—near catalytic loop (a)	1	15	1
SH30—near catalytic loop (b)	9	1	1
SH1—near catalytic loop	2	3	9
SH2—near catalytic loop	5	35	1
SH30—near metal ion	2	37	7
SH1—near metal ion	1	46	2
SH2—near metal ion	1	30	1

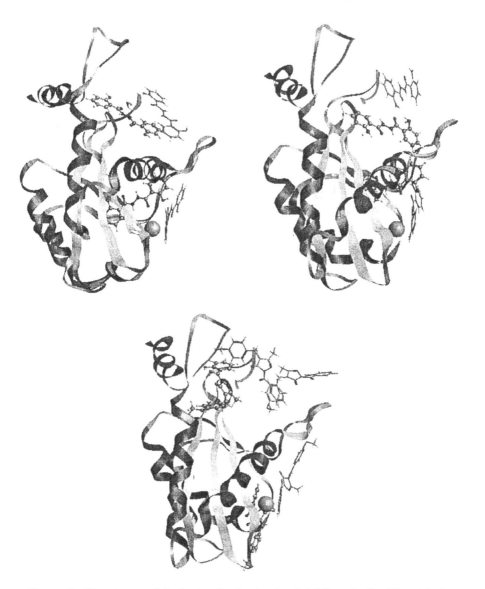

FIGURE 8 Structures of docked salicylhydrazine inhibitors in the IN catalytic core. Salicylhydrazines positioned by the MOE program are shown as stick models, those positioned by the fast Fourier transform method are shown as ball and stick models. SH1, SH2, and SH30 are shown on the top left, top right, and bottom, respectively.

cates that the choice of focus areas in the vicinity of the metal ion and above the catalytic loop was appropriate. The relative rankings of inhibitor interactions in these two sites of interest are shown in Table 3. The docked complexes are shown in Figure 8.

These results demonstrate that a better consensus among the evaluation methods is achieved for the two active inhibitors, SH30 and SH1, than for the inactive compound, SH2. The evaluation methods rate complexes having the inhibitor above the catalytic loop more consistently and as better ligands than those having the inhibitor near the metal ion. The ability of both docking methods to populate the site near the metal ion as well as above the catalytic loop does indicate that different inhibitors may show a preference for different sites based

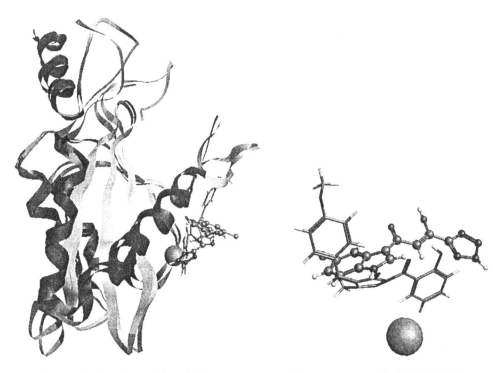

Figure 9 Overlay of the HIV IN crystallographic complex with CITEP (PDB entry 1QS4) on the highest-ranking docked complex of SH30 in the IN active site. Left: Ribbon model of the protein, with the metal ion shown as a spacefilling model, CITEP shown as a ball-and-stick model, and SH30 shown as a stick model. Right: Close view of the metal ion (spacefilling), CITEP (ball and stick), and SH30 (stick).

on their structure. In fact, crystallographic evidence points to this possibility, for the crystal structure of ASV IN with 4-acetylamino-5-hydroxynaphthalene-2,7-disulfonic acid (Y3) (41) shows the inhibitor populating the site above the catalytic loop, whereas the crystal structure of HIV IN with 1-(5-chloroindol-3-yl)-3-hydroxy-3-(2H-tetrazol-5-yl)-propenone (ClTEP) (16) shows the inhibitor populating the site near the metal ion.

Comparison of our docking results for the salicylhydrazines against the crystal structure of the HIV IN catalytic core with ClTEP shows some interesting correspondences. Figure 9 shows an overlay of this structure on our highest-ranking complex of SH30 in the metal ion site. The two inhibitors occupy overlapping regions of space and have weak π interactions with the metal ion. The distances observed for these π interactions are 4.4 Å for our SH30 complex and 5.7 Å in the ClTEP crystal structure. These are essentially the same, due to the resolution of the experimental structure. These π interactions are not optimal. Our optimizations of metal ion π complexes with both the thiazolothiazepines and the salicylhydrazines result in distances of less than 3 Å. The apparent weakness of this interaction may demonstrate why our studies of inhibitor complexes with metal ions did not support a direct metal ion interaction. Studies are under way to dock the ClTEP inhibitor to the integrase catalytic core to evaluate whether the scoring methods used to rank the salicylhydrazine complexes are able to accurately provide a consensus that the metal ion site is more favorable for this inhibitor. The result of this evaluation will determine whether our apparent consensus favoring the site above the catalytic loop for the salicylhydrazines does indeed represent the preferred binding site for this class of inhibitors.

5. CONCLUSIONS

The multipronged approach toward elucidating the role of divalent metal cations in HIV IN inhibition has generated preliminary data that lend support to an allosteric involvement of the metal ion in inhibition by two chemical classes of HIV IN inhibitors, the salicylhydrazines and the thiazolothiazepines. First, both the metal-ion-sensitive salicylhydrazines and the metal-ion-insensitive thiazolothiazepines show different location preferences for chelation to Mg^{2+} and Mn^{2+}. Second, a series of 10 salicylhydrazines show a similar correlation between Mg^{2+} or Mn^{2+} binding affinity and biological activity determined in the presence of Mn^{2+}. Third, different metal ions do confer minor structural differences on a model of the IN active site that may propagate into the surrounding region of the enzyme. Finally, docking results with the salicylhydrazines slightly favor salicylhydrazine binding above the catalytic loop rather than in the vicinity of the metal ion, although evaluation of our scoring methods on the recently crystallized HIV IN complex with ClTEP is still needed.

ACKNOWLEDGMENTS

The generous support of the National Science Foundation (STI-9602656 for laboratory renovation and CHE-9708517 for computational resources) and the National Institute of Allergy and Infectious Diseases (1 R15 AI45984-01) are gratefully acknowledged. The support of the Chemical Computing Group in the form of the MOE program is also appreciated.

REFERENCES

1. MD Andrake, AM Skalka. J Biol Chem 271:19633–19636, 1996.
2. BG Turner, MF Summers. J Mol Biol 285:1–32, 1999.
3. JK Rockstroh, M Altfeld, B Kupfer, R Kaiser, G Fatkenheuer, B Salzberger, KE Schneweis, U Spengler. Eur J Med Res 4:271–274, 1999.
4. BA Larder, G Darby, DD Richman. Science 243:1731–1734, 1989.
5. WEJ Robinson. Infections Medicine 15:129–137, 1998.
6. JA Este, C Cabrera, D Schols, P Cherepanov, A Gutierrez, M Witvrouw, C Pannecouque, Z Debyser, RF Rando, B Clotet, J Desmyter, E De Clercq. Mol Pharmacol 53:340–345, 1998.
7. J Sakai, M Kawamura, J-I Sakuragi, S Sakuragi, R Shibata, A Ishimoto, N Ono, S Ueda, A Adachi. J Virol 67:1169–1174, 1993.
8. PM Cannon, ED Byles, SM Kingsman, A Kingsman. J Virol. 70:651–657, 1996.
9. F Dyda, AB Hickman, TM Jenkins, A Engelman, R Craigie, DR Davies. Science 266:1981–1986, 1994.
10. P Rice, R Craigie, DR Davies. Curr Opin Struct Biol 6:76–83, 1996.
11. G Bujacz, J Alexandratos, Q Zhou-Liu, C Clément-Mella, A Wlodawer. FEBS Lett 398:175–178, 1996.
12. Y Goldgur, F Dyda, AB Hickman, TM Jenkins, R Craigie, DR Davies. Proc Natl Acad Sci USA 95:9150–9154, 1998.
13. M Cai, R Zheng, M Caffrey, M Craigie, R Craigie, GM Clore, AM Gronenborn. Nature Struct Biol 4:567–577, 1997.
14. M Cai, R Zheng, M Caffrey, R Craigie, GM Clore, AM Gronenborn. Nature Struct Biol 4:839–840, 1997.
15. PJ Lodi, JA Ernst, J Kuszewski, AM Hickman, A Engelman, R Craigie, GM Clore, AM Gronenborn. Biochemistry 34:9826–9833, 1995.
16. Y Goldgur, R Craigie, GH Cohen, T Fujiwara, T Fujishita, H Sugimoto, T Endo, H Murai, DR Davies. Proc Natl Acad Sci USA 96:13040–13043, 1999.
17. E Asante-Appiah, AM Skalka. J Biol Chem 272:16196–16205, 1997.
18. MR Fesen, KW Kohn, F Leteurtre, Y Pommier. Proc Natl Acad Sci USA 90:2399–2403, 1993.
19. TRJ Burke, MR Fesen, A Mazumder, J Wang, AM Carothers, D Grunberger, J Driscoll, K Kohn, Y Pommier. J Med Chem 38:4171–4178, 1995.
20. MR Fesen, Y Pommier, F Leteurtre, S Hiroguchi, J Yung, KW Kohn. Biochem Pharm 48:595–608, 1994.

21. A Mazumder, M Gupta, DM Perrin, DS Sigman, M Rabinovitz, Y Pommier. AIDS Res Hum Retroviruses 11:115–125, 1995.
22. H Zhao, N Neamati, A Mazumder, S Sunder, Y Pommier, TRJ Burke. J Med Chem 40:1186–1194, 1997.
23. A Mazumder, A Gazit, A Levitzki, M Nicklaus, J Yung, G Kohlhagen, Y Pommier. Biochemistry 34:15111–15122, 1995.
24. K Mekouar, J-F Mouscadet, D Desmaële, F Subra, H Leh, D Savouré, C Auclair, J d'Angelo. J Med Chem 41:2846–2857, 1998.
25. MC Nicklaus, N Neamati, H Hong, A Mazumder, S Sunder, J Chen, GWA Milne, Y Pommier. J Med Chem 40:920–929, 1997.
26. DJ Hazuda, PJ Felock, JC Hastings, B Pramanik, AL Wolfe. J Virol 71:7005–7011, 1997.
27. N Neamati, H Hong, S Sunder, GWA Milne, Y Pommier. Mol Pharm 52:1041–1055, 1997.
28. N Neamati, H Hong, JM Owen, S Sunder, HE Winslow, JL Christensen, H Zhao, TRJ Burke, GWA Milne, Y Pommier. J Med Chem 41:3202–3209, 1998.
29. N Neamati, JA Turpin, HE Winslow, JL Christensen, K Williamson, A Orr, WG Rice, Y Pommier, A Garofalo, A Brizzi, G Campiani. J Med Chem 42:3334–3341, 1999.
30. Spartan, Wavefunction, Inc, Irvine, 1993.
31. S Maignan, J-P Guilloteau, Q Zhou-Liu, C Clément-Mella, V Mikol. J Mol Biol 282, 359–368, 1998.
32. MOE, Chemical Computing Group, Montreal, 1998.
33. WJ Stevens, H Basch, M Krauss. J Chem Phys 81:6026–6033, 1984.
34. TR Cundari, WJ Stevens. J Chem Phys 98:5555–5565, 1993.
35. WJ Stevens, H Basch, M Krauss, P Jasien. Can J Chem 70:612–630, 1992.
36. M Krauss, WJ Stevens. Ann Rev Phys Chem 35:357–385, 1985.
37. General Atomic and Molecular Electronic Structure System: GAMESS, Gordon Research Group, Iowa State University, Ames, IA, 1998.
38. G Bujacz, M Jaskolski, J Alexandratos, A Wlodawer, G Merkel, RA Katz, AM Skalka. Structures 4:89–96, 1996.
39. G Bujacz, J Alexandratos, A Wlodawer, G Merkel, M Andrake, RA Katz, AM Skalka. J Biol Chem 272:18161–18168, 1997.
40. TR Cundari, J Deng. J Chem Inf Comput Sci 39:376–381, 1999.
41. J Lubkowski, F Yang, J Alexandratos, A Wlodawer, H Zhao, TRJ Burke, N Neamati, Y Pommier, G Merkel, AM Skalka. Proc Natl Acad Sci USA 95:4831–4836, 1998.
42. AA Bliznyuk, JE Gready. J Comput Chem 20:938–988, 1999.
43. R Wang, L Liu, L Lai, Y Tang. J Mol Model 4:379–394, 1998.

8

Cyclometallation of a Computationally Designed Diene: Synthesis of (−)-Androst-4-ene-3,16-dione

Douglass F. Taber
University of Delaware, Newark, Delaware

James P. Louey
Sacred Heart University, Fairfield, Connecticut

Yanong Wang
American Home Products, Pearl River, New York

Wei Zhang
Bristol-Myers Squibb Pharmaceutical Research Institute, New Brunswick, New Jersey

1. INTRODUCTION

We recently reported (1) a new approach to the stereoselective construction of polycyclic systems, illustrated by the cyclozirconation/carbonylation (2) of a *computationally* designed diene **1** (Scheme 1) to give the tetracyclic ketone **2**. Ketone **2** was converted over several steps to (−)-androst-4-ene-3,16-dione (**3**). (3) We describe here the history of this work and the development of the computational approach that led to the design of diene **1**. We also lay out further lines of exploration that will be interesting to pursue.

2. INTRAMOLECULAR DIENE CYCLOZIRCONATION

When we began this work in 1988 (4), substantial work on the intramolecular cyclometallation of eneynes and diynes had been reported (5). It was, therefore, surprising

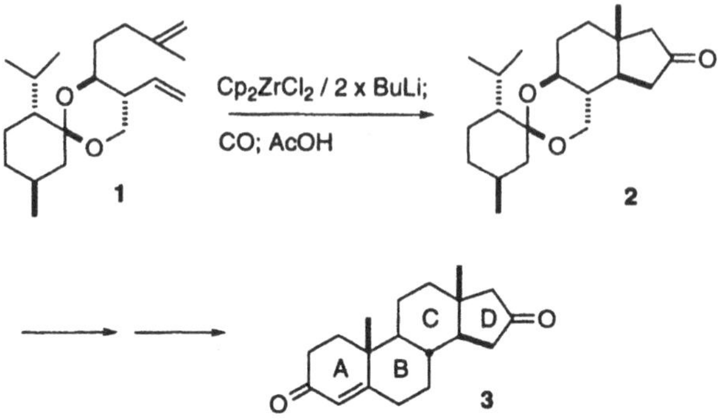

Scheme 1

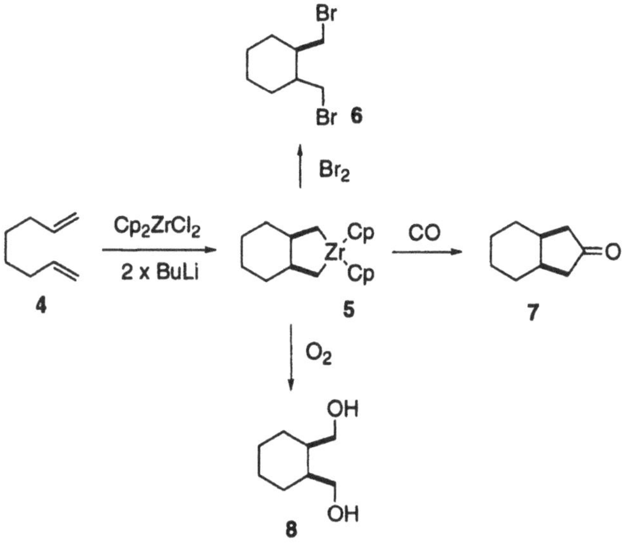

Scheme 2

that a general procedure for the intramolecular cyclometallation of *dienes* (Scheme 2) had not yet been developed (6). We were delighted to find that reduction of $(Cp)_2ZrCl_2$ with BuLi (7) in the presence of a diene **4** gave the zirconacycle **5** and that the metallacycle would react with a variety of interesting electrophiles.

3. THERMAL REVERSIBILITY OF DIENE CYCLOZIRCONATION

We discovered the thermal reversibility of the diene cyclometallation when we attempted cyclization of the diene **9** (Scheme 3) (8,9). The reaction proceeded to completion after two hours at room temperature, but rather than the expected tricyclic alcohol **13**, the product was the dimer **11**, from the reaction of two monosubstituted alkenes. We repeated the cyclozirconation, but let it proceed for a longer time. After 18 hours at room temperature, a new product had appeared (TLC analysis). After 1.5 hours at 75°C, the reaction was complete. The tricyclic alcohol **13** was isolated in 63% yield from **9**.

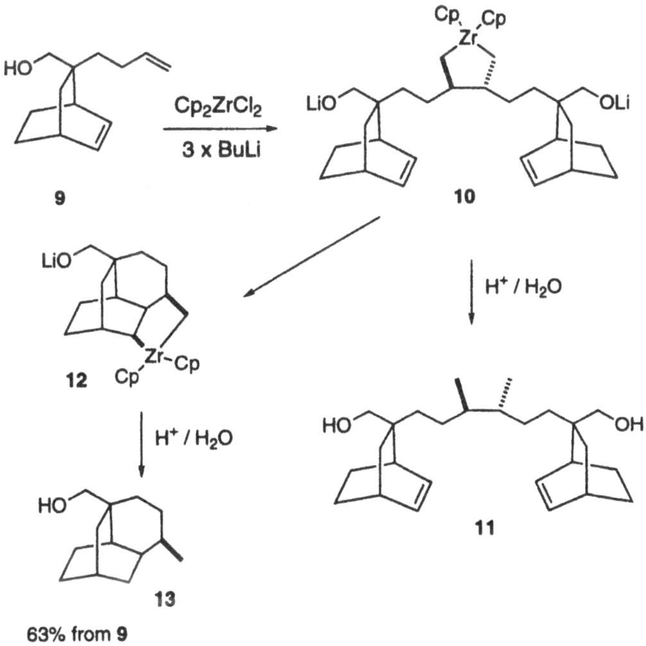

Scheme 3

4. COMPUTATIONAL ANALYSIS OF THE RELATIVE STABILITY OF ZIRCONACYCLES: ZINDO

This observation of the thermal reversibility of diene cyclozirconation led us to reinvestigate the earlier systems (10). Indeed, while cyclozirconation of 1,6-heptadiene **14** at room temperature for 1 h, followed by bromination, gave predominantly the trans-fused dibromide **15** (Scheme 4), as reported, cyclozirconation at low temperature (−78°C to 0°C) followed by bromination gave predominantly the cis dibromide **16**. We hypothesized that diene **14** under kinetic conditions gave the cis zirconacycle, which then equilibrated to the more stable trans zirconacycle on reaching room temperature. *Qualitatively*, the preference for the trans-fused metallacycle **17** is understandable, since the cis-fused metallacycle **18** is folded, with one of the cyclopentadienyl rings extending toward the concave face of the ring system. The trans-fused metallacycle **17** is extended and so is not destabilized by such a steric interaction.

While this qualitative picture was interesting, we needed a method that would give us a *quantitative* assessment of the relative stability of equilibrating zirconacycles. A survey of different computational methods (ZINDO (11,12), molecular mechanics, and density functional theory) led to the conclusion that ZINDO (Scheme 5) was the most reliable for calculating the relative stabilities of the diastereomeric zirconacycles. Using ZINDO, the *trans*-5,5-zirconacycle **17** was calculated to be more stable than the *cis*-5,5-zirconacycle **18** by 2.5 kcal/mol.

We had observed that cyclozirconation of diene **4** at room temperature (Scheme 6), even overnight, gave after oxygenation predominantly the cis diol **8**. Yet, ZINDO calculations (Scheme 5) indicated that the trans zirconacycle **19** should be more stable than the cis zirconacycle **20**. Given this evidence, we re-

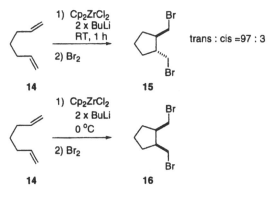

Scheme 4

ZINDO

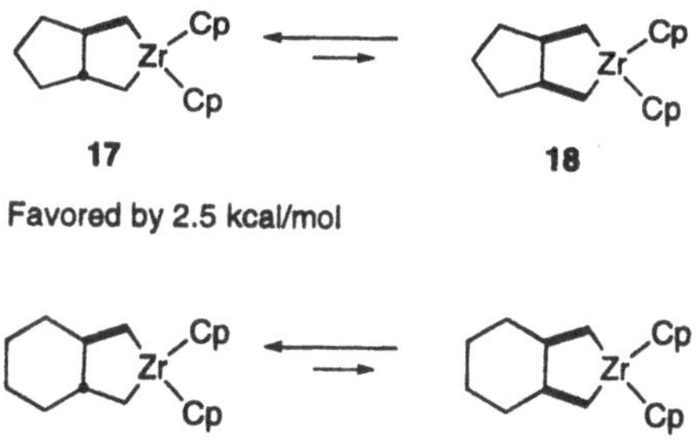

17   **18**

Favored by 2.5 kcal/mol

19   **20**

Favored by 2.8 kcal/mol

Scheme 5

peated the cyclozirconation at higher temperature and found that we did receive, after oxygenation, the expected trans diol **21**. We hypothesize that the 6/5 zirconacycle (**19/20**) is more stable than the 5/5 zirconacycle (**17/18**), so the activation energy for equilibration is higher for the 6/5 zirconacycle.

5. NATURAL PRODUCT SYNTHESIS: (−)-HALICLONADIAMINE

Carbonylation of the equilibrated zirconacycle **22** gave the cyclopentanone **23** (Scheme 7), a valuable synthon for the construction of natural products. This simple, one-step procedure for the preparation of the trans-fused cyclopentanone

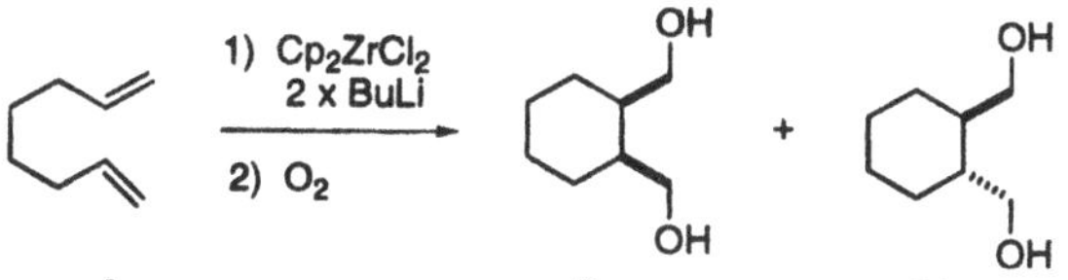

4	**8**	**21**
A: RT, 22 h 72%	75	25
B: 75 °C, 4 h 87%	1	99

Scheme 6

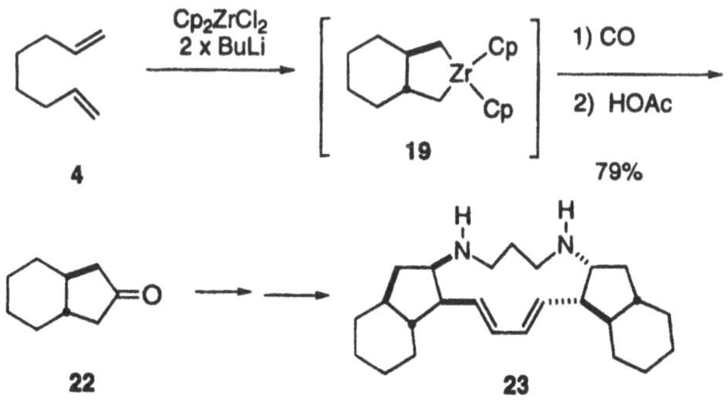

Scheme 7

22 laid the foundation for our recent enantioselective synthesis of (−)-haliclonad-iamine **23** (13).

6. GLOBAL EQUILIBRATION OF DIASTEREOMERIC ZIRCONACYCLES: (+)-ELEMOL

The equilibration of **17** and **18** or of **19** and **20** (Scheme 5) can be accomplished by exchanging the π face of just one of the two alkenes. The next question to address was whether *global* equilibration, that is, sequential dissociation and re-addition of Zr to each face of each alkene, could be achieved under the conditions of cyclozirconation. We therefore investigated the cyclozirconation of diene **24** (Scheme 8), readily prepared from α-terpineol (14). There are four diastereomeric

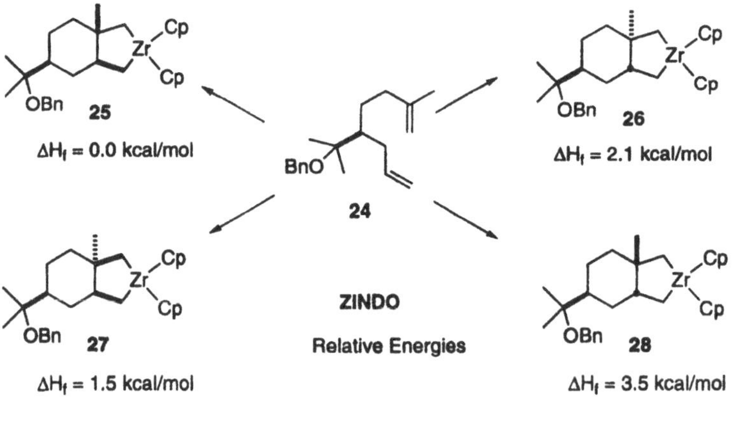

Scheme 8

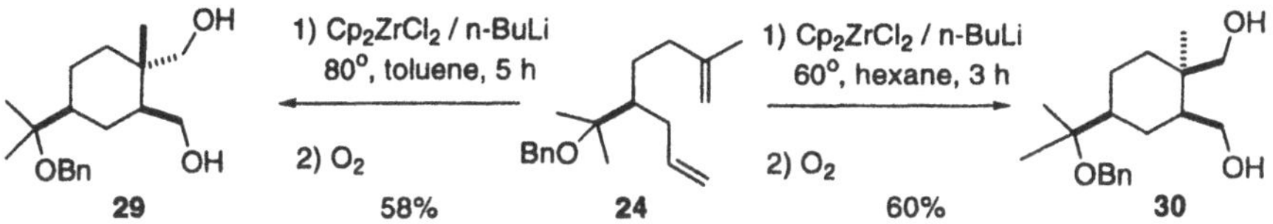

Scheme 9

zirconacycles **25–28** that could be formed from **24**. On the basis of the relative stabilities (ZINDO) of these diastereomers (Scheme 8), we predicted that global equilibration of the zirconacycles followed by oxygenation should give **29**. In the event, cyclozirconation of **24** at 80°C for 5 hours, followed by oxygenation, gave predominantly diol **29**, as predicted (Scheme 9).

It was equally striking that cyclozirconation under kinetic conditions (60°C, 3 hours), followed by oxygenation, gave only one of the two possible cis diastereomers of the product diol. Diol **30** derives from the more stable of the two possible cis zirconacycles, **27** and **28**. This suggests that ZINDO may possibly also prove useful for predicting the *kinetic* products from such diene cyclozirconations.

7. COMPUTATIONAL DIENE DESIGN: (–)-ANDROST-4-ENE-3,16-DIONE

Having established that intramolecular diene cyclozirconation can be carried out under conditions of either kinetic or thermodynamic control, and having shown that semiempirical calculations (ZINDO) can be used to predict the relative stabilities of diastereomeric zirconacycles, we next undertook the computationally based design of a diene such that cyclozirconation would be *directed* toward a *desired* diastereomer.

Our initial objective was the construction of the steroid skeleton (e.g., 3, Scheme 10) with control of both relative and absolute configuration. We first considered a B → BCD construction, starting with diene **31** (Scheme 11). Unfortunately, computational analysis (ZINDO) predicted that the undesired cis-fused zirconacycle **33** would be *more* stable than the desired trans-fused zirconacycle **32**. The prospects did not improve with the acetonide **34**. Again (Scheme 11), computational analysis (ZINDO) predicted that the cis-fused **36** would be more stable than the desired trans-fused **35**.

It was clear that the protecting group on the diol had to introduce steric bulk underneath the ring system of the tricyclic zirconacycle, to destabilize the cis diastereomer. After considering several other alternatives, we settled on the

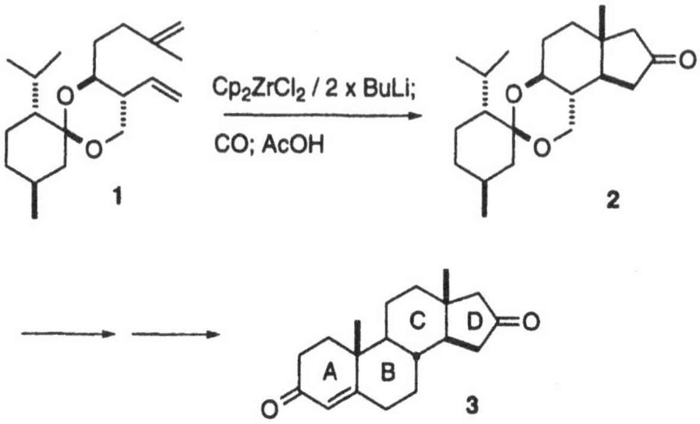

Scheme 10

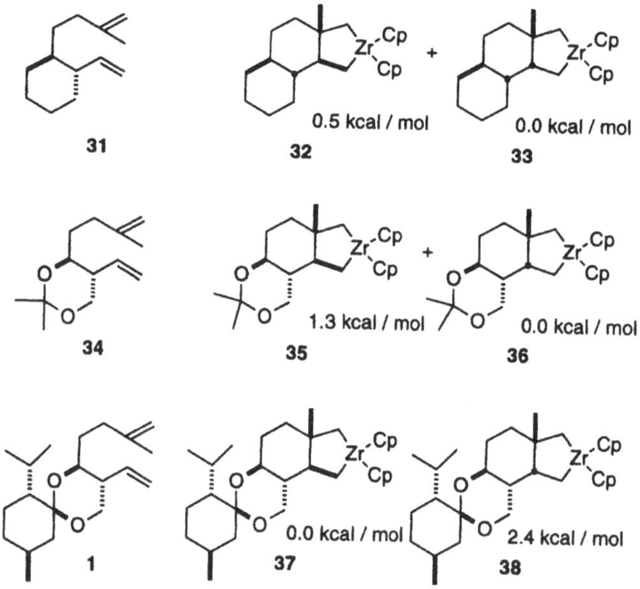

Scheme 11

TABLE 1 Cyclozirconation/Carbonylation of 1

	T (°C)	t (h)	% yield	2	δ 72.3	δ 63.2	δ 61.6
1	80	5	28	49	20	10	21
2	80	12	42	58	10	7	25
3	90	2	48	47	20	12	21
4	100	1	19	60	14	13	13
5	80	10	63	52	21	8	19
6	80	14	26	47	19	10	24
7	80	24	10	56	8	5	31

menthonide **1**. This introduced steric interactions such the desired trans-fused **37** was predicted to be *more* stable than the cis-fused **38**. For each of these three dienes (**31**, **34**, and **1**), the other two diastereomeric zirconacycles were predicted to be significantly less stable (from **31**, the other trans-fused diastereomer (**39**) was calculated at 9.6 kcal/mol, while the other cis-fused diastereomer (**40**) was calculated at 9.9 kcal/mol, compared to **33**).

Cyclozirconation conditions were varied (Table 1) to optimize the yield of **2**. In each case, the crude diastereomeric mixture of zirconacycles was carbonylated, and the yield and the ratios of the mixture of four product ketones (easily discerned by their oxygenated methines, ^{13}C NMR δ 73.1 (**2**), δ 72.3, δ 63.2, δ 61.6) were recorded. At temperatures in excess of 80°C (entries 1–4), substantial thermal degradation set in. Returning to 80°C (entries 2, 5–7), it was apparent that while the proportion of **2** was still increasing at 12 h, thermal degradation was again competing, lowering the overall yield. Pure **2** was isolated from the mixture of four product ketones by crystallization, and the structure was established by X-ray crystallography. Ketone **2** was carried over several steps to (−)-androst-4-ene-1,16-dione **3**.

To assure ourselves of the role of the menthone ketal in the cyclozirconation of **1**, we also effected cyclozirconation (80°C, 10 h) of the diol **40** (Scheme 12),

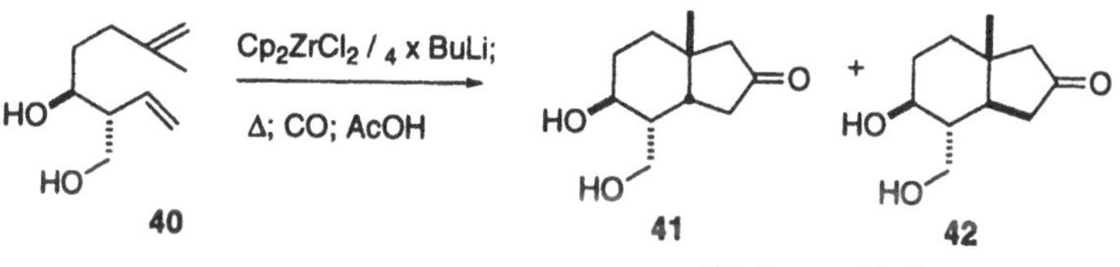

cis : trans = 5 : 1

Scheme 12

using an extra two equivalents of *n*-BuLi to deprotonate the alcohols. Cyclozirconation and carbonylation proceeded smoothly, but as would be expected from the calculations (Scheme 11) the cis-fused product **41** was dominant (65%), and the trans-fused ketone **42** was only 13% of the mixture of product ketones.

8. DIRECTIONS FOR THE FUTURE

These exciting results established the validity of this computational approach. This is, however, just the beginning. Although we could push the proportion of **2** (Scheme 10) to 5:1 and even higher, with longer times and/or higher temperatures for the cyclometallation step, the yields of product dropped off, due to thermal degradation of the metallacycle. In addition to extending the cyclozirconation to more challenging dienes, we are therefore also exploring other metal and ligand combinations to effect intramolecular diene cyclometallation. Our objective is to establish a metal/ligand combination such that full equilibration of the cyclometallation products can be achieved efficiently.

Rothwell (15) recently reported that reduction of (ArO)₂TiCl₂ **43** (ArOH = 2,6-diphenyl phenol) (Scheme 13) in the presence of 1,7-octadiene **4**, with warming only to room temperature, led to the trans-fused titanacycle **45**. We prepared **43** from the commercially available 2,6-diphenylphenol and repeated this cyclotitanation, oxygenating the intermediate titanacycles to give diols **8** and **21**. It is apparent that the 6/5 titanacycle is equilibrating much more rapidly (4 h, rt) than the 6/5 bis-Cp zirconacycle (3 h, 70°). If ZINDO calculations are valid with the titanacycles (we have not yet established this), it is also apparent that

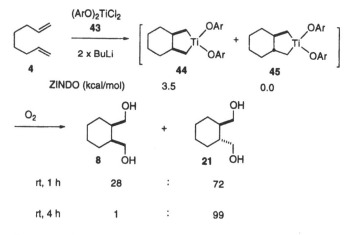

Scheme 13

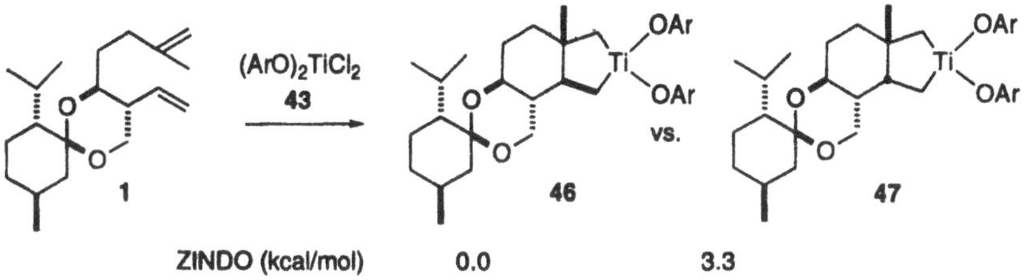

Scheme 14

the much more sterically demanding $(ArO)_2Ti$ favors the trans ring fusion by a somewhat larger margin than does the Cp_2Zr (for Cp_2Zr, trans is favored by 2.8 kcal/mol).

We prepared and purified the air- and moisture-sensitive complex **43** to carry out these studies. We have also found that it is possible to generate this complex in situ, with the starting ArOH (two equivalents) and two equivalents of BuLi, followed, at rt, by $TiCl_4$, and then at $-78°$ by the diene and two more equivalents of BuLi. The cyclization results are the same as with the preprepared complex **43**. We will use this latter approach to quickly screen a variety of other alcohols and diols, including enantiomerically pure diols such as BINOL (1,1'-bi-2-naphthol) and Taddol ($\alpha,\alpha,\alpha',\alpha'$-tetraaryl-1,3-dioxolane-4,5-dimethanol), in this cyclization.

Our primary interest with **43** is to develop an alternative procedure for diene cyclometallation such that the intermediate metallacycles will equilibrate at lower temperature and more efficiently than is observed with Cp_2ZrCl_2. It is also striking that with more substituted dienes, the trans metallacycles derived from **43** are also favored over the cis by a more substantial margin than with Cp_2ZrCl_2. For diene **1** (Scheme 14), for instance, the trans titanacycle **46** is calculated to be more stable than the cis titanacycle **47** by 3.3 kcal/mol.

REFERENCES

1. DF Taber, W Zhang, CL Campbell, AL Rheingold, CD Incarvito. J Am Chem Soc 122:4813–4814, 2000.
2. (a) For an overview of carbonylative diene and enyne cyclometallation, see LS Hegedus, Transition Metals in the Synthesis of Complex Organic Molecules. 2nd ed. Sausalito, CA: University Science Books, 1999. For other recent references to carbonylative cyclometallation, see: (b) Z Zhao, Y Ding, G Zhao. J Org Chem 63: 9285–9291, 1998. (c) E-I Negishi, J-L Montchamp, L Anastasia, A Elizarov, D Choueiry. Tetrahedron Lett 39:2503–2506, 1998. (d) Y-T Shiu, RJ Madhushaw, W-T Li, Y-C Lin, G-H Lee, S-M Peng, F-L Liao, S-L Wang, R-S Liu. J Am Chem

Soc 121:4066–4077, 1999. (e) M Murakami, K Itami, Y Ito. J Am Chem Soc 121: 4130–4135, 1999. (f) FA Hicks, NM Kablaoui, SA Buchwald. J Am Chem Soc 121: 5881–5898, 1999.

3. For recent examples of steroid total synthesis, see: (a) M Kurosu, LR Marcin, TJ Grinsteiner, Y Kishi. J Am Chem Soc 120:6627, 1998. (b) PA Grieco, SA May, MD Kaufman. Tetrahedron Lett 39:7047–7050, 1998. (c) PA Zoretic, H Fang, A Ribeiro. J Org Chem 63:7213–7217, 1998. (d) C Heinemann, M Demuth. J Am Chem Soc 121:4894–4895, 1999, and references cited therein.

4. WA Nugent, DF Taber. J Am Chem Soc 111:6435–6437, 1989.

5. SL Buchwald, RB Nielsen. Chem Rev 88:1047–1058, 1988.

6. The literature contained several reports of cyclization of 1,7-octadiene by early transition metal complexes to isomeric mixtures of metallaindanes: (a) JX McDermott, ME Wilson, GM Whitesides. J Am Chem Soc 98:6529–6536, 1976. (b) RH Grubbs, A Miyashita. J Chem Soc, Chem Commun 864–865, 1977. (c) SJ McLain, CD Wood, RR Schrock. J Am Chem Soc 101:4558–4570, 1979. (d) KI Gell, J Schwartz. J Chem Soc, Chem Commun 244–246, 1979.

7. E-I Negishi, FE Cederbaum, T Takahashi. Tetrahedron Lett 27:2829–32, 1986.

8. DF Taber, JP Louey, JA Lim. Tetrahedron Lett 34:2243–46, 1993.

9. Negishi independently observed the reversibility of intermolecular diene cyclozirconation. T Takahashi, T Fujimori, S Takashi, M Saburi, Y Uchida, CJ Rousset, E-I Negishi. J Chem Soc, Chem Commun 182–183, 1990.

10. DF Taber, JP Louey, Y Wang, WA Nugent, DA Dixon, RL Harlow. J Am Chem Soc 116:9457–9463, 1994.

11. Both ZINDO and molecular mechanics were used as implemented on the Tektronix CAChe workstation. For leading references to ZINDO, a semiempirical program that has been parameterized for the first two rows of transition metals, see: (a) MC Zerner, GW Loew, RF Kirchner, UT Mueller-Westerhoff. J Am Chem Soc 102: 589–599, 1980. (b) WP Anderson, TR Cundari, RS Drago, MC Zerner. Inorg Chem 29:1–5, 1990.

12. Although ZINDO was originally parameterized to give good spectroscopic results, it had also been used in studies of the energetics and structures of transition metal-based catalytic systems: (a) GL Estiu, MC Zerner. J Phys Chem 97:13720–13729, 1993. (b) GL Estiu, MC Zerner. Int J Quantum Chem 26:587, 1992.

13. DF Taber, Y Wang. J Am Chem Soc 119:22–26, 1997.

14. DF Taber, Y Wang. Tetrahedron Lett 36:6639–42, 1995.

15. MG Thorn, JE Hill, SA Waratuke, ES Johnson, PE Fanwick, IP Rothwell. J Am Chem Soc 119:8630–8641, 1997.

9

Rhodium-Mediated Intramolecular C–H Insertion: Probing the Geometry of the Transition State

Douglass F. Taber
University of Delaware, Newark, Delaware

Pascual Lahuerta and Salah-eddine Stiriba
University of Valencia, Valencia, Spain

James P. Louey
Sacred Heart University, Fairfield, Connecticut

Scott C. Malcolm
Harvard University, Cambridge, Massachusetts

Robert P. Meagley
Intel Corporation, Hillsboro, Oregon

Kimberly K. You
BASF Corporation, Wyandotte, Michigan

1. INTRODUCTION

The power of Rh-mediated intramolecular C–H insertion can be seen in the cyclization of the α-diazo ester **1** (Scheme 1). Although four diastereomers could have been formed from this cyclization, only **2**, the key intermediate for the synthesis of the dendrobatid alkaloid 251F **3**, was in fact observed. This outcome, as explained in detail shortly, had first been predicted computationally. This chapter summarizes our computational approach toward understanding the transition state ("point of commitment") for these Rh-mediated cyclizations. As we discuss at the end of this chapter, there is yet much left to be learned.

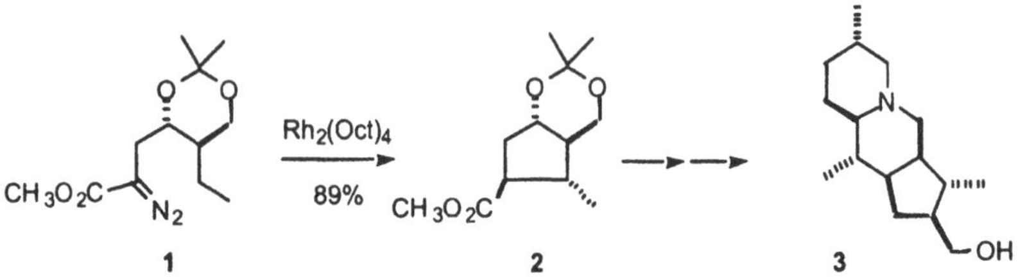

Scheme 1

2. CYCLIZATION VERSUS ELIMINATION

We observed (1,2) that an α-diazo β-keto ester **5** (Scheme 2) would, on exposure to a catalytic amount of $Rh_2(OAc)_4$, undergo smooth cyclization to the cyclopentane derivative **6**. Since the α-diazo β-keto ester **5** was readily prepared by diazo transfer (3) to the corresponding β-keto ester **4**, this established a general route to highly substituted cyclopentanes. The subsequent observation (4) that use of $Rh_2(O_2CR)_4$ catalysts derived from more electron-donating carboxylic acids al-

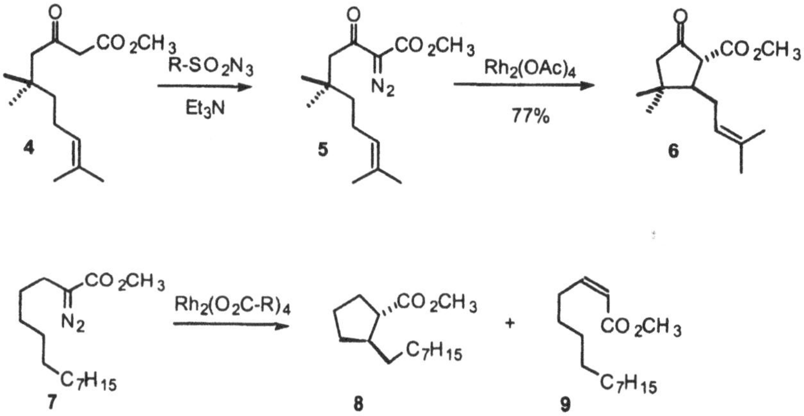

R	8	9
-CH$_3$	66	34
-CF$_3$	52	48
-C(CH$_3$)$_3$	85	15

Scheme 2

lowed the efficient cyclization of simple α-diazo esters (5) such as **7** (Scheme 2) to the corresponding cyclopentanes with high diastereocontrol set the stage for the work described here.

3. DEVELOPMENT OF THE COMPUTATIONAL APPROACH

Any or all of products **11–14** (Scheme 3) could have been formed by cyclization of the α-diazo ester **10**. In the event, only **12** was observed (6). In an attempt to rationalize this result, we developed a computationally based model for the transition state for C–H insertion.

An understanding of the mechanism (7) for Rh-mediated intramolecular C–H insertion begins with the recognition that these α-diazo carbonyl derivatives can also be seen as stabilized ylides, such as **15** (Scheme 4). The catalytic Rh(II) carboxylate **16** is Lewis acidic, with vacant coordination sites at the apical positions, as shown. The first step in the mechanism, carbene transfer from the diazo ester to the Rh, begins with complexation of the electron density at the diazo carbon with an open Rh coordination site, to give **17**. Back-donation of electron density from the proximal Rh to the carbene carbon with concomitant loss of N_2 then gives the intermediate Rh carbene complex **18**.

The mechanism by which this intermediate Rh carbene complex **18** reacts can be more easily understood if it is written as the inverted ylide **19**. This species would clearly be electrophilic at carbon. We hypothesized that for bond formation to proceed, a transition state (**20**) in which the C–Rh bond is aligned with the target C–H bond would be required (8). As the C–H insertion reaction proceeded, the electron pair in the C–H bond would drop down to form the new C–C bond,

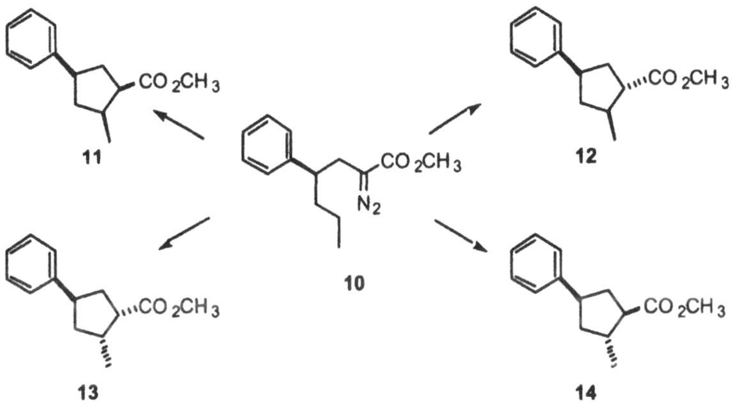

Scheme 3

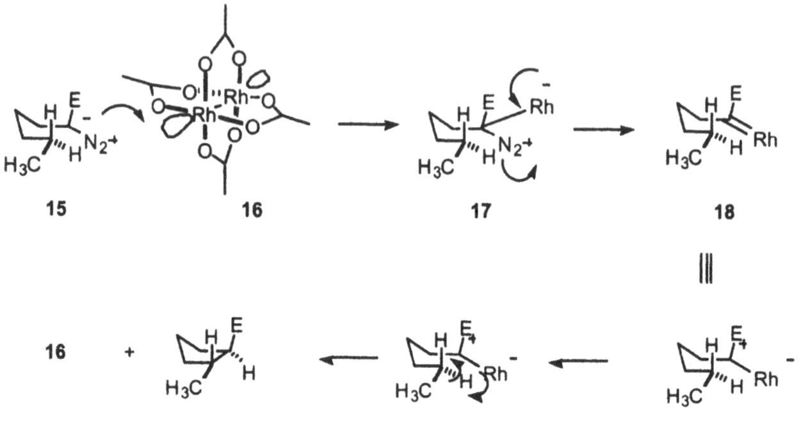

Scheme 4

and at the same time the electron pair in the C–Rh bond would slide over to form the new C–H bond. This would give the product (21) and release the initial Rh species 16, completing the catalytic cycle.

A central assumption of this mechanism is that the actual C–H insertion is concerted and that it proceeds with retention of absolute configuration. We had already, in a related case (9), demonstrated that Rh-mediated C–H insertion indeed proceeded with retention of absolute configuration.

The actual product from a cyclization will be determined as the intermediate carbene *commits* to a particular diastereomeric transition state. If these diastereomeric transition states are indeed in full thermal equilibrium (10), then computational modeling of the diastereomeric transition states (20) could allow us to predict which would be favored and thus which diastereomeric product would be formed.

4. DEVELOPMENT OF THE COMPUTATIONAL MODEL

To construct the transition state 20, we locked (Scheme 5) the Rh–Rh–C bond angle at 180°. To secure overlap between the C–Rh bond and the target C–H bond, we established weak bonds [meaningful in mechanics (11)] between the two incipiently bonding carbons and between the target H and the proximal Rh (10). As mechanics tends to rehybridize weakly bonded carbons, we also found it necessary to lock the H–C–C–C dihedral angle of the target C–H at 60° (or, in the inverted transition state, −60°), to maintain sp^3 geometry.

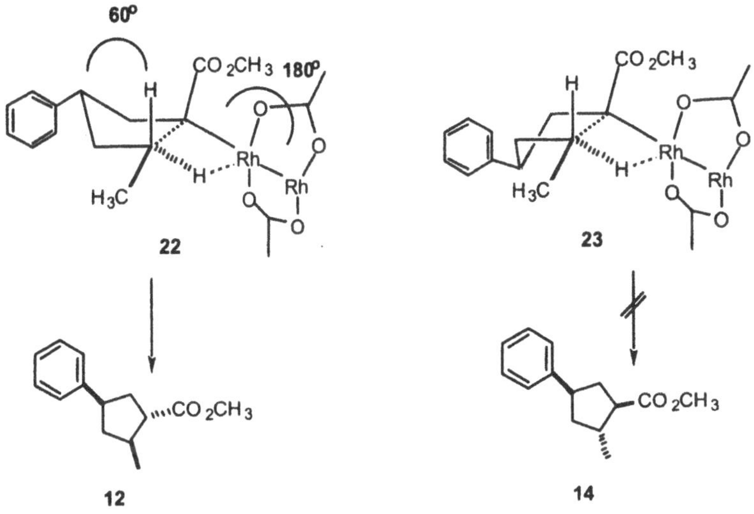

Scheme 5

There are still two possibilities for the transition state, **22** and **23**. In transition state **22**, the Rh carbene is pointed away from the flip of the incipient cyclopentane ring (a "chairlike" transition state, counting the five carbons and the Rh in the six-membered ring), whereas in **23** the Rh carbene is pointed toward the flip of the incipient cyclopentane ring (a "boatlike" transition state). As **10** (Scheme 3) cyclizes to **12**, in which the methyl and the phenyl are on the same face of the cyclopentane, we concluded that at the point of commitment to product formation, the transition state leading to cyclization must be chairlike (**22**) rather than boatlike (**23**).

Sterically (mechanics), there is no significant energy difference between the competing transition states **22** and **23**. We assume that the difference is electronic, that the conformation **22** makes electron density more readily available from the target C–H bond than does conformation **23**. This interplay between steric and electronic effects will be important throughout this discussion of Rh-mediated intramolecular C–H insertion.

5. APPLICATION OF THE COMPUTATIONAL MODEL

For the cyclization of **10**, there are four diastereomeric chairlike transition states, **22**, **24**, **25**, and **27** (Table 1), each leading to one of the four possible diastereomeric products. With the angles and bonds as stated, we minimized each of the four transition states with mechanics (11). Transition state **22** was found to be

TABLE 1 Calculated Relative Energies of Transition States and Products

Entry	TS	Product	TSΔE, kcal/mol	Product ΔE, kcal/mol
1	**22** (Ph, CO_2CH_3, H_3C, H, Rh)	**12** (Ph, CO_2CH_3)	0.0	0.0
2	**24** (Ph, CO_2CH_3, H_3C, H, Rh)	**14** (Ph, CO_2CH_3)	5.3	0.1
3	**25** (Ph, H_3C, CO_2CH_3, H, Rh)	**26** (Ph, CO_2CH_3)	6.1	0.2
4	**27** (Ph, H_3C, CO_2CH_3, H, Rh)	**28** (Ph, CO_2CH_3)	7.4	0.9

the lowest in energy, by 5.3 kcal/mol compared to the next most stable. This contrasts with the relative stability of the four diastereomeric *products*, **12**, **14**, **26**, and **28** (Table 1), which are quite comparable one to another.

Using this approach, we have successfully predicted the major product from the cyclization of more than 30 α-diazo esters and α-diazo β-keto esters (6), including the cyclization (12) of **1** to **2** (Scheme 1). Not all Rh-mediated intramolecular C–H insertion reactions will proceed to give a single dominant diastereomer. Our interest in this initial investigation has been to develop a model for the transition state that will allow us to discern those cyclizations that *will* proceed with high diastereoselectivity.

6. CHIRAL AUXILIARY CONTROL

Returning to the cyclization of a simple α-diazo ester (Scheme 6), we wanted to design (13) a chiral ester that would direct the cyclization preferentially to one absolute configuration of the product cyclopentane. In attempting to extend our computational approach to the design of such a chiral auxiliary, we found that we were missing a key piece of data, the dihedral angle between the ester carbonyl and the rhodium carbenoid at the point of commitment to cyclization (*30-syn* vs. *30-anti*). We and others have, in the past, speculated that the ester carbonyl and the rhodium carbenoid could be *syn* (14), *anti* (15), or *orthogonal* (15a), but no experimental or computational evidence in favor of any of these had been put forward. Since our computational approach did not allow us to answer this question directly, we devised an indirect approach based on the cyclization of α-diazoester **29**, derived from the naphthylborneol **33** (16). Our conclusion from this study is that the ester carbonyl and the rhodium carbenoid are *syn* in the transition state leading to the cyclization of esters such as **29**.

The chiral diazo ester **29** was cyclized with four commonly used rhodium carboxylate catalysts (Table 2). It was found as before that rhodium pivalate (17) (entry 4) was most efficient for forming the cyclopentanes and that rhodium trifluoroacetate (entry 1) was best for forming the alkenes (18). For the pivalate, both the yield of the cyclization and the diastereoselectivity improved at lower temperature (entry 5).

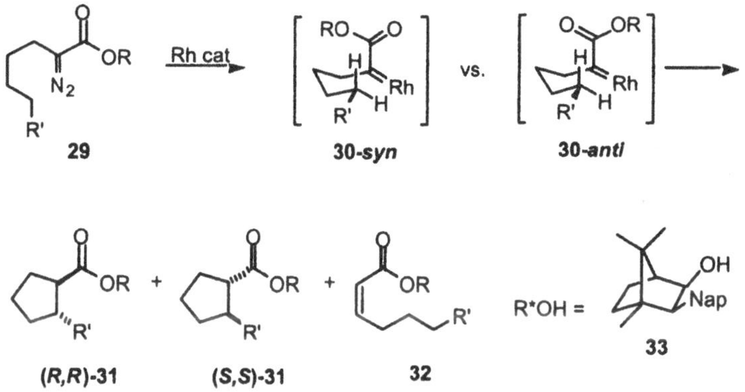

Scheme 6

TABLE 2   Influence of the Ligand Bound to Rhodium on the
Diastereoselectivity of the Cyclization of Ester **29**

Entry	Ligand	Reaction temperature	Yield	(*R,R*)-31		(*S,S*)-31		32
1	CF$_3$CO$_2$	18°C	89%	1.5	:	1.0	:	4.2
2	CH$_3$CO$_2$	18°C	92%	2.6	:	1.0	:	0.9
3	*n*-C$_7$H$_{15}$CO$_2$	18°C	82%	3.8	:	1.0	:	0.7
4	(CH$_3$)$_3$CCO$_2$	18°C	99%	8.4	:	1.0	:	1.0
5	(CH$_3$)$_3$CCO$_2$	−78°C	99%	14	:	1.0	:	1.0

7. COMPUTATIONAL ANALYSIS OF THE NAPHTHYLBORNYL-DERIVED ESTER

Assuming that the rhodium carbenoid and the ester carbonyl should be coplanar, the critical question as we extended our computational analysis to the naphthylbornyl-derived ester **29** was whether the rhodium carbenoid and the ester carbonyl would be *syn* or *anti* at the point of commitment to product formation. As illustrated in Table 3, there are four possible products, each of which could have come via either a *syn* or an *anti* transition state. Thus, there could be eight competing transition states leading to cyclization. We carried out, as already outlined, the minimizations for each of the corresponding eight "points of commitment." The minima, summarized in Table 3, were established using a meticulous grid search.

7.1. Analysis

The *syn* and the *anti* conformations leading to (*R,R*)-**31**, illustrated in Table 3, are calculated (mechanics) to be the two lowest-energy transition states for the cyclization of **29**. Of the two, the *anti* conformation (Rh carbene and carbonyl coplanar but pointing in opposite directions) is the more stable, by 3.37 kcal/mol. If steric factors alone governed the outcome of these cyclizations, we would expect that the *anti* transition state leading to (*R,R*)-**31** would be competing with the *syn* transition state leading to (*S,S*)-**31**. The former would be favored by 4.35 kcal/mol. We have found that if the difference in calculated transition-state energies is greater than 2 kcal/mol, a single product is always formed in high (>95%) diastereomeric excess. We do not observe such high diastereoselectivity in the cyclization of **29**, so we conclude that steric factors *alone* do not govern the stereochemical outcome of this cyclization.

We propose that there is in fact a substantial electronic preference, not reflected in the mechanics calculations, for the ester carbonyl and the C=Rh

TABLE 3 Transition States and Products Resulting from Cyclization of Naphthylbornyl 2-Diazoheptanoate (29)

Possible diastereomeric T.S.	Possible diastereomer	Relative energy of the transition state in kcal/mol	
		syn	anti
(structure)	(R,R)-31	3.37	0.00
(structure)	(R,S)-31	10.02	18.64
(structure)	(S,S)-31	4.35	14.38
(structure)	(S,R)-31	8.30	9.97

E* = CO₂R*

bond to be *syn* at the point of commitment to cyclization. This preference is strong enough to overcome the calculated steric preference (3.37 kcal/mol) for the *anti* transition state. The competition then is between the *syn* transition state leading to **(R,R)-31**, and the *syn* transition state leading to **(S,S)-31**. The relative energies of these two transition states differ by less than 1 kcal/mol, so we predict, and observe, low diastereoselectivity.

We have posed this *syn/anti* question using a sterically demanding ester for which there is a significant conformational bias in favor of the *anti* transition state. We therefore believe that the conclusion, that there is a substantial preference for the ester carbonyl and the C=Rh bond to be *syn* at the point of commitment to cyclization, is general and is not limited to this particular case.

While we have had some success, we are aware of the limitations inherent in a transition-state model for rhodium-mediated C–H insertion that attempts to predict product ratios on the basis of mechanics calculations. Arbitrary decisions limiting the several degrees of freedom possible in the transition state could lead one to a model for the "point of commitment" to cyclization that would be far from reality. The work described here is important because it offers *experimental* evidence for a key rotational degree of freedom, the dihedral angle between the ester carbonyl and the rhodium carbenoid.

Our initial objective, in this investigation, had been to design a useful chiral auxiliary. We were pleased to find that naphthylborneol **33** itself, on optimization of the catalyst and the reaction temperature, served effectively. Until effective chiral *catalysts* are developed, naphthylborneol **33** will be of significant practical value for directing the absolute course of rhodium-mediated intramolecular C–H insertion reactions.

8. IMPLICATIONS FOR CHIRAL CATALYST DESIGN

It was striking (Table 2) how much changing the ligand on the rhodium carboxylate changed the product distribution from the cyclization of **29**. It has been consistently observed by us and by others that, electronically, the ligands exert substantial control over the reactivity of the intermediate carbenoid. It is apparent (**20**, Scheme 4) that a strongly electron-withdrawing ligand will result in a more reactive carbenoid and that, with such a ligand, commitment to product formation will occur while the carbenoid carbon and the target C–H are still some distance apart. By changing the ligand from octanoate to pivalate, the reactivity of the carbenoid is apparently attenuated, resulting in a tighter transition state. The distance between the carbenoid carbon and the target C–H is then smaller at the point of commitment, bringing the chiral ester in closer proximity to the reaction center, where it can better influence the product distribution by the handedness of its steric bulk. An effective chiral *catalyst*, then, will have to direct the reaction sterically, and at the same time be electron-donating enough to have a late, *tight* "point of commitment."

9. COMPUTATIONAL DESIGN OF A RHODIUM CATALYST: BRIDGING THE TETRAKISCARBOXYLATODIRHODIUM CORE

In approaching the design of a chiral catalyst, the first question was whether or not our computational approach would allow us to predict the conformation of ligands around the Rh–Rh core. In particular, it seemed important to us, if we were to effectively control the three-dimensional space surrounding the Rh carbene/C–H interaction, to design (19) a family of dicarboxylate ligands that could occupy four of the eight O–Rh sites on the dirhodium tetracarboxylate (35, Scheme 7).

Although several hundred tetrakiscarboxylato metal–metal dimers were known (20), there had been no report of a dicarboxylic acid that would bridge two positions on such a dimer (21). We reasoned that the best chance for success would be with a dicarboxylic acid that was specifically designed to fit across the 5.4-Å gap between the carboxylate ligands.

To approach the design of a dicarboxylate ligand that would effectively bridge the Rh–Rh core, we first optimized the parent $Rh_2(CF_3CO_2)_4$ using ZINDO (22,23). We locked this structure, bridged two of the acetate methyl groups with an increasing number of methylenes, and evaluated the strain energy of the resulting (hypothetical) complexes using mechanics (11). As expected, the initial very high strain energy for a zero-methylene bridge decreased rapidly with increasing bridge length, until the bridge reached four carbons. After that, the strain energy did not significantly decrease with increasing bridge length. Recognition that entropy considerations would favor bridging by a *convergent* bidentate ligand then led us to *m*-benzenedipropanoic acid 35 as a likely candidate (24,25).

A priori, there was cause to be concerned that exchange of a tetrakiscarboxylato metal–metal dimer with a dicarboxylic acid would lead only to oligomers. In that event (Scheme 7), heating 35 in dichloroethane with tetrakis(trifluoroacetato)dirhodium 34 led to smooth exchange to give the emerald green complex 36, which was easily purified by silica gel chromatography. Prolonged heating of the reaction mixture led to more polar materials. The structure of 36 was confirmed by X-ray diffraction of the derived bis-acetone adduct. Our calculated structure for the bis-acetone adduct was exactly superimposable (19) on the X-ray structure.

A key question was whether or not the bridged dimer would effectively catalyze the C–H insertion reaction. We were pleased to observe (Scheme 8) that

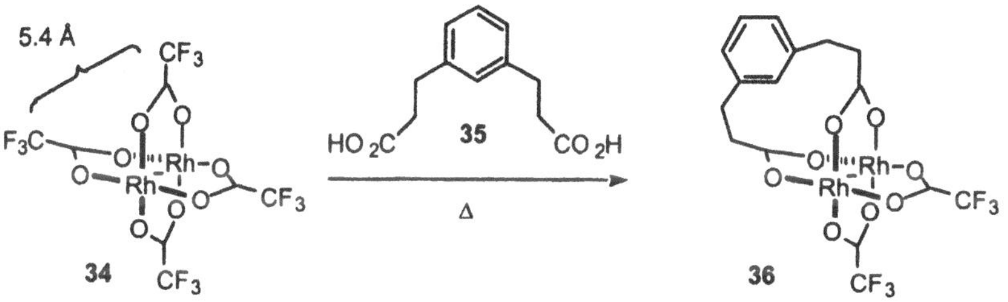

Scheme 7

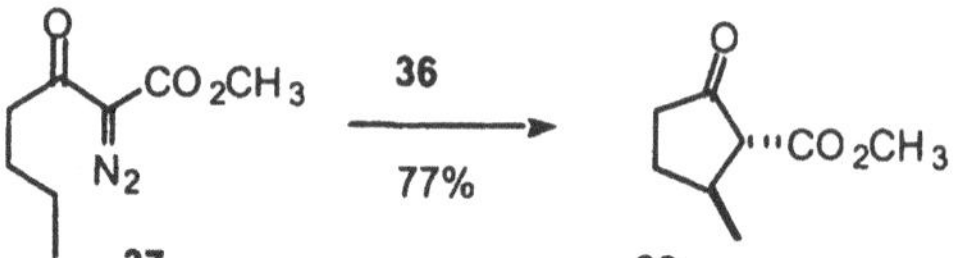

Scheme 8

complex **36** is in fact an efficient catalyst (1610 turnovers) for the cyclization of **37** to **38**.

10. DESIGN OF AN ENANTIOSELECTIVE CATALYST

With the assurance that our computational approach permitted the reliable prediction of the three-dimensional shape of the carboxylate ligands around the Rh–Rh complex, we turned our attention to the rational design of chirally substituted analogs (27,28) that might direct the *absolute* sense of the cyclization of **37** to **38** (Scheme 8). The best catalyst reported to date for the cyclization of **37** is that of Hashimoto (29), which effects (Scheme 8) C–H insertion with 27% ee.

There are two competing transition states for C–H insertion, **39** and **40** (Scheme 9). In transition state **39**, insertion is taking place into H_A. In transition state **40**, insertion is taking place into the enantiotopic H_B. The challenge is to design a chiral rhodium catalyst such that transition state **39** is favored over transition state **40** by *at least* the 2.5 kcal/mol we have observed is necessary to expect substantial diastereoselectivity in the C–H insertion reaction.

In cartoon form, what is needed is a carboxylate that will extend sterically to set up the three-dimensional environment around the apical position of the Rh, where the carbene binds and where the C–H insertion reaction is taking place. This is depicted schematically in Scheme 10. The challenge, then, is to design a ligand such that the resulting chiral environment favors transition state

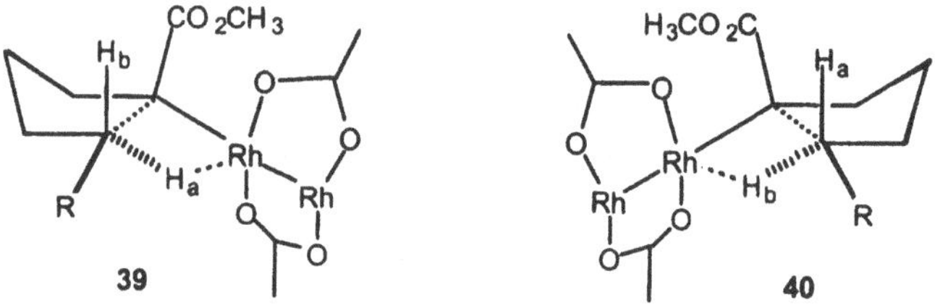

Scheme 9

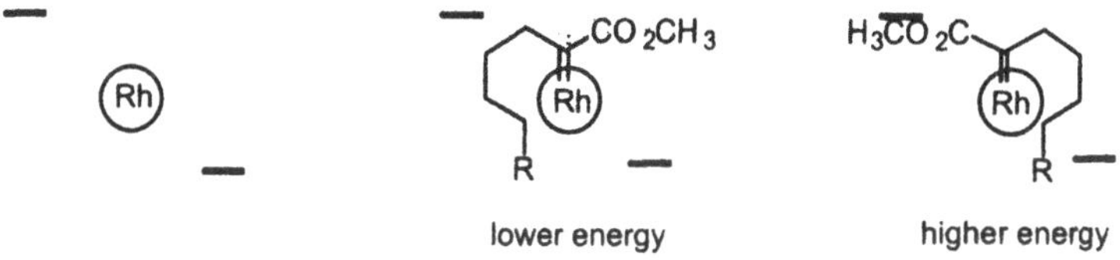

lower energy   higher energy

41   42   43

Scheme 10

41, leading to one enantiomer, over transition state **42**, leading to the competing enantiomer.

We used our computationally based model to design and assess a series of chiral Rh(II) carboxylates. It was quickly apparent that designs based on simple mono carboxylates were too flexible—there was never an unequivocal energy difference between the two competing diastereomeric transition states. We are therefore pursuing two complementary strategies: the use of ortho-metalated head-to-tail triarylphosphine complexes, and the use of diacids that can bridge two sites on the dirhodium core.

11. TRIARYLPHOSPHINE-DERIVED CATALYSTS

This part of the work (30) was carried out in collaboration with Professor Pascual Lahuerta of the University of Valencia, Spain. Most of the work was done by Salah Stiriba, a Ph.D. student from Valencia who also spent three months in our laboratory.

All approaches to the design of enantiomerically pure Rh(II) catalysts (28–30) had depended on the attachment of enantiomerically pure ligands to the rhodium core. We undertook a complementary strategy (Scheme 11), the preparation

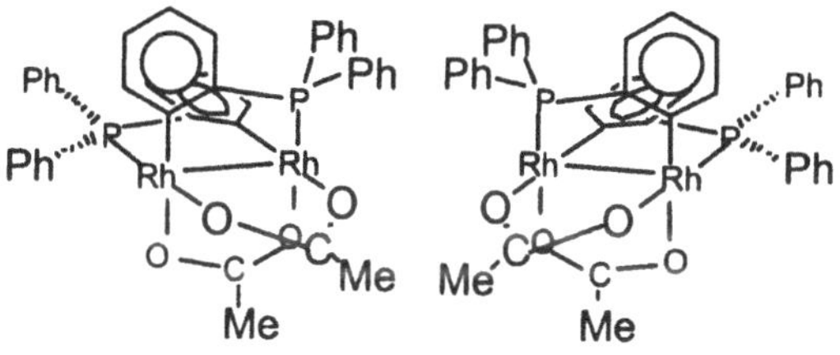

(P)-**44**   (M)-**44**

Scheme 11

of Rh(II)-dimer (P)-**44** and its enantiomer (M)-**44** having backbone chirality (31,32).

Our proposed transition state for Rh-mediated C–H insertion seemed to fit the chiral twist of complexes (P)-**44** and (M)-**44** particularly well. In fact, using the approach outlined earlier we calculated that the transition state **46a** (Scheme 12) should be sterically favored over the transition state **46b** by 4.2 kcal/mol.

Motivated by this possibility, we considered strategies by which complexes such as (P)-**44** and (M)-**44** might be obtained as single enantiomers. Our first approach, separation of the diastereoisomers resulting from addition of chiral ligands to the axial positions of the dimers [to make $Rh_2(PC)_2(O_2CR)_2L_2^*$], turned out not to be practical, due to the high kinetic lability of those ligands. We therefore turned to an alternate possibility, separation of the diastereomers derived from the attachment of chiral carboxylate groups [$Rh_2(PC)_2(O_2CR^*)_2L_2$].

As a chiral auxiliary (Scheme 13) we used the inexpensive N-(4-methylphenylsulfonyl-(L)-prolinate), (Protos, **48**). Replacement of acetate by Protos in the orthometalated acetate mixture (P)-**44** and (M)-**44** yielded the expected 1:1 mixture of the desired diastereomers **49a** and **49b**. These were separable by silica gel chromatography (10% Et_2O/CH_2Cl_2).

The two enantiomerically enriched complexes (P)-**50** and (M)-**50** were obtained via ligand exchange of **49a** and **49b** (separately) with trifluoroacetic acid. The enantiomeric purities of (P)-**50** and (M)-**50** (>98% ee) were checked by ^{31}P NMR in the presence of (−)-1-1(1-naphthyl)ethylamine. The absolute configurations of (M)-**50** and (P)-**50** were established by X-ray diffraction. Further exchange with pivalic acid then gave (M)-**51** and (P)-**51**.

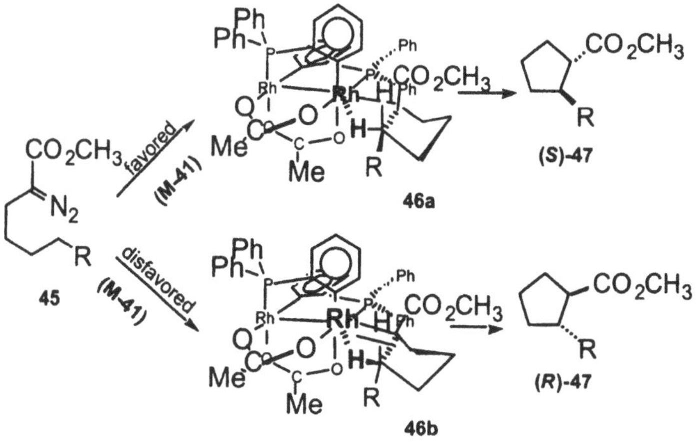

Scheme 12

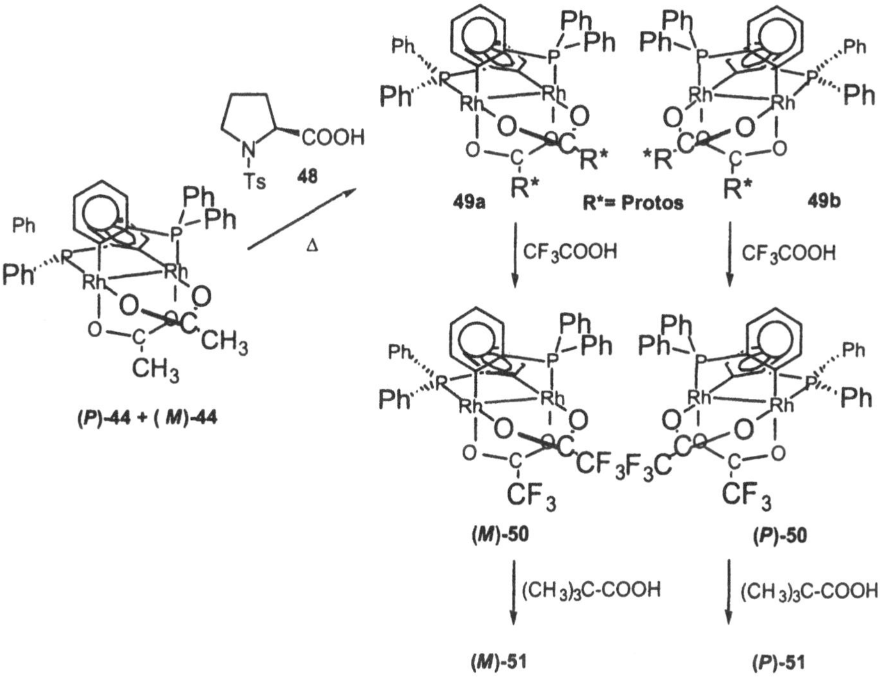

Ph, Ph, P–Ph, Ph,,,R, Ph'' Rh

N–Ts, COOH

48

Ph, Ph, P–Ph, Ph'''R, Ph' Rh, O–C–C, CH₃, CH₃

(P)-44 + (M)-44

Δ

Ph, Ph P–Ph, Ph''' R, Ph' Rh—Rh, O, O–C–C–R*, R*

49a R* = Protos

Ph Ph–P, Ph, Rh—Rh, O, O, *R–C–C–O, R*

49b

CF₃COOH | CF₃COOH

Ph,, Ph P–Ph, Ph''' R, Rh—Rh, O, O–C–C–CF₃, CF₃

(M)-50

Ph Ph–P, Ph, Rh—Rh, O, F₃C–C–C–O, CF₃

(P)-50

(CH₃)₃C-COOH | (CH₃)₃C-COOH

(M)-51 | **(P)-51**

Scheme 13

11.1. Assessment of Catalyst Reactivity

It is not likely that a highly reactive catalyst will be highly selective. Electron donation from the target C–H bond, and concomitant commitment to bond formation will be too early, when the target C–H is at too great a distance to feel the chirality of the ligands on Rh. We therefore needed a method to establish the *relative reactivity* of a series of Rh catalysts. We have developed ester **52** (Scheme 14) as our standard substrate. We have observed (4) that on exposure to rhodium [tetrakis]trifluoroacetate, **52** gave only the eliminated product **54**. On exposure to rhodium [tetrakis]pivalate, on the other hand (pivalate is the most electron-donating ligand we have yet found), **52** gave an 8:1 ratio of **53** to **54**. Rhodium [tetrakis]acetate gave about a 2.3:1 ratio of **53** to **54**, and rhodium [tetrakis]octanoate gave a 4:1 ratio. We have therefore taken the **53/54** ratio to be a useful measure of the reactivity (correlating with the electrophilicity and thus with the length of the incipient C–C bond at the point of commitment to cyclization) of a rhodium complex. Unfortunately, the chiral Rh complexes prepared by Doyle (27) gave only elimination from **52**, with no **53** being observed at all. By

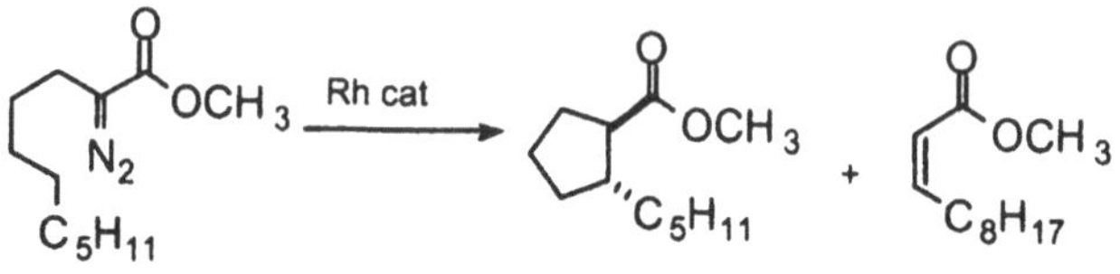

52   **53**   **54**

Catalyst

(M)-50   0   :   100

(M)-51   53   :   47

Scheme 14

this same analysis, it was apparent that the orthometallated triphenylphosphine catalysts are somewhat more reactive than the [tetrakis]carboxylates.

11.2. Enantiomeric Excess

We knew from the cyclization/elimination ratio that the Rh carbenes derived from (P)-50 and (P)-51 were very reactive, and so we did not expect them to be highly selective. We were delighted to observe that even these very reactive catalysts, with commitment to bond formation occurring far from the chiral environment of the ligands, still gave significant enantiomeric excess. It is noteworthy that each of the three substrate types, **52**, **55**, and **57** (Scheme 15), showed about

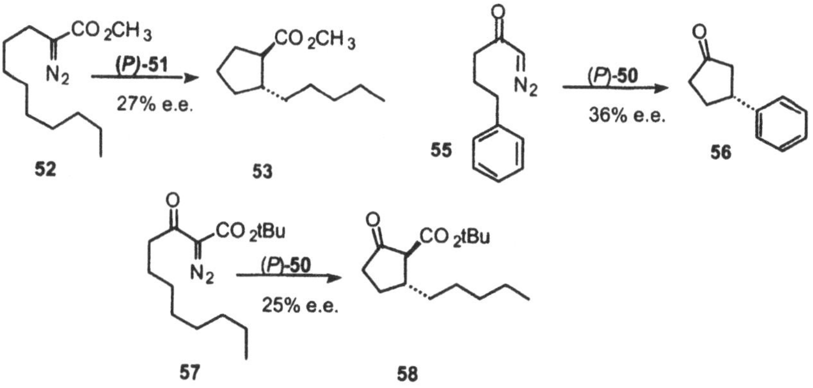

Scheme 15

the same degree of enantioselectivity. It is especially encouraging that the major enantiomer observed is in each case the one predicted by our computational model.

12. DIRECTIONS FOR THE FUTURE

At 27% enantiomeric excess, we are observing a $\Delta\Delta G$ of about 0.4 kcal/mol, or about 10% of that estimated computationally. Our hypothesis is that the enantiomeric excess is low because commitment to bond formation with the reactive carbene is occurring very early. With an early, open transition state, the substrate is not feeling the full influence of the chiral ligands on rhodium. We propose to test this hypothesis by preparing Rh complexes that will give less reactive carbenes and assessing their catalytic activity. We are pursuing the following two complementary strategies.

12.1. More Electron-Donating Ligands

It is apparent from the results with the cyclization of **52** to **53** vs. **54** (Scheme 14) that more electron-withdrawing ligands on the Rh make the derived carbene more reactive. Thus, we should be using more electron-donating phosphines to prepare analogs of (*M*)-**50** and (*M*)-**51**. So far, attempts to prepare such analogs have failed at the orthometallation stage.

12.2. Chiral Analogs of Pivalate

Electronically, it is important that the carboxylate ligand on Rh be as electron-donating as possible. In practice, this means that α,α,α-trialkylated carboxylates are going to be the most effective. Combination of this concept with the bridged design **59** (Scheme 16) and the need to make the ligand usefully chiral led to the ligand **60**. It was envisioned that as **60** wrapped equatorially around the Rh–Rh core, the cyclohexyl rings would extend outward. The phenyl substituents would then project upward and downward, creating a chiral space around the apical position of the Rh, where the carbene would be located.

We have not yet prepared **60**, but we have calculated that one of the two diastereomeric transition states for C–H insertion (Scheme 17) would be favored over the other by 8.2 kcal/mol. This suggests that **61** and **62** could be highly selective catalysts for C–H insertion.

One advantage of this approach is that we plan to assemble **60** in a modular fashion (Scheme 18). By systematically varying the pendant arene, the cycloalkane with its substituents, and the group that bridges the two carboxylates, we should be able to prepare a combinatorial family of catalysts.

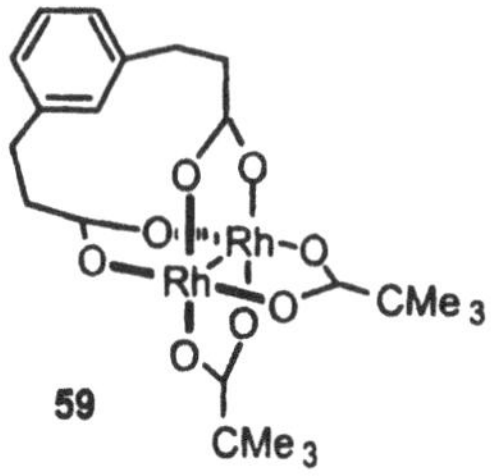

59

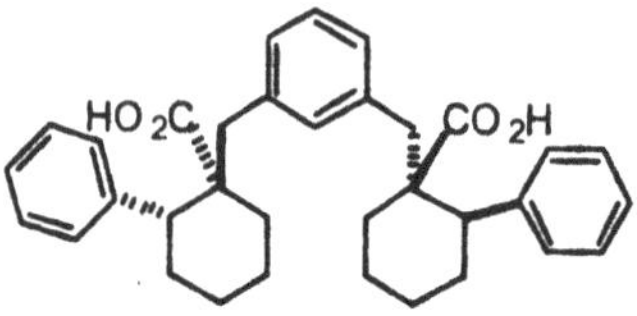

60

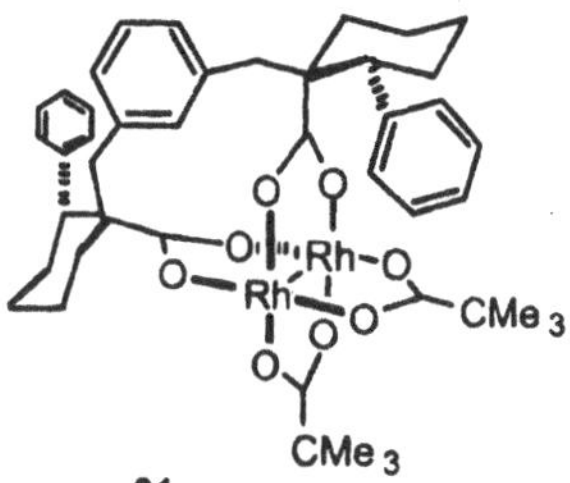

61

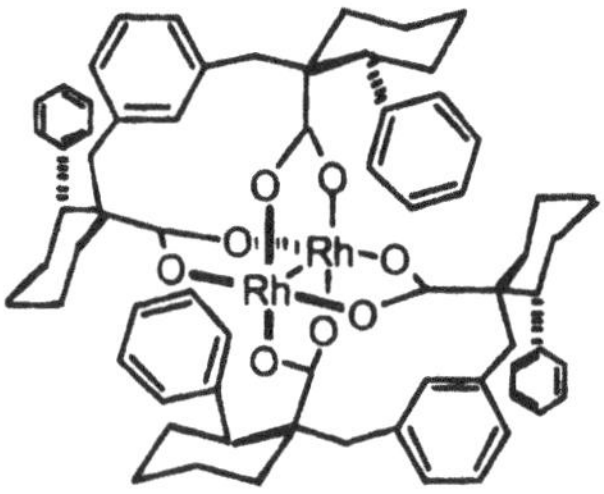

62

Scheme 16

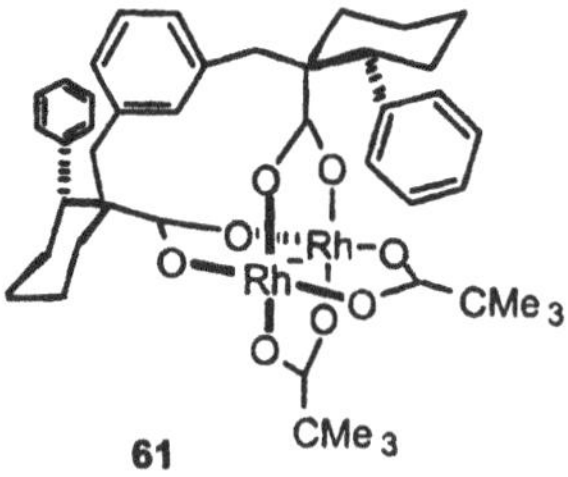

61

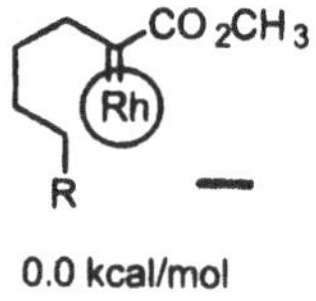

62
0.0 kcal/mol

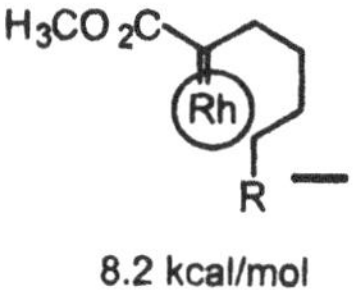

63
8.2 kcal/mol

Scheme 17

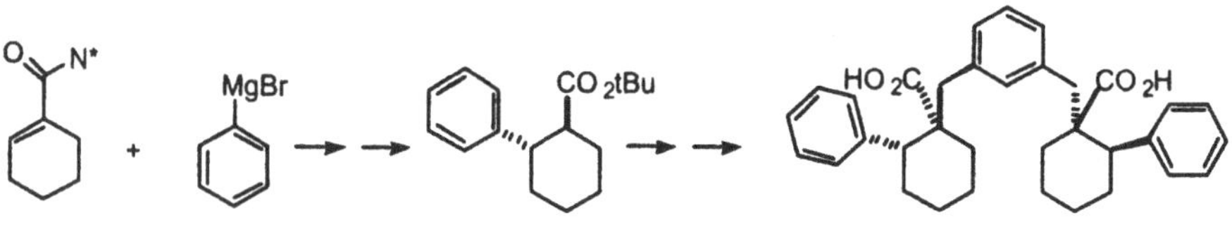

64   **65**   **66**   **60**

Scheme 18

REFERENCES

1. (a) DF Taber, EH Petty. J Org Chem 47:4808–4809, 1982. (b) DF Taber, RE Ruckle Jr. J Am Chem Soc 108:7686–7693, 1986.
2. For general reviews of rhodium mediated C–H insertions see: (a) DF Taber. Comprehensive Organic Synthesis. In G Pattenden, ed. Oxford: Pergamon Press, 1991, Vol 3, pp 1045–1062. (b) A Padwa, DJ Austin. Angew Chem Int Ed Engl 33:1797–1815, 1994. (c) T Ye, MA McKervey. Chem Rev 94:1091–1160, 1994. (d) MP Doyle, MA McKervey, Y Tao. Modern Catalytic Methods for Organic Synthesis with Diazo Compounds, New York: Wiley, 1998.
3. DF Taber, RE Ruckle, Jr., MJ Hennessy. J Org Chem 51:4077–4084, 1986.
4. DF Taber, MJ Hennessy, JP Louey. J Org Chem 57:436–441, 1986.
5. DF Taber, K You, Y Song. J Org Chem 60:1093–1094, 1995.
6. DF Taber, KK You, AL Rheingold. J Am Chem Soc 118:547–556, 1996.
7. For related analyses of transition states for Rh carbene insertions, see: (a) MP Doyle, LJ Westrum, WNE Wolthius, MM See, WP Boone, V Bagheri, MM Pearson. J Am Chem Soc 115:958–964, 1993. (b) KC Brown, T Kodadek. J Am Chem Soc 114:8336–8338, 1992. (c) MC Pirrung, AT Morehead Jr. J Am Chem Soc 116:8991–9000, 1994. (d) HML Davies, NJS Huby, WR Cantrell Jr, JL Olive. J Am Chem Soc 115:9468–9479, 1993.
8. Alternatively, the initial complex of the electron-deficient carbon with the electron density in the target C–H could be depicted as a three-center, two-electron bond (Ref. 7a). We initially took this approach computationally, but found that the results did not correlate with the diastereoselectivity observed for the reaction.
9. DF Taber, EH Petty, EHK Raman. J Am Chem Soc 107:196–199, 1985.
10. For reversible Rh-complexation with a C–H bond, see: (a) BH Weiller, EP Wasserman, RG Bergman, CB Moore, GC Pimentel. J Am Chem Soc 111:8288–8290, 1989. (b) EP Wasserman, CB Moore, RG Bergman. Science 255:315–318, 1992.
11. Mechanics was used as implemented on the Tektronix CAChe workstation. Although our initial work (Ref. 6) included minimizing the Rh–Rh core with ZINDO, we have subsequently found that this approach works just as well with mechanics alone. The CAChe workstation is particularly well suited to the sort of analysis outlined here, for its superb three-dimensional visualization facilitates understanding of the competing transition states.
12. DF Taber, KK You. J Am Chem Soc 117:5757–5762, 1995.
13. DF Taber, SL Malcolm. J Org Chem 63:3717–3721, 1998.
14. MP Doyle. Chem Rev 86:919–939, 1986.
15. (a) MP Doyle, LJ Westrum, WNE Wolthuis, MM See, WP Boone, V Bagheri, MM Pearson. J Am Chem Soc 115:958–964, 1993. (b) MP Doyle, Recl Trav Chim Pays-Bas 110:305–316, 1991. (c) HML Davies, NJS Huby, WR Cantrell Jr, JL Olive. J Am Chem Soc 115:9468–9479.
16. DF Taber, K Raman, MD Gaul. J Org Chem 52:28–34, 1987.
17. Rhodium pivalate {dirhodium tetrakis[μ-(2,2-dimethylpropanato O:O′)]} was synthesized by refluxing commercially available rhodium trifluoroacetate in eight equivalents of pivalic acid for 24 hours followed by removal of excess acid under vacuum. The crude catalyst was purified by flash chromatography using an MTBE:petroleum

ether gradient. TLC R_f (10% MTBE/petroleum ether) = 0.52. For leading references to the preparation of other rhodium carboxylates see: (a) TR Felthouse. Prog Inor Chem 29:73–166, 1982. (b) FH Jardine, PS Sheridan. In: G Wilkinson, ed. Comprehensive Coordination Chemistry. Vol IV. New York: Pergamon Press, 1987, pp 901–1096.

18. DF Taber, RJ Herr, SK Pack, JM Geremia. J Org Chem 61:2908–2910, 1996.

19. DF Taber, RP Meagley, JP Louey, AL Rheingold. Inorg Chim Acta 239:25–28, 1995.

20. Tetrakis(carboxylato)dimetal complexes have been prepared from Cr, Cu, Mo, Re, Rh, Ru, Tc, and W. For leading references, see: FA Cotton, RA Walton. Multiple Bonds Between Metal Atoms. New York: Wiley, 1982.

21. Chisholm has reported diacids that can connect two tungsten (or molybdenum) dimers, to make tetramers. RH Cayton, MH Chisholm, JC Huffman, EB Lobkovsky. J Am Chem Soc 113:8709–8721, 1991.

22. ZINDO was used as implemented on the Tektronix CAChe workstation. For leading references to ZINDO, a semiempirical program that has been paramaterized for the first two rows of transition metals, see: (a) MC Zerner, GW Loew, RF Kirchner, UT Mueller-Westerhoff. J Am Chem Soc 102:589–599, 1980. (b) WP Anderson, TR Cundari, RS Drago, MC Zerner. Inorg Chem 29:1–3, 1990.

23. Although ZINDO was originally parameterized to give good spectroscopic results, it had also been used in studies of the energetics and structures of transition metal-based catalytic systems. (a) GL Estiu, MC Zerner. J Phys Chem 97:13720–13729, 1993. (b) GL Estiu, MC Zerner. Int J Quantum Chem 26:587, 1992.

24. Diacid **35** was prepared by coupling α,α′-dibromo *m*-xylene with allyl magnesium chloride, followed by RuO$_4$-mediated cleavage of the resultant diene (26).

25. Diacid **35**, prepared by an alternative route, was already a known compound. P Ruggli, P Bucheler. Helv Chim Acta 30:2048–2057, 1947.

26. (a) PHJ Carlsen, T Katsuki, VS Martin, KB Sharpless. J Org Chem 46:3936–3938, 1981. (b) M Caron, PR Carlier, KB Sharpless. J Org Chem 53:5185–5187, 1983. (c) LM Stock, K W-T Tse. Fuel 62:974–976, 1983.

27. MP Doyle, WR Winchester, JAA Hoorn, V Lynch, SH Simonsen, R Ghosh. J Am Chem Soc 115:9968, 1993.

28. (a) HML Davies, PR Bruzinski, DH Lake, N Kong, MJ Fall. J Am Chem Soc 118: 6897–6907, 1996. (b) For a very effective chiral modification of **36**, see HML Davies, N Kong. Tetrahedron Lett 38:4203–4206, 1997.

29. S-i Hashimoto, N Watanabe, T Sato, M Shiro, S Ikegami. Tetrahedron Lett 34:5109–5112, 1993. This catalyst works much better with other substrates, giving ee's approaching 90% for insertion into benzylic C–H.

30. DF Taber, SC Malcolm, K Bieger, P Lahuerta, M Sanau, S-E Stiriba, J Perez-Prieto, MA Monge. J Am Chem Soc 121:860–861, 1999.

31. (a) FA Cotton, RA Walton. Multiple Bonds Between Metal Atoms. 2nd ed. Oxford: Oxford University Press, 1993. (b) AR Chakravarty, FA Cotton, DA Tocher, JH Tocher. Organometallics 4:8–13 1985. (c) F Estevan, P Lahuerta, J Perez-Prieto, M Sanau, S-E Stiriba, MA Ubeda. Organometallics 16:880–886, 1997.

32. For a description of the use of "P-" and "M-" designators for helical molecules, see EL Eliel, SH Wilen, LN Mander. Stereochemistry of Organic Compounds. New York: Wiley Interscience, 1994.

10

Molecular Mechanics Modeling of Organometallic Catalysts

David P. White and Warthen Douglass
University of North Carolina at Wilmington,
Wilmington, North Carolina

1. INTRODUCTION

Organometallic chemistry is interesting in part because it has applications to catalytic processes. Since the discovery of C–H bond activation and the homogeneous hydrogenation of olefins, the importance of organometallic complexes has been undisputed. Many experimental studies of organometallic catalysis have focused on catalyst and substrate structure, kinetics of transformations, mechanisms, thermochemical properties, turnover, selectivity, etc., and a massive quantity of experimental data has been accumulated. Molecular mechanics can be used to compile and analyze these data in order to direct the design of novel catalytic systems.

Organometallic chemists have long attempted to employ molecular mechanics to the rational design of catalysts. However, molecular mechanics was developed in order to study organic molecules, whose structures are well defined and show easily predicted trends in structure. Organometallic complexes, on the other hand, exhibit a wide variety of different structures, most of which are specific to the metal under investigation (see Chapter 2) (1). This diversity of both coordination number and geometry has resulted in individual workers developing

237

specific molecular mechanics models for a single class of catalyst, often confining the study to one reaction type containing a single type of catalyst (2). We have four goals for this chapter: 1) present an overview of the steps commonly employed to study organometallic catalysis, 2) show how the principles underlying molecular mechanics methods are applied to three specific examples (stereoselectivity in asymmetric hydrogenation, olefin polymerization, and host/guest interactions in zeolites), 3) briefly illustrate the practical applications of molecular modeling to catalysts used in industry, and 4) present a limited survey of the literature to illustrate how different workers have applied molecular mechanics to the study of properties of catalysts of importance to organometallic chemists.

2. WHERE TO BEGIN

2.1. Force Field Parameterization

Before we can model any catalytic process, we need to have at our disposal some molecular mechanics code and a well-parameterized force field. A general overview of molecular mechanics is presented in Chapters 2 and 3. In essence, molecular mechanics computes the energy required to deform a molecule from its ideal, "strain-free" geometry. Broadly speaking, there are two different types of molecular mechanics code: programs that are based on empirically assigned parameters for each type of bond and bond angle, and programs that assign molecular mechanics parameters based on rules. Parameter-based code, for example, MM2 (3), explicitly assigns a force constant and equilibrium value to all bond lengths, angles, torsion angles, and van der Waals interactions in the molecule. Rule-based code, for example, the Universal Force Field, UFF (4), derives these parameters from rules based on "normal" distances and angles. For example, a normal bond distance is the sum of covalent radii of the connected atoms. In the UFF a strain-free bond distance, r_{ij} (Eq. 1), between atoms i and j is given by the sum of covalent radii, $r_i + r_j$, with corrections for electronegativity, r_{EN} (Eq. 2) and bond order r_{BO} (Eq. 3):

$$r_{ij} = r_i + r_j + r_{EN} + r_{BO} \tag{1}$$

$$r_{EN} = \frac{r_i r_j (\sqrt{\chi_i} - \sqrt{\chi_j})^2}{(\chi_i r_i + \chi_j r_j)} \tag{2}$$

$$r_{BO} = -\lambda(r_i + r_j) \ln(n) \tag{3}$$

In Eqs. (2) and (3), χ is electronegativity, n is the bond order, and λ is a parameter. Another example of a rule-based force field is VALBOND, from the Landis group.

There are many different pieces of code available for molecular mechanics, ranging from the simple, such as MM2, to the elaborate, such as Cerius^2, SYBYL, Spartan, and HyperChem. The code chosen for a particular model of catalytic processes depends on two factors: (1) the complexity of the system that is to be studied, and (2) the amount of computer expertise available. Complicated structures, such as surfaces and zeolites, generally require specialized software packages for their visualization; typically workers use commercial code with perhaps minor modifications. Simpler systems, such as modeling vanadium oxo species, are amenable to study using simpler codes, such as MM2, that are customized to suit the specific needs of the research group. It should be noted that the various available packages employ different force assumptions and some force fields are more suitable to one kind of application than to another (see Chapter 2).

A good molecular mechanics model is only as good as the parameters or theory upon which it is based (5). Equilibrium geometrical parameters, such as distances, angles, and torsion angles, are usually found from crystallographic data. Traditionally, force constants are found from either spectroscopic data or quantum mechanical calculations (see Chapter 2). It is customary to assume that the metal-independent parameters for the organic portion of the organometallic complex are the same as the parameters found in any molecular mechanics code optimized for organic compounds. Once we have a set of parameters, we generally compute a structure and then carry out a point-by-point comparison between the computed and experimentally determined structure, usually an X-ray crystal structure. An alternative to the point-by-point comparison of computed and experimental structures, recently proposed by Cundari, is to use genetic algorithms or neural networks to compare the computed structure with many crystal structures in order to optimize the parameter set (6,7).

2.2. The Mechanism

Mechanisms in organometallic chemistry can be quite complicated and are often matters of considerable controversy (8–10). However, there is usually one step in the mechanism that enables us to get a handle on the problem we wish to solve. For example, when we look at Ziegler–Natta polymerization we shall see that the face of the olefin that coordinates to the metal determines the stereochemistry of the polymer. In this case, the molecular mechanics model focuses on the differences in energy between the two different coordination modes of the olefin. In general, a molecular mechanics model needs to focus on the step in the mechanism that gives rise to the interesting, or surprising, chemistry. Most often, we model a single step in a mechanism to determine the outcome of the reaction. The issue is to determine the rate-determining step in the mechanism and model that step. However, there may be pre-equilibria that also play a role in the reaction, so we tend to model the rate-determining step along with any pre-equilibria that may be important. Since mechanisms are often not known with certainty, we must be careful in our interpretation of results.

2.3. Conformers and Conformational Searching

In the experimental laboratory, we deal with moles of substances. On the computer, we often look at a single structure or a small set of structures. It is important to realize that there may be an entire ensemble of conformers that can participate in a reaction. Computational chemists take one of two approaches to the problem of multiple conformations: (1) carry out conformational searches as efficiently and exhaustively as feasible, or (2) study a model of the system in which there are as few conformational degrees of freedom as possible.

There are two general approaches to the search of the conformational space of a molecule: systematic and stochastic. *Systematic* approaches generate a conformational grid in which all torsion angles in the molecule are varied yielding many conformers. Consequently, systematic searches are feasible only with molecules that contain a few rotatable bonds. *Random*, or *stochastic*, searches often use Monte Carlo–type algorithms in which all torsion angles in the molecule are varied in a random manner, usually simultaneously. Efficient conformational searching is essential to developing a reliable computational model of any system. The topic of conformational searching is usually discussed in most texts on molecular modeling (11–13). In addition to these "traditional approaches," molecular dynamics and genetic algorithms are currently being used to search the conformational space of molecules (11–13).

Once we have established the focus question we wish to address and have the appropriate molecular mechanics code, force field, and parameter set, we can begin our computations. In the next three sections we look, in detail, at two homogeneous systems and one heterogeneous system. We begin with homogeneous asymmetric hydrogenation in which molecular mechanics addresses the question of how stereochemistry is transferred from the ligand to the substrate. Then we look at homogeneous Ziegler–Natta olefin polymerization to examine the use of molecular mechanics in determining the stereochemistry of a growing polymer chain. Finally, we look at the shape selectivity of zeolites as an example of heterogenous catalysis. At the end of this chapter, we present two tables summarizing other applications of molecular mechanics to organometallic catalysis. For convenience and ease of use, we include the software and force field used as well as the location of parameters and the problem studied.

3. HYDROGENATION

Olefin hydrogenation has been known since 1966, when Wilkinson and coworkers reported the homogeneous hydrogenation of olefins by rhodium catalysts (see

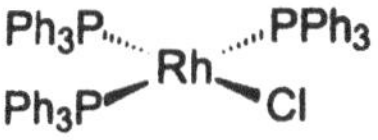

Ph₃P⠀⠀⠀⠀⠀⠀⠀⠀⠀⠀⠀⠀⠀⠀⠀⠀⠀⠀⠀⠀⠀⠀⠀

$Ph_3P_{\cdots\cdots}$ ⠀⠀$_{\cdots\cdots}PPh_3$
⠀⠀⠀⠀⠀Rh
Ph_3P ⠀⠀⠀⠀Cl

FIGURE 1 Wilkinson's catalyst, [Rh(PPh₃)₃Cl].

Fig. 1 for catalyst structure) (14–16). Since its discovery, homogeneous hydrogenation has grown and is not limited to olefins as substrates. Shortly after the discovery of homogeneous olefin hydrogenation, it became apparent that by modifying the ligands from PPh₃ to a chiral ligand, stereoselective homogeneous hydrogenation became possible. Only low enantiomeric excesses, ee's,* were achievable in the initial studies. However, moving to bidentate chiral ligands (Fig. 2) resulted in a dramatic increase in ee.

Molecular mechanics modeling of the asymmetric hydrogenation must begin with the mechanism of the reaction. When the prochiral olefin binds to the catalyst containing chiral bidentate phosphine, two possible diastereomers result: one with the *re* face and one with the *si* face of the olefin coordinated to the metal (Fig. 3). Work in the Halpern and Brown laboratories has shown that the observed enantiomeric product cannot result from the diastereomer observed in solution (17–20). Thus, the minor diastereomer, which cannot be observed, must be responsible for the dominant chiral product. Any molecular mechanics model of the asymmetric hydrogenation reaction must explain how the minor diastereomer reacts faster than the major.

To effectively model the asymmetric hydrogenation reaction, we must look at the mechanism carefully. The first step involves the displacement of solvent and the coordination of the enamide to produce the two diastereomers (Fig. 3) (17–20). It appears as though the enamide-coordinated diastereomers are in rapid equilibrium with each other through the solvento species (Fig. 4). This square planar rhodium(I) cation is then attacked by dihydrogen to form an octahedral rhodium(III) complex (Fig. 4). Hydrogen then inserts into the Rh–C bonds, and the product is reductively eliminated (Fig. 4). From a molecular mechanics standpoint we have three entities to model: the square planar rhodium(I) solvento species and the two intermediates (square pyramidal dihydrogen complex and the octahedral dihydride).

In order to model the square planar rhodium(I) complex we need to realize that the positions trans to the diphosphine may not be equivalent, since the diphosphine is chiral. Consider the [(diphosphine)Rh(norbornadiene)]⁺ as a model for the solvento species. In order to distinguish between the nonequivalent phosphorus atoms, we label them P_a and P_b. Each olefin is 90° from one phosphorus atom

---

* Enantiomeric excess is defined as % R enantiomer − % S.

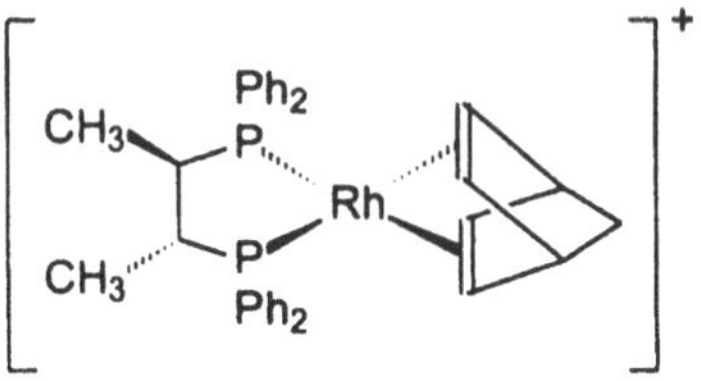

FIGURE 2   Example of an asymmetric hydrogenation catalyst, [(S,S-CHIRA-PHOS)Rh(nobornadiene)]⁺. The norbornadiene ligand is used to represent the coordination of solvent molecules.

and 180° from the other. It is very difficult to model a structure in which one interaction has two different equilibrium bond angles. There are two approaches in the literature to the problem: 1) assign the P atoms different labels (P_a and P_b) and then define each interaction uniquely (this results in a significant increase in the number of parameters) (21–23), and 2) redefine the potentials, creating a more general force field (24). Once we have decided upon an appropriate force field, we need to turn our attention to modeling an η^2-bonded olefin.

In molecular mechanics a chemical bond is considered to be composed of two spheres attached by a spring. Modeling of M-olefin systems presents a simple problem: Where do we anchor the metal? (Strictly speaking, the metal should be anchored to the center of the olefin C=C bond, but there is no atom at the C=C centroid to anchor the metal.) One approach is to bond the metal to both carbon atoms in the olefin. This creates a metallocycle, which is not a realistic model for olefin binding. An alternate approach is to define a pseudoatom (an atom with

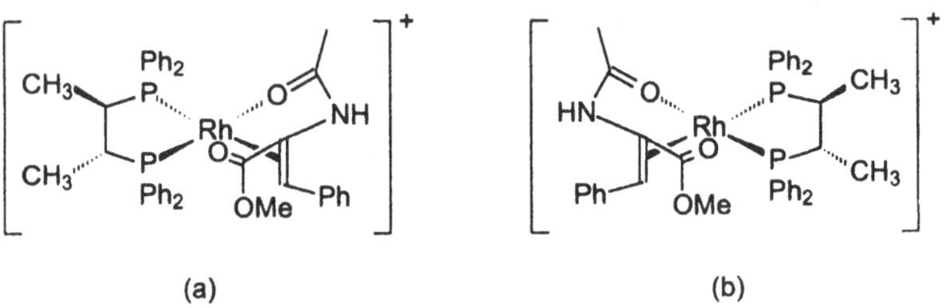

(a)   (b)

FIGURE 3   Structures of [(S,S-CHIRAPHOS)Rh(MAC)]⁺ (MAC is methyl (Z)-α-acetamidocinnamate; see Fig. 6). Notice that (a) has the *re* face of the olefin coordinated to the Rh, whereas (b) has the *si* olefin face coordinated.

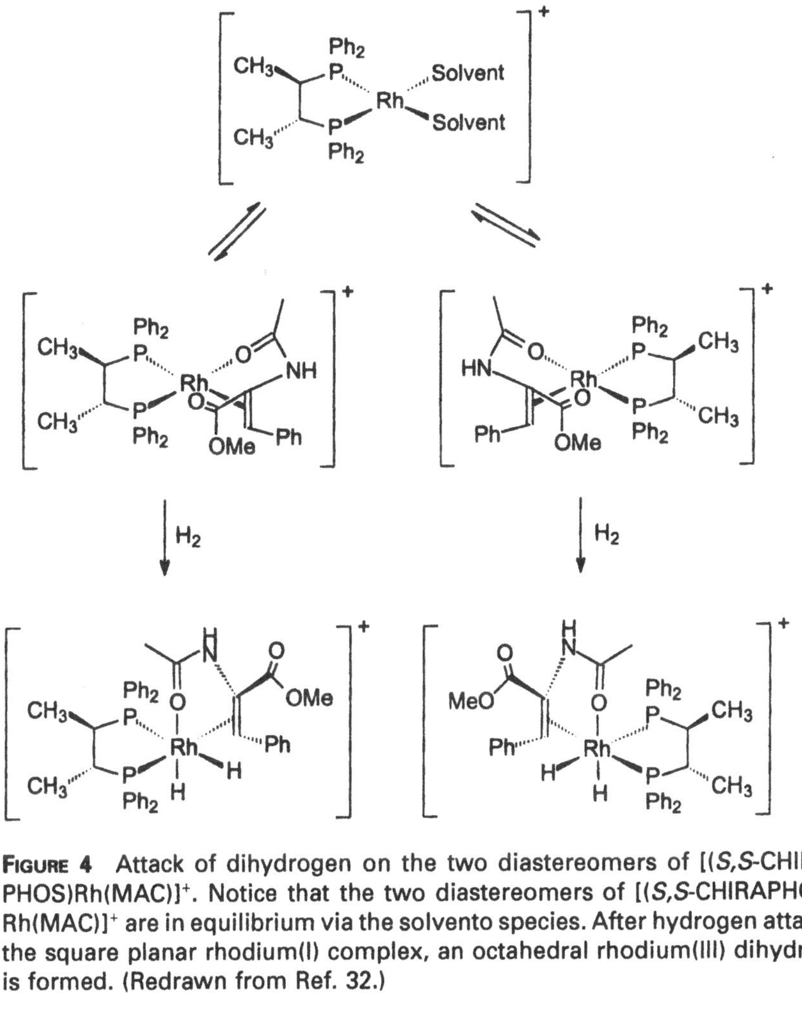

FIGURE 4 Attack of dihydrogen on the two diastereomers of [(*S,S*-CHIRA-PHOS)Rh(MAC)]⁺. Notice that the two diastereomers of [(*S,S*-CHIRAPHOS) Rh(MAC)]⁺ are in equilibrium via the solvento species. After hydrogen attacks the square planar rhodium(I) complex, an octahedral rhodium(III) dihydride is formed. (Redrawn from Ref. 32.)

zero van der Waals radius and zero force constants) in the C=C centroid and then bond the pseudoatom to the metal. However, if the pseudoatom interrupts the bonding, then the two halves of the olefin rotate with respect to each other and physically unrealistic results emerge (Fig. 5). One resolution is to place a pseudoatom at the centroid of the C=C bond but to leave the C–C bond intact (23). Thus, we generate a single point of attachment of the olefin to the metal, but we do not interrupt the C–C connectivity. A similar approach has been used for the molecular mechanics modeling of cyclopentadienyl ligands (25–28).

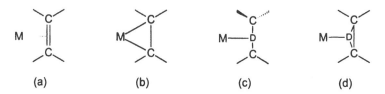

FIGURE 5   Bonding models for an η^2-olefin interaction. (a) Shows the actual bonding in the complex, (b) a molecular mechanics model of the metallocycle, (c) shows how the two halves of the olefin can rotate relative to each other if a pseudoatom, D, interrupts the bonding, and (d) shows a molecular mechanics model that is used in the literature. (From Ref. 23.)

    Now that we have the ability to bond olefins and enamides to the rhodium, we are in a position to be able to model asymmetric hydrogenation. Using CHEMX,* Brown and Evans modeled a series of [(diphosphine)Rh(dehydroamino acid)]$^+$ complexes (29). CHIRAPHOS was used as the chiral diphosphine with ethyl (Z)-α-acetamidocinnamate (EAC) as the substrate (Fig. 6). The structures of interest were assembled from fragments derived from X-ray crystal structures. Hydrogens were added at standard 1.08-Å distances. Only the van der Waals energy was minimized, using TORMIN in COSMIC molecular modeling package,† assuming the P–Ph torsional barriers were insignificant. Brown and Evans noted that the energy difference between the *re* and *si* diastereomers was small enough to lie within the computational limits of accuracy. However, they concluded that the main difference between diastereomers occurs in the nonbonded interaction between α-ester group and the aryl groups on the phosphorus atoms of the diphosphine.

    With the geometries of the diastereomers established, Brown and Evans then modeled the addition of H$_2$ to Rh (29). Four different pathways for hydrogen attack were considered, two for each diastereomer. The hydrogen was placed 1.60 Å from the metal and the van der Waals energy minimized. Sterically impossible structures were eliminated, and the resulting two diastereomers showed a large energy difference (42.2 kcal/mol). The high-energy diastereomer contained a significant nonbonded interaction between ester and P–Ph group, whereas the low-energy diastereomer did not. Finally, these workers calculated an energy surface for the attack of dihydrogen on the metal. The major and minor diastereomers were found to respond quite differently to the addition of dihydrogen. Sub-

---

* CHEMX was reported to be available from Dr. K. Davies and associates, Molecular Design, Oxford, UK. See Ref. 29.

† COSMIC and TORMIN were reported to be available from Dr. J. G. Venter, Smith Kline and French, Welwyn Garden City, See Ref. 29.

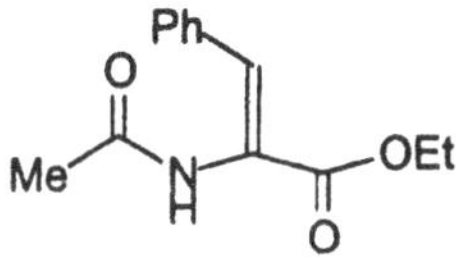

FIGURE 6 Structure of ethyl (Z)-α-acetamidocinnamate (EAC).

sequent to this paper, Landis and coworkers have approached the same problem from a quantum mechanical perspective, yielding results that are more sound from a theoretical perspective.

Bosnich and coworkers analyzed asymmetric hydrogenation using molecular graphics with MODEL-MMX* (30). Dihydrogen addition to both major and minor diastereomers was analyzed for the [(S,S-CHIRAPHOS)Rh(EAC)]+ complex. (EAC is ethyl-N-acetyl-α-aminocinnamate.) As with the Brown approach (29), Bosnich considered only van der Waals terms in the computation of energies and the partial minimization of the complexes using the method reported by Davies and Murrall (31). The crystal structure of the major diastereomer was used for input, and eight dihydride structures were analyzed (Fig. 7). Only two of the eight possible trajectories gave feasible energies for dihydrogen attack at the metal. Calculations agreed with experiment in that the computed low-van-der-Waals-energy structures contained the correct alignment about the M–H and M–olefin bonds for product formation.

In 1993 Landis began a detailed study of the asymmetric hydrogenation reaction (32). In this work, he analyzed the structural features of the catalyst that give rise to high enantioselectivity. In particular, he focused on methyl (Z)-α-acetamidocinnamate (MAC) as the substrate with DIPAMP, CHIRAPHOS, and DIPH as phosphines (Fig. 8). In addition to the prochiral MAC substrate, Landis also included norbornadiene as a test substrate to develop the methodology. In the molecular mechanics calculations, Landis used a modified SHAPES force field (24) within CHARMM (33). Electrostatics were included using Rappé and Goddard's QEq method (34). Inclusion of electrostatics was found not to alter results significantly. Finally, the conformational space of the molecule was sampled by using a constrained grid-search technique.

The differential equilibrium constant for the binding to form the two diastereomers (using MAC as the substrate) is reported to decrease from about 30

---

* MODEL-MMX was reported to be obtained from Clark Still modified by K. Steliou, University of Montreal. MMX, developed by K. Gilbert and J. J. Gajewski, Indiana University, was obtained through Serena Software.

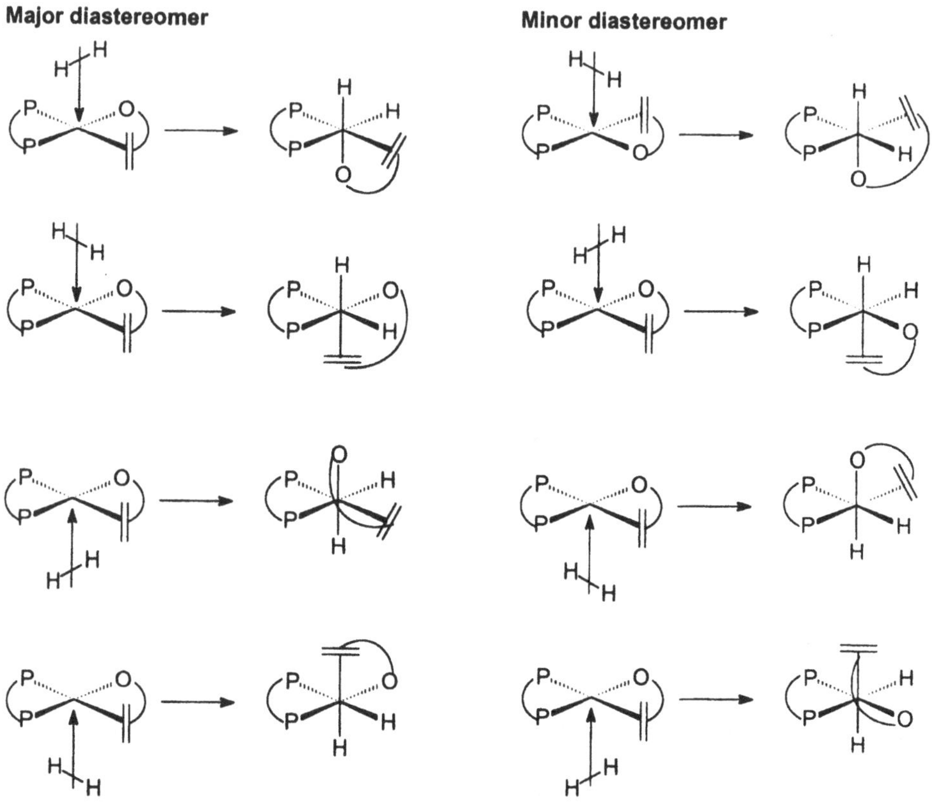

Major diastereomer   **Minor diastereomer**

FIGURE 7   Dihydrogen attack trajectories for the major and minor diastereomers of [(diphosphine)Rh(enamide)]+ studied by Bosnich and coworkers (30).
In this figure the enamide is represented by coordinated O and olefin and the
diphosphine by the coordinated phosphorus atoms. (Redrawn from Ref. 30.)

to 10 to 1 on moving from CHIRAPHOS to DIPAMP to DIPH (18,35). For
[(CHIRAPHOS)Rh(MAC)]+, Landis found a diastereomeric energy difference
of 2.3 kcal/mol using molecular mechanics (32). Changing the diphosphine from
CHIRAPHOS to DIPH, a 1.3-kcal/mol energy difference between diastereomers
was computed (as compared with ±0.3 kcal/mol reported in the literature). These
results suggest that the margin of error involved in this methodology is about
1 kcal/mol. Finally for the DIPAMP complex, a 0.4-kcal/mol energy difference
between diastereomers was computed (as opposed to the experimental ΔH of 1.4
kcal/mol).

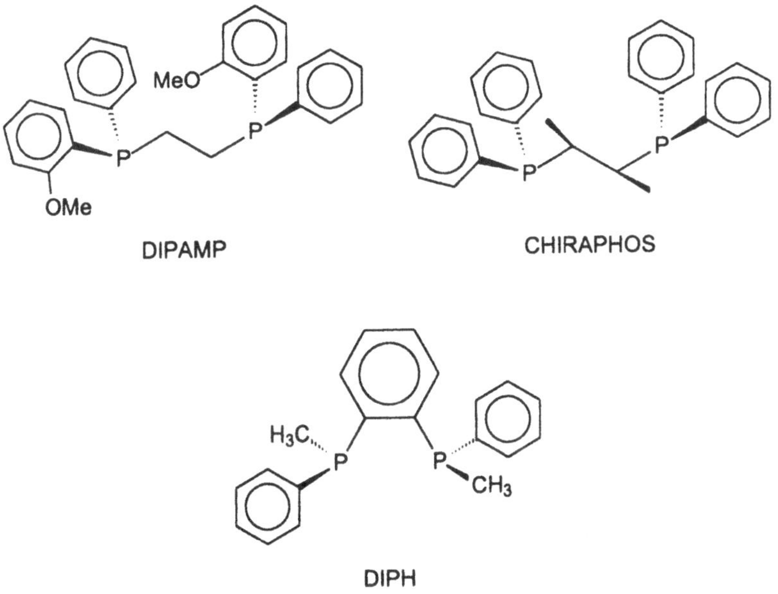

DIPAMP   CHIRAPHOS

DIPH

FIGURE 8 Structures of DIPMAMP, CHIRAPHOS, and DIPH. (Redrawn from Ref. 32.)

Since enantioselectivity in this reaction is a result of the energy difference between the diastereomeric transition states after H_2 is added, Landis modeled the addition of H_2 to the diastereomers of the CHIRAPHOS and DIPAMP complexes with MAC as the substrate. Landis posed a simple question: Is there a significant barrier to hydrogen attack at the Rh center that can be modeled by molecular mechanics? In the first study Landis found that all possible attack trajectories allowed almost strain-free attack of dihydrogen (molecular mechanics barriers were less than 3 kcal/mol) (32). In a subsequent study, a better picture of the reaction coordinate was generated using DFT and quantum mechanical models, which are outside the scope of this chapter.

Several other workers have used different force fields to model enantioselective hydrogenation to different substrates (see Table 1). For example, Schwalm and coworkers approached the enantioselective hydrogenation of α-ketoesters using AMBER (36) from within MACROMODEL (37,38). Mortreux and coworkers have used CAChe (39) augmented with MM2 (3) to model asymmetric hydrogenation of ketopantolactone (40). Ruiz has used Cerius[2] (41) with the Universal Force Field (4) to model hydrogenation using Rh and Ir complexes of rigid dithioethers as ligands and acrylic acids (42).

TABLE 1 Summary of the Application of Molecular Mechanics to Organometallic Catalysis, with Examples

Reaction	Catalyst	Program	Force field	Added parameters	Purpose	Ref.
Allylation	η³-Allyl palladium complexes with chiral phenanthroline ligands	MacMimic	MM2	η³-Allyl-Pd interaction (91,92)	Model the stereoselectivity in complexes with conformationally flexible ligands in asymmetric allylation	93
Allylation	Chiral η³-allyl palladium(II) catalysts	MacroModel	MM2	Derived from crystal structure data; available as supplementary material	Determine the factors that govern stereodifferentiation in [(chiral diphosphine)Pd(η³-allyl)]⁺ complexes	94
Allylation	η³-Allyl palladium(II) catalysts	MacMimic	MM2	Obtained from the literature (91)	Quantify the steric interaction between an incoming nucleophile and η³-allyl palladium complex during allylation	95
Dihydroxylation	Osmium tetraoxide with various other ligands	MacroModel, MacMimic	Modified MM2, MM3	NH₃-type nitrogen parameters taken from MM2(91); osmium parameters taken from the literature (96)	Explain enantiofacial selectivities and selectivity trends observed for the various olefin classes	97
Epoxidation	Mn(salen) complexes	MacroModel MM3	MM3	Parameters for Mn obtained from crystal structure data (98)	Probe mechanism of epoxidation by Mn(salen) complexes via metallaoxetane intermediates	99
Force field development	Metallocenes (M = Fe, Ru, Os, V, Cr, Co, Fe, Ni)	CHEM-X	CHARMM	Vibrational data	Develop a self-consistent molecular mechanics force field, based on spectroscopic data, for linear metallocenes	25
Force field development	Bent Ti, Zr, and Hf metallocenes	CHARMM	Modified CHARMM (27)	Derived from vibrational data of [Cp₂MCl₂] complexes	Generate a self-consistent, accurate force field for bent metallocenes	27

Application	System			Data used	Comments/Goal	Ref.
Force field development	Inorganic Fe^{2+} and Ni^{2+} complexes with N-donor ligands	CHARMM	CHARMM	Augmented with values for pyridine and metal	Generate parameters appropriate for modeling the selectivity of the macrocyclic reagents to the size of the metal atom	2
Force field development	Vanadium-oxo complexes	MM2	MM2	X-ray data and quantum calculations to determine metal-dependent parameters	Develop force field for vanadium-oxos; compare MM and SEQC methods	100
Force field development	Low-spin Ni(II) complexes with tetraaza macrocycles	MOLMEC	MM2	Extensions for aliphatic amines and aromatic systems from crystal structure data	Use trial-and-error process to derive force field parameters for the Ni(II) part of the molecule that gave the best fit with X-ray data	101
Force field development	Allylic nickel phosphine complexes	PCMODEL	MM2	Crystallographic data and ab initio calculations	Develop force field to predict diastereo-induction in intramolecular Ni-catalyzed [4 + 4] cycloadditions	102
Force field development	Vanadium peroxides ($L_nV(O_2)_m$; $m = 0–4$)	MM2	MM2	Metal-dependent parameters derived from quantum mechanical calculations and crystallography	Demonstrate utility of estimating metal-dependent molecular mechanics parameters from quantum calculations	103
Force field development	Transition metal carbonyl clusters	Custom	MM2	Presented in paper	Develop a new force field for the molecular mechanics simulation of ligand structures in transition metal carbonyl clusters	104
Force field development	WCl_xCHR (R = H, CH_3, CH_2CH_3)	PCMODEL	METMOD1	Crystal structure data	Develop and evaluate tungsten carbene parameters	105
Force field development	Cobalt(II), nickel(II), and copper(II) complexes with amine and imine ligands	DOMMINO	CLFSE MM (106)	Crystal structure data	Extend molecular mechanics scheme with cellular ligand field ligand stabilization energy (CLFSE) terms that explicitly treat the electronic effects arising from changes in the d-orbital energies	107

TABLE 1 Continued

Reaction	Catalyst	Program	Force field	Added parameters	Purpose	Ref.
Force field development	Layered α- and γ-zirconium phosphates	Cerius²	UFF	Derived from AIQC on model compounds	Molecular mechanics parameters derivation; compare molecular mechanics results to crystal data and AIQC (CRYSTAL95 program)	108
HIV-1 Protese	HIV-1 Protease	FRODO, CHAIN, AMMP (109)	UFF	Protein crystal structure and IR data, used to improve parameters for proteins and nucleic acids (110)	Calculate protease/peptide interaction energies	111
HIV-1 Protease	HIV-1 Protease	Insight II Discover	CVFF	Crystallography (112)	Calculate inhibitor binding energies	113
Hydrocyanation	[NiCl$_2$L$_2$], L$_2$ = electronically tuned Thixantphos diphosphines	SYBYL	TRIPOS	Reported in paper	Study effect of ligand bite angle and backbone rigidity on hydrocyanation selectivity	114
Hydrodesulfurization	MoS$_2$ slabs on γ-Al$_2$O$_3$ support	Cerius²	Dreiding	Listed in paper	Model nonbonded interactions of MoS$_2$ with Al$_2$O$_3$	115
Hydrodesulfurization	MoS$_2$ slabs supported on γ-alumina and β-quartz	Cerius²	Dreiding	MoS$_2$ parameters from Ref. 115; other parameters listed in paper	Model free MoS$_2$ clusters and nonbonded interactions between MoS$_2$ sheets and planes of γ-alumina or β-quartz for hydrodesulfurization catalysts	116
Hydroformylation	Rhodium complexes containing BINAPHOS	Insight II Discover	Extended cff91	Parameters based on DFT calculations	Use molecular mechanics to include steric effects in DFT calculations on the stereoselectivity of hydroformylation	117
Hydroformylation	[Pt(CO)XL$_2$] complexes (L = diphosphine, X = halide)	Polygraf	Dreiding	Parameters included in paper	Evaluate importance of steric factors in determining regioselectivity	118

Hydroformylation	Rhodium diphosphine complexes	MacroModel	Amber	Augmented with values for tertiary phosphines (19)	Probe the different aldehyde regioselectivity of phosphine ligands	120
Hydroformylation	[Rh(modified xanthene)H(CO)L] and [Rh(diphosphine)(H)(CO)$_2$] complexes	SYBYL	TRIPOS	Crystal structure data and Ref. 121	Develop new bidentate diphosphines by modeling the effect of bite angle on regioselectivity	122
Hydrogenation	Rh diphosphine complexes	CHEM-X COSMIC	COSMIC	Crystal structure fragments assembled in COSMIC	Define the source of stereoselectivity in binding of prochiral enamide to the chiral Rh diphosphine fragment	29
Hydrogenation	[Rh(chiral bisphosphine)(MAC)]$^+$ complexes	QUANTA CHARMM	SHAPES (24)	Presented in paper	Probe structural features that give rise to the enantioselectivity observed in the hydrogenation of the substrates	32
Hydrogenation	Cinchona-modified Pt/alumina catalysts	AMBER	MacroModel	Unknown origin of parameters	Rationalize interaction between chiral modifier and substrate	38
Hydrogenation	Rh(I) aminophosphine-phosphinite complexes	CAChe	MM2	Presented in paper	Support thermodynamically controlled asymmetric hydrogenation of ketopantolactone	40
Hydrogenation	Ir complexes with chiral dithioether ligands that form seven-membered rings	Cerius^2	UFF	UFF parameters only	Study the relative stability of possible isomers	42
Hydrogenation	[Rh(S,S-CHIRAPHOS)]$^+$ enamide complexes	CHEM-X PCMODEL	MMX	Crystallography and within MMX	Analyze the addition of H$_2$ to the major and minor diasteremeric [Rh(S,S)-CHIRAPHOS)]$^+$ fragments	30
Hydroxylation	Osmium tetraoxide	MacroModel	MM2	Used values for RuO$_4$	Explained selectivity trends for various olefins	97

TABLE 1 Continued

Reaction	Catalyst	Program	Force field	Added parameters	Purpose	Ref.
Hydroxylation	Iron and manganese prophyrins	TOPO, MMID	MM2	Referenced in paper	Model catalyzed saturated alkane hydroxylation	123
Insertion	[(SiH$_2$-C$_5$H$_4$-NH)MCH$_3$]$^+$ (M = Ti, Zr, Hf), [(SiH$_2$-C$_5$H$_4$-NH)TiCH$_3$]	POLYGRAF	MM2	Unknowns approximated	Probe mechanism of chain propogation using constrained catalyst geometries	124
Insertion	Cp$_2$*U(H)[(1s)-endol-bornoxide] (Cp* = η^5-C$_5$Me$_5$)	BIOGRAF	Dreiding and MMP2	Crystallographic data	Search for most sterically favorable approach of olefin toward actinide center	125
Insertion	Rh$_2$(5S-MEPY)$_4$, Rh$_2$(5R-MEPY)$_4$, Rh$_2$(4R-BNOX)$_4$, Rh$_2$(4S-BNOX)$_4$	CAChe	Augmented MM2	Crystallographic data	Determine preferred conformation of intermediate metal carbenes, and measure steric effect of chiral ligands	126
Insertion	Dirhodium carboxylates and carboxamides	CAChe	Augmented MM2	Estimated for atom types not in MM2	Model pseudotransition-state structures to identify steric factors that control regioselectivity	127
Ligand design	Cyclophosphazenic polypodands and glymes and their complexes with ion pairs M$^+$I$^-$, M = Li, Na, K, and Rb	MM2I (128)	MM2	Derived from crystallographic data (129)	Investigate the catalytic activity of these ligands in solid–liquid phase transfer reactions	130
Ligand design	Ni(II) with tetraaza macrocyclic ligands	ALCHEMY and BOYD (131)	Modified TRIPOS (132)	Derived from crystallographic data and reported in paper	Predict the steric strain in the higher-field-strength complexes that could not be synthesized	133
Ligand design	Chiral crown ethers derived from camphor	Hyperchem	AMBER	From Hyperchem, with additions listed in the paper	Study mechanism of ionic reactions catalyzed by chiral crown ethers; model stereoselectivity of catalysts	134

Metathesis	Supported tungsten phenoxides	Insight II	ESFF	Crystal structure data	Model the surface structure to determine the preferred arrangement of the tungsten diphenoxide species on the hydroxylated support	135
Metathesis	Tungsta-carbenes	PCMODEL	METMOD1 MMX	MMX parameters and crystal structure data	Create an adequate model to study the catalysts, the intermediates, and products of this reaction type	136
Organic	TADDOL–TiCl$_2$	Chem3D	MM2	None specified	Measure the influence that substituents on the dioxolane ring exert on stereoselectivity	137
Organic	Co$_2$(CO)$_8$	PCMODEL	MMX	None specified	Model proposed cobaltacycle intermediates of bicyclization of substrate	138
Organic	Redox active cavitand ligands with ferrocenyl redox centers	CHEMMOD	CHEMMIN	Taken from CHEMMIN	Find the minimum-energy position of the ligand within the cavitand	139
Organic	Bleomycin (BLM) bound to Fe(III)	MM2MX	1.5MM2/MX	General metal complex values used	Make a qualitative study of Fe(III)BLM bound to HOOH	140
Polymerization	Bridged zirconocene dichlorides	Discover	Modified cff91	Unpublished data	Use molecular mechanics to study equilibration between conformers of the catalysts	141
Polymerization	Cu(I) carboxylates	MOPAC PCMODEL	PM3 and MMX	None specified	Propose possible mechanism of the dimerization of 1-alkynes	142
Polymerization	Ansa zirconocenes with chiral ethylene bridges	Not specified	Not specified	None specified	Rationalize stereoregularity of polypropene samples in the presence of diastereomeric complexes with different bridges	143

TABLE 1 Continued

Reaction	Catalyst	Program	Force field	Added parameters	Purpose	Ref.
Polymerization	[WCl$_6$]/[SnMe$_4$]	METMOD	MM2	Collected from literature or spectroscopic data	Create a model for studying the steric effects governing the stereoselectivity of olefin metathesis in the elementary steps of olefin polymerization	144
Polymerization	Silylene-bridged zirconocene	MM2	MM2	None reported (Zr treated as a pseudoatom)	Investigate the effects of alkyl substituents on the Cp rings and the olefin substrate	55
Polymerization	Zirconocenes	Custom code	Modified CHARMM (27,33) AMBER	From the literature (69)	Determine relationship between regiospecificity and type of stereoselectivity in propene polymerization	59
Polymerization	Cyclopentadienyl complexes of Ti and Zr	Custom code	CHARMM MM2	None specified	Present a geometrical and nonbonded energy analysis on possible catalytic intermediates	60
Polymerization	Bis-(2-phenylindenyl) zirconium chloride precursor	Custom code	CHARMM (27,33)	From the literature (69)	Analyze the stereospecificity and enantioselectivity caused by the isomerization of the catalyst	61
Shape selectivity	Fe(II)/Fe(III) complexes of DIPIC, 2PP$_6$C, and CHOX	SYBYL	TAFF (145)	Added parameters to describe Fe(II) and Fe(III) with these ligands	Model the formation of the complexes of Fe(II) and Fe(III) of linear NO$_2$-donor set ligands	146
Shape selectivity	Bis(pentamethylcyclopentadienyl) complexes of Ca, Sr, Ba, Sm, Eu, and Yb	CHARMM	CHARMM	Parameters obtained from the literature (27)	Understand why these metallocenes are bent	26

Shape selectivity	Vanadium oxo complexes	Custom code	MM2	Metal-dependent parameters were derived as outlined in paper	Compare quantum and molecular mechanics methods in analysis of the steric and electronic energy differences between isomers	100
Surface study	Rh/SiO_2 and Rh[Sn(n-C_4H_9)$_x$]$_x$/SiO_2	SYBYL	TRIPOS	Listed in paper, obtained from Refs. 147–149	Calculate steric hindrance for Sn complexes grafted on the metallic surface	150
Surface study	Pt/SiO_2 and Sn(n-C_4H_9)	SYBYL	TRIPOS	Listed in paper and Ref. 149	Calculate steric hindrance for Sn complexes grafted on the metallic surface	151
Surface study	MoS_2 slab	PC-CHEMMOD	Dreiding in CHEMMIN	Used values from literature (152)	Model the active site of MoS_2 and binding of thiophene	153
Zeolites	Zeolite L	Chemgraph	Only van der Waals terms	Parameters taken from the literature (154)	Calculate minimum-energy position of pyridine in zeolite L	155
Zeolites	Silicalite and zeolite NaY	ZEOLITHEN-ERGIE	Only van der Waals terms	Reported in papers	Use molecular mechanics to develop a molecular dynamics approach to studying small molecules in zeolites	156–159
Zeolites	Zeolite NaY	ZEOLITHEN-ERGIE	Reported in paper	Crystal structure data	Determine adsorption sites and locations of aromatic molecules in zeolite NaY under catalytic conditions	74
Zeolites	H_x · ZSM-5/benzene complexes (x ≤ 4)	Custom method (160,161)	Only van der Waals terms	Listed in paper and modified from literature for sorbate (162,163)	Predict preferred proton locations in zeolite	76
Zeolites	Mordenite, zeolite L	Insight II Discover	CVFF	Obtained from the literature (78)	Obtain the minimum-energy profiles for the selective isopropylation of naphthalene	77
Zeolites	FER, ZSM-48, EUO, MFI zeolites (small-, medium-, and large-pore zeolites)	CATALYSIS	CVFF and cff91	Within CATALYSIS and computed structures compared with crystal structure data	Determine the minimum-energy pathway for the diffusion of the substrates through the zeolites	80

TABLE 1 Continued

Reaction	Catalyst	Program	Force field	Added parameters	Purpose	Ref.
Zeolites	Y, mordenite, ZSM-5 and beta zeolites	Cerius²	Burchart (zeolite), Dreiding (sorbate)	Cerius²	Estimate adsorption strength of all the carbenium ion isomers derived from the olefins on the zeolites	81
Zeolites	Zeolite-Y with lanthanide and actinide ions	QUANTA/CHARMM	Reported in paper	From crystal structure data	Study C–S bond cleavage in N-substituted carbonimido dithiolates	82
Zeolites	HZSM-5	ZEOLITHEN-ERGIE		Parameters taken from the literature (164)	Determine the location of naphthalene and 2-methylnaphthalene in the HZSM-5	83
Zeolites	Faujasite-type zeolites	Dizzy (79)	Reported in paper	Vibrational spectra	Develop a force field to explicitly distinguish between Al and Si in FAU-type zeolites	79
Ziegler–Natta	Ansa zirconocenes and hafnocenes	None specified	Modified CHARMM	None specified	Rationalize dependence of the stereoselectivity of the catalyst on the π-ligand alkyl substitutions	64
Ziegler–Natta	Meso- and rac-bis(2-penylindenyl)zirconium dichloride	MCM	UFF	No parameters needed	Model the source of barrier between meso and rac forms of the catalyst; determine the role of π-stacking on polymer formation	165

Ziegler–Natta	rac-(1,2-Ethylenebis(η⁵-indenyl))Zr(IV) catalysts	BIOGRAF	Dreiding	Ab initio calculations for activated complex; molecular mechanics parameters reported in paper	Determine effect of catalyst substituents on the tacticity of polypropylene	166
Ziegler–Natta	Ansa metallocenes	Cerius²	UFF and VALBOND	No additional parameters necesary	Develop a force field capable of calculating transition states and kinetic isotope effects for the systems of interest	167
Ziegler–Natta	Ansa zirconocenes	BIOGRAF, POLYGRAF	Dreiding	Dreiding extended to include tetrahedral Zr and pseudoatoms for Cp, allyl, and olefin centroids; parameters reported in paper	Evaluate ability of the force field to predict tacticity of known and unknown zirconocene catalysts	54
Ziegler–Natta	Zirconocene-based and titanocene-based catalysts	None given	Modified CHARMM	None specified	Rationalize the probability distributions of stereochemical configurations of the regioirregular units in isotactic polymer units	57
Ziegler–Natta	$Cp_2ZrC_2H_5^+$ and $Cp_2ZrC_3H_9^+$ with the various bridging ligands	CHARMM	Unspecified	DFT calculations; additional parameters from the literature (33,65–67)	Evaluate steric effects due to bulky substituents	62

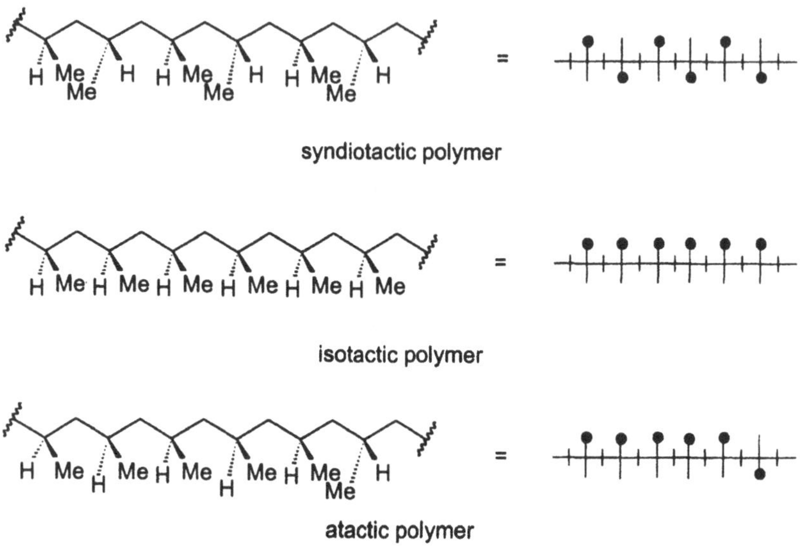

syndiotactic polymer

isotactic polymer

atactic polymer

FIGURE 9 Structures of syndiotactic (regular alternating of stereochemistry along the polymer chain), isotactic (same stereochemistry across the polymer chain), and atactic (random stereochemistry along the polymer chain) polymers.

4. HOMOGENEOUS ZIEGLER–NATTA POLYMERIZATION

Polymerization of α-olefins is one of the cornerstone reactions in organometallic catalysis. Ziegler–Natta olefin polymerization is an important chemical process (43,44). A typical catalyst for the heterogeneous polymerization reactions is $TiCl_3/Et_2AlCl$ (45). However, the homogeneous system, on which much of the mechanistic work has been carried out, uses $Cp_2ZrCl_2/AlCl_3$ (46–52).

Three different types of polymer can be produced using Ziegler–Natta catalysis: syndiotactic (regular alternating of stereochemistry along the polymer chain), isotactic (same relative configuration along the polymer chain), and atactic (random stereochemistry along the polymer chain) polymers (Fig. 9). In the literature, two methods are proposed to control stereochemistry: (1) chirality of catalyst (largely responsible for isotactic polymers, the catalysts are usually group 4B metallocenes with MAO*), and (2) control by chirality of the last monomer unit inserted (responsible for syndiotactic polymers, called the *syndiotactic defect*). Most workers in this area attempt to apply molecular mechanics (or combined

---

* MAO is methylaluminoxane, $[Al(CH_3)_3\text{-}O]_n$, used as a cocatalyst.

quantum mechanics/molecular mechanics methods) to understand the origins of stereoselectivity in polymer production.

In this section, we shall work through a few different approaches to the molecular mechanics modeling of Ziegler–Natta catalysis. Other approaches used to model this system are listed in Tables 1 and 2.

Rappé and coworkers used the Dreiding force field (53) to model the isotactic catalyst (S,S)-$C_2H_4(4,5,6,7$-tetrahydro-1-indenyl)$_2$ZrCl$_2$ (Fig. 10a) and the atactic catalyst (S,S)-C_2H_4(indenyl)$_2$ZrCl$_2$ (Fig. 10b) (54). Within the Dreiding force field, Rappé added a tetrahedral Zr atom (covalent radius 1.54 Å) and pseudoatoms for the cyclopentadienyl ring centroid, the η^3-allyl centroid, and the olefin centroid, all with covalent radii of 0.73 Å (54). All the necessary force field parameters were adjusted to give realistic geometries compared to the crystal structures.* To sample the conformational space of the complexes, the workers used molecular dynamics with the positions of the pseudoatoms constrained. (Pseudoatoms are used to bind the cyclopentadienyl, allyl, and olefin ligands to the metal. In molecular dynamics, atomic velocity is related to the mass of the atoms. The zero-mass psuedoatoms move too much if their positions are not constrained.)

A model of the transition state was generated using ab initio methods (54). This transition state was transferred into Dreiding and a conformational search carried out using anneal dynamics. Once the lowest-energy conformation of the transition state was determined, molecular mechanics energy differences between transition state and structures leading to (1) syndiotactic and (2) isotactic chain extensions were calculated. Rappé found that the isotactic polymer was favored over the syndiotactic polymer by less than 3 kcal/mol for the (S,S)-$C_2H_4(4,5,6,7$-tetrahydro-1-indenyl)$_2$ZrCl$_2$ (Fig. 10a) catalyst (54).

In addition to studies dealing with selectivity, Rappé also looked at polymer chain length versus energy differential (54). In other words, the work examines the energy difference between transition state and structures leading to the second and third insertions. Again, the isotactic polymer was favored over the syndiotactic polymer by 6 kcal/mol (for the second insertion) and 5 kcal/mol (for the third insertion) for the (S,S)-$C_2H_4(4,5,6,7$-tetrahydro-1-indenyl)$_2$ZrCl$_2$ (Fig. 10a) catalyst. Steric congestion was found to cause the energy difference between structures.

Turning attention to the (S,S)-C_2H_4(indenyl)$_2$ZrCl$_2$ (Fig. 10b) catalyst, Rappé found that isotactic polymer formation was also favored for the first insertion by 1 kcal/mol, which was reported to be significant (54). Using meso-$C_2H_4(4,5,6,7$-tetrahydro-1-indenyl)$_2$ZrCl$_2$, which gives rise to the atactic polymer experimentally, no preference for the two enantiotopic faces of the olefin was

---

* Reference 54 lists all the additional parameters.

TABLE 2 Summary of Reviews and Perspectives About Molecular Mechanics Modeling of Catalysts

Subject area	Catalysts	Literature covered	Ref.
Perspective	Transition metal *d*- and *f*-block complexes	Overview of the challenges in the application of molecular mechanics to *d*- and *f*-block complexes	1
Perspective	Homogeneous *d*-block transition metal catalysts	Challenges in and requirements for molecular modeling of catalysts and catalytic reactions	168
Review		Recent literature regarding application of molecular mechanics methods to inorganic compounds	28
Review		Development of molecular mechanics models for complexes of a single metal ion with organic ligands	5
Zeolites	Zeolites and other uniform catalysts	Application of molecular mechanics to zeolites and other uniform catalysts, including sorption sites, energetics of sorption, dynamics of diffusion, and reaction mechanisms	169
Ziegler–Natta	Various types of Ziegler–Natta catalysts	Polymerization mechanism and applications of molecular mechanics to the study of enantioselectivity of some stereospecific catalytic systems having chiral site stereocontrol	56
Ziegler–Natta	Homo- and heterogeneous Ziegler–Natta catalysts	Mechanism of enantioselectivity and its relevance to catalytic systems	58
Ziegler–Natta	Homo- and heterogeneous Ziegler–Natta catalysts	Compares stereospecificity of polymerization for homo- and heterogeneous systems; develops a model for the catalytic sites	63

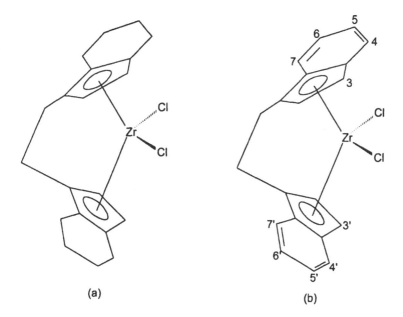

(a)   (b)

FIGURE 10   Structures of (a) (S,S)-$C_2H_4(4,5,6,7$-tetrahydro-1-indenyl)$_2$ZrCl$_2$ and (b) (S,S)-$C_2H_4($indenyl)$_2$ZrCl$_2$. (Redrawn from Ref. 166.)

found using molecular mechanics. This lack of energy discrimination can be understood by realizing that the catalyst has no symmetry because of the puckering of the indenyl rings. Since there is no ability for the catalyst to discriminate between the two different faces of the olefin, an atactic polymer should, and does, result.

Modifications to (S,S)-$C_2H_4(4,5,6,7$-tetrahydro-1-idenyl)$_2$ZrCl$_2$ were undertaken in order to probe the catalyst for those structural features that give rise to good differentiation between olefin faces (54). Rappé and coworkers found that a methyl group in position 4 of the indenyl ring (see Fig. 10) should increase stereoselectivity by inducing unfavorable steric interactions. From molecular mechanics calculations, this structure was found to increase the energy difference between transition states by 1 kcal/mol for the third insertion (as opposed to the 5 kcal/mol calculated earlier) (54). Rappé concluded that steric congestion occurs in the space that leads to the syndiotactic defect. However, if a methyl group was placed in position 3 of the indenyl ring, an atactic polymer was produced, because the energy difference between transition states for the third insertion was almost zero (54).

Morokuma and coworkers attempted a similar approach to understand both regio- and stereoselectivity of *ansa* metallocene catalysts (55). Ab initio methods were again used to determine the structure and energies of the transition states to ethylene insertion. Molecular mechanics was used to determine the effect of methyl substituents on olefins and cyclopentadienyl rings on the regio- and stereo-selectivity of Ziegler–Natta polymerization. Molecular mechanics using MM2 was carried out using a pseudoatom for Zr. Minimal conformational searching was employed, and only the substrate was energy-minimized (55). From this limited model, the authors found that primary insertion with one methyl group on the cyclopentadienyl ring was 3 kcal/mol lower than secondary insertion. The authors concluded that steric effects were responsible for the unfavorable energy.

A slightly different approach to modeling *ansa* metallocenes is to use an iso or *sec*-butyl group to represent the growing polymer chain (56–64). Guerra and coworkers modeled both Ti and Zr complexes using methods and parameters obtained from the literature (65–72). Crystal structures for the titanium-based systems were used to set up the zirconium systems. The Zr–C(alkyl) bond length was set to 2.28 Å and the Zr–C(olefin) bond length to 2.30 Å (by analogy with the Ti complexes) and all aromatic hydrogens were allowed to bend 10° out of plane. Energy maps were calculated as a function of M–C(alkyl) and M–C(olefin) bond rotations. Relative energies between complexes with *re* and *si* coordinated faces were generated. Molecular mechanics models were found to agree with experimental findings on the racemic system. In addition, secondary insertions were directed by interactions between the methyl group on the olefin and the cyclopentadienyl and indenyl ligands.

A number of other specific studies were found in the literature concerning Ziegler–Natta polymerization (see Table 1). In essence, all studies followed paths analogous to those illustrated in this section. To summarize: The discriminating step in the mechanism is the orientation of olefin insertion. This orientation can be analyzed by modeling simple complexes, with both *re* and *si* faces of the olefin coordinated, and then looking for energy differences between the *re* and *si* complexes. By looking at energy differences between *re* and *si* complexes, any errors introduced in the method are averaged out.

5. MODELING OF ZEOLITES

Thus far in this chapter we have concentrated on homogeneously catalyzed transformations, which have grown in popularity in industry over the past decade (73). Heterogeneously catalyzed transformations are still very popular in industry and deserve attention (44). Interest in zeolites as heterogeneous catalysts has grown recently, largely because of their ability to discriminate between different molecular shapes. Molecular mechanics has been used largely to identify specific sites for guest molecule adsorption in the zeolite channels. In this section, we shall consider some of the challenges involved in the molecular mechanics modeling of host/guest interactions in zeolites.

Location of aromatic molecules in zeolites has attracted much interest (74). Using a customized modeling code, ZEOLITHENERGIE, Fuess and coworkers have calculated the positions of aromatics in zeolite Y (75). More importantly, the methodology outlined has been adapted to many different hosts and guests (see Table 1). Their calculation takes place in seven steps:

1. Interaction energies are calculated between one guest molecule and the zeolite lattice. Considering only a single guest molecule in the zeolite models the case of infinite dilution.
2. No distinction is made between the zeolite T-sites (Si and Al). All are treated as 75% Si and 25% Al. For example, zeolite NaY is modeled with a Si/Al ratio of 3.0.
3. Because the zeolite system is so large, it is very difficult to model all interactions using a conventional force field such as MM2. Instead, the zeolite cage and guest molecule are held rigid, and the structure is minimized considering only nonbonded interactions.
4. Nonbonded energy, ϕ_{tot}, is calculated as the sum of Leonnard–Jones and electrostatic interactions:

$$\phi_{tot} = \sum_i \sum_j \left(B_{\alpha\beta} r_{ij}^{-12} - C_{\alpha\beta} r_{ij}^{-6} + \frac{q_\alpha q_\beta}{4\pi\varepsilon_0 r_{ij}} \right) \tag{4}$$

In this equation, $B_{\alpha\beta}$ characterizes short-range repulsion between atoms α and β, $C_{\alpha\beta}$ characterizes the dispersive interaction between α and β, and r is the internuclear distance. All parameters used in the study of Fuess and coworkers are reported in Ref. 75.

5. Hydrogen bonding is ignored.
6. Internal adsorption energy, ΔU_{ads}, is calculated by a Boltzmann weight of ϕ_{tot} summed at constant temperature, T:

$$\Delta U_{ads} = \frac{\sum_v \phi_{tot} e^{-\phi_{tot}/RT}}{\sum_v e^{-\phi_{tot}/RT}} \tag{5}$$

7. A three-dimensional grid is defined in an asymmetric zeolite unit, and the guest molecule is placed in the center of the grid. The orientation of the molecule is optimized by a Monte Carlo procedure: First a 0.25-Å grid is used and ϕ_{tot} is calculated over 1000 orientations per grid point. Then a 0.1-Å grid is defined and ϕ_{tot} is calculated over 5000 orientations per grid point.

Using the method just outlined, it has been argued that preferred adsorption sites depend on the nature of cations and hydrogen bonding possibilities in zeolite Y (76). The adsorption of benzene in H_x·ZSM-5, $H_xAl_xSi_{96-x}O_{192}$, has been studied because of the importance of ZSM-5 as a shape-selective commercial catalyst (76). Molecular mechanics has revealed that there are four different orientations of the benzene molecule in the zeolite channel: one with the aromatic plane perpendicular to the [010] channel axis, two with the aromatic plane parallel to the channel axis, and one intermediate. To come to these conclusions Mentzen and coworkers used a Buckingham potential to minimize the energy of the benzene molecule in the zeolite (76). Mentzen and coworkers also noted the importance of starting with a good set of positional parameters for the zeolite, which are usually obtained from X-ray or neutron diffraction studies.

One important goal of molecular modeling is to be able to predict molecular behavior in order to save time and money. In zeolite chemistry, for example, it would be advantageous to be able to predict which zeolite would be optimal for a given transformation. Consider 2,6-diisopropyl naphthalene (DIPN) and 2,7-diisopropyl naphthalene: solid acid catalysts tend to produce a mixture of these two isomers, which requires costly separation. If a zeolite catalyst with appropriate channel size were used, then only the desired 2,6-isomer could be formed. Workers have turned to molecular mechanics to determine which zeolite will best catalyze the selective production of 2,6-DIPN (77). Using Insight II, Horsley and coworkers used molecular graphics to determine that zeolite L has pores so large that they cannot distinguish between the 2,6- and 2,7-isomers of DIPN. On the other hand, mordenite selectively adsorbs 2,6-DIPN (77). With parameters from the literature (78), Horsley and coworkers modeled mordenite and zeolite L using Discover.

To determine which of the two isomers, 2,6-DIPN or 2,7-DIPN, was better adsorbed by the zeolite, the interaction energy along the general diffusion pathway was calculated (77). The diffusion path was defined by a series of points along the channel axis. Using a strong harmonic potential, the sorbate was constrained to lie a fixed distance from the extremes, and the zeolite lattice was fixed in the crystallographically determined structure. The sorbate was then energy-minimized, moved by 0.2 Å, and reminimized. This process was repeated until the sorbate was at the end of the section of channel. A plot of energy as a function of distance was used to determine the energy barrier as the sorbate moved through the zeolite. For mordenite, 2,6-DIPN showed an energy barrier of 4 kcal/mol, whereas the 2,7-isomer showed 18 kcal/mol (77). These results were in agreement with the molecular graphics visualizations.

Simple energy minimization did not result in reasonable results for the zeolite L system: areas of severe strain were noted, which were not visible in the molecular graphics analysis. As the sorbate moved through the channels in zeolite L, the isopropyl groups were caught on bits of zeolite. Horsley and coworkers added a Monte Carlo step to their procedure: the sorbate was added to the channel, a Monte Carlo conformational search was carried out, then the lowest energy structure was used for the energy calculation. The results showed that zeolite L could not distinguish between 2,6- and 2,7-diisopropyl naphthalene.

Recently a new force field for the modeling of Faujasite-type zeolites has been reported (79). This force field was developed to address a deficiency in the approaches mentioned earlier: the new force field models the Si and Al atoms explicitly, as opposed to the T-model, in which the Si/Al ratio is held constant. Jaramillo and Auerbach note that the T-model is acceptable when the guest molecule is reasonably far from the zeolite T-sites. However, when the guest molecule is an ion, for example, Na^+, then there is a small distance between guest and T-sites, requiring the Si and Al sites to be modeled explicitly. Partial charges, which are reported in the paper, were used for the zeolite potential (on Si, Al, both types of oxygen, and Na) (79). Both flexible and rigid frameworks were used to model the positions of sodium cations in the zeolite using a Buckingham potential (Eq. 4). Jaramillo and Auerbach developed a program, Clazyx, to convert computed coordinated into cationic sites, which can be compared with experiment. Barriers between sites were modeled using molecular dynamics at 1000 K with 1-fs time steps. The new force field was able to reproduce experimental cation positions, site occupancies, and vibrational frequencies.

Molecular mechanics has also been used to study skeletal isomerization of 1-butene to isobutene (80), olefin selectivity in fluid catalytic cracking using ZSM-5, zeolite Y, mordenite and β (81), carbon–sulfur bond cleavage over zeolite Y (82), and the location of naphthalene and 2-methylnaphthalene in HZSM-5 (83). In all cases, a methodology similar to those described earlier were adopted (75,77).

6. APPLICATIONS OF MOLECULAR MODELING TO CATALYSTS IN INDUSTRY

Several companies make use of different levels of molecular modeling of catalytic processes. In this section we present some of the work by Shell Chemical Company and BP Amoco Chemicals. Because of propriety issues, we present an overview of some of the work that pertains to this chapter.

6.1. Shell Higher Olefin Process (SHOP)

Shell Chemical Company has developed a method in which ethylene is oligomerized to higher α-olefins using a nickel-based catalyst (84). The ligands, shown in Figure 11, are chelating O- and P-donors.

Shell uses computational chemistry because ligand synthesis is costly (in terms of both time and money), experimental results can be ambiguous, and the

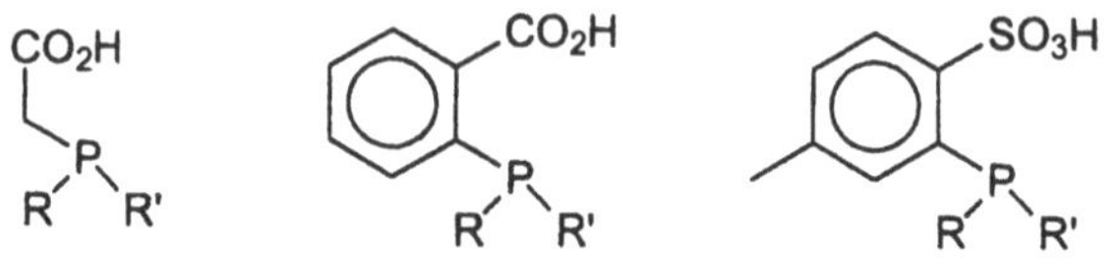

FIGURE 11   Examples of the O-, N-donor bidentate ligands used in the Shell higher olefin process. (Redrawn from Ref. 87.)

computational resources are readily available within the company. Their goal is to use molecular modeling to direct the experimental program.

Two levels of theory are commonly used in the design of the nickel-based catalysts shown in Figure 11: Density Functional Theory (B3LYP functional used with effective core potentials for Ni and 6-31G* for everything else in the complex) and molecular mechanics (both the UFF (4) and reaction force field, RFF (85,86) are used) (87). All these methods are complementary, and the experiments are guided from the results of several calculations using different molecular modeling techniques.

As with many computational models of catalysis, we need a key step in the mechanism to model. Since chain growth occurs with the olefin trans to the coordinated oxygen, workers at Shell have modeled the cis/trans ratio in order to determine how to modify the ligand to force trans coordination (Fig. 12).

When 1-butene inserts into the growing oligomeric chain there are two possible products: a branched and an internal olefin. The product olefin, branched or internal, is determined by the manner in which the butene coordinates to the nickel (Fig. 13). The reason for the interest in the branched/internal ratio is the fact that internal olefins are more desirable than branched ones. Shell's scientists have successfully used the RFF to determine both the cis/trans ratio (Fig. 12) and the branched/internal ratio. In addition, workers found that the stereochemistry of butene coordination is determined by steric effects (Fig. 13). In particular, the substituents on phosphorus were found to exert a steric control over the branched/internal ratio resulting from butene insertion.

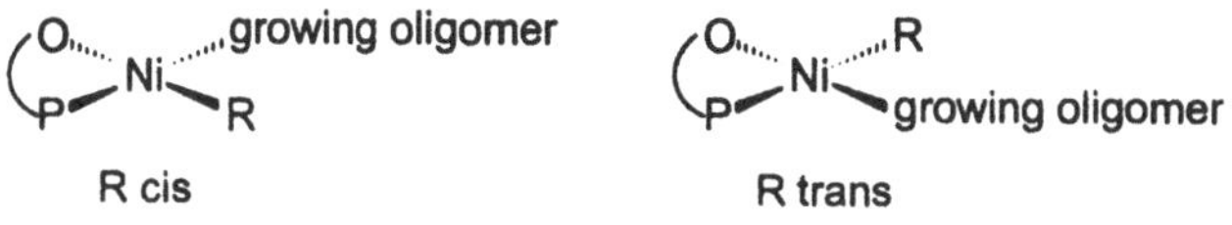

FIGURE 12   Structures of the nickel-based catalysts used in the SHOP. The chelating bidentate ligand is represented by the arc connecting O and P. (Redrawn from Ref. 87.)

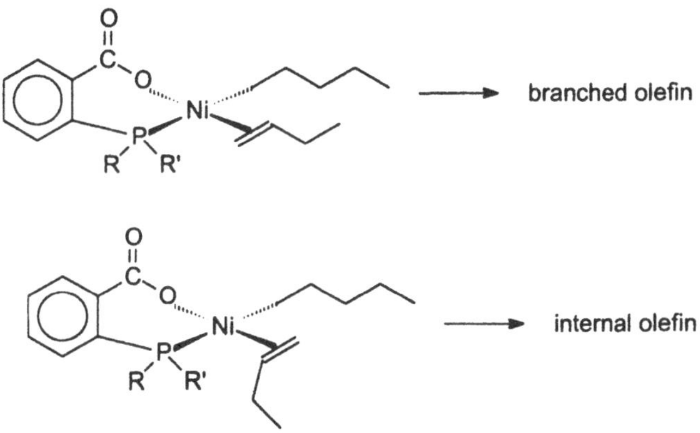

FIGURE 13 Formation of branched versus internal olefins from the insertion of 1-butene into the growing polymer chain. Notice that the branched/internal ratio depends on how the olefin binds to the metal. (Redrawn from Ref. 87.)

6.2. Molecular Modeling at BP Amoco Chemicals

Work at BP Amoco Chemicals is based on the philosophy that the design and improvement of products is practically achieved in three steps: 1) reading the literature, garnering ideas, and brainstorming, 2) testing the best ideas using molecular modeling, and then 3) confirming the computational results with laboratory work. BP Amoco Chemicals' scientists have successfully used computational chemistry in many projects, one of which is the single-site catalyst model (88).

In the polypropylene catalyst shown in Figure 14, an isotactic-atactic block copolymer can be formed by rotation of one ring relative to the Zr-centroid axis. (For descriptions of polymer stereochemistry, see Fig. 9.) The isospecific *rac* rotamer of the catalyst gives rise to the isotactic block, while the aspecific *meso* form gives rise to the atactic block (Fig. 14). Using UFF (4) and RFF (85,86) (as well as ab initio methods and DFT), the workers were able to confirm experimental evidence (89,90) that the indenyl substituent, R in Figure 14, could influence the equilibrium between the *rac* and *meso* rotomers. Using RFF the workers were able to successfully predict the relative amount of isotactic and atactic blocks in the polymer and to correlate that with R.

In addition to modeling polypropylene formation, scientists at BP Amoco Chemicals have successfully used computations to model desulfurization of light naphtha, flue gas multicomponent equilibria, methane-to-methanol conversion, and oxygen scavenging films.

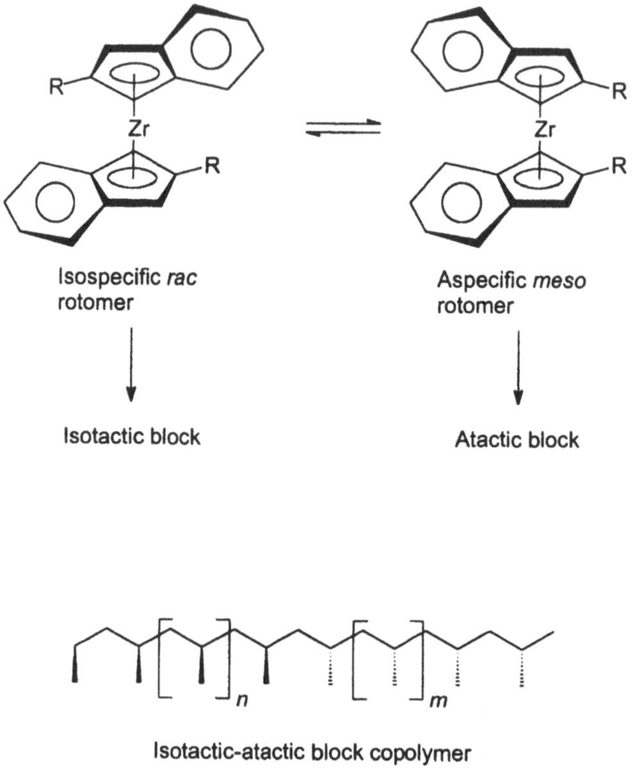

Isospecific *rac*   Aspecific *meso*
rotomer   rotomer

Isotactic block   Atactic block

Isotactic-atactic block copolymer

FIGURE 14   Structures of the *rac* and *meso* rotomers that lead to isotactic and atactic polymer blocks in the polymerization of polypropylene. An example of an isotactic-atactic block copolymer is shown. The composition of the block copolymer is determined by the equilibrium constant for the interconversion between *rac* and *meso* forms of the catalyst. (Redrawn from Ref. 88.)

7. SUMMARY

Molecular mechanics has been used to model a variety of different catalytic processes. Almost all the major catalytic transformations in organometallic catalysis have been studied using some form of molecular modeling, and most with molecular mechanics. Workers have successfully built models that offer new insights into how a catalyst interacts with the substrate. In particular, molecular modeling has found a home within the chemical industry. Molecular modeling is used in both fundamental and applied research because it is more efficient to guide experiments using ligands designed via computational methods than by trial and error in the laboratory. When we develop a molecular mechanics model of a reaction, we have to carefully examine the mechanism for that reaction in order to build the model. In the process of building the molecular mechanics model, we develop an ability to visualize reaction centers, which leads to new insights into catalytic reactivity. In this manner, we begin to allow our imaginations to guide the design of novel catalysts. These new designs can be tested using computational models far faster, and more efficiently, than by carrying out experiments in the laboratory.

ACKNOWLEDGMENTS

We thank Professor Theodore L. Brown, University of Illinois, for many helpful suggestions during manuscript preparation. We thank Ms. Rebecca Simon, University of Illinois, Thomas M. Thorpe, Procter and Gamble, David S. Brown, Shell Chemical Company, and Joseph T. Golab, BP Amoco Chemicals, for help with the section on industrial applications.

REFERENCES

1. TR Cundari. J Chem Soc Dalton Trans 2771–2776, 1998.
2. IV Pletnev, VL Mel'nikov. Russ J Coordination Chem 23:188–196, 1997.
3. NL Allinger. J Am Chem Soc 99:8127–8134, 1977.
4. AK Rappé, CJ Casewit, KS Colwell, WA Goddard, WM Skiff. J Am Chem Soc 114:10024, 1992.
5. BP Hay. Coord Chem Rev 126:177–236, 1993.
6. J Hunger, G Huttner. J Comput Chem 20:445–471, 1999.
7. TR Cundari, WF Fu. Inorg Chim Acta, in press, 1999.
8. JD Atwood. Inorganic and Organometallic Reaction Mechanisms. New York: Wiley, 1997.
9. RB Jordan. Reaction Mechanisms of Inorganic and Organometallic Systems. 2nd ed. New York: Oxford University Press, 1998.
10. ML Tobe, J Burgess. Inorganic Reaction Mechanisms. Essex, UK: Longman, 1999.
11. P Comba, TW Hambley. Molecular Modeling of Inorganic Compounds. Wenheim, Germany: VCH, 1995.
12. AR Leach. Molecular Modelling Principles and Applications. Essex, UK: Longman, 1996.
13. AK Rappé, CJ Casewit. Molecular Mechanics Across Chemistry. Sausalito, CA: University Science Books, 1997.
14. JA Osborne, FH Jardine, JF Young, G Wilkinson. J Chem Soc A:1711, 1966.
15. FH Jardine, JA Osborne, G Wilkinson. J Chem Soc A:1574, 1967.
16. S Montelatici, A vanderEnt, JA Osborne, G Wilkinson. J Chem Soc A:1054, 1968.
17. ACS Chan, JJ Pluth, J Halpern. J Am Chem Soc 102:5952, 1980.
18. CR Landis, J Halpern. J Am Chem Soc 109:1746–1754, 1987.
19. JM Brown, PA Chalconer. J Chem Soc Chem Commun, 344, 1980.

20. JM Brown. Chem Soc Revs 22:25–47, 1993.
21. ML Caffery, TL Brown. Inorg Chem 30:3907–3914, 1991.
22. KJ Lee, TL Brown. Inorg Chem 31:289–294, 1992.
23. DP White, TL Brown. Inorg Chem 34:2718–2724, 1995.
24. VS Allured, CM Kelly, CR Landis. J Am Chem Soc 113:9493, 1991.
25. TN Doman, CR Landis, B Bosnich. J Am Chem Soc 114:7264–7272, 1992.
26. TK Hollis, JK Burdett, B Bosnich. Organometallics 12:3385–3386, 1993.
27. TN Doman, TK Hollis, B Bosnich. J Am Chem Soc 117:1352–1368, 1995.
28. CR Landis, DM Root, T Cleveland. In: DB Boyd, KB Lipkowitz, eds. Reviews in
     Computational Chemistry. New York: VCH, 1995, pp 73–148.
29. JM Brown, PL Evans. Tetrahedron 44:4905–4916, 1988.
30. PL Bogdan, JJ Irwin, B Bosnich. Organometallics 8:1450–1453, 1989.
31. EK Davies, NW Murrall. J Comput Chem 13:149–156, 1988.
32. JS Giovannetti, CM Kelly, CR Landis. J Am Chem Soc 115:5889, 1993.
33. R Brooks, RE Bruccoleri, BD Olafson, DJ States, S Swaminathan, M Karplus. J
     Comput Chem 4:187, 1983.
34. AK Rappè, WA Goddard. J Chem Phys 95:3358, 1991.
35. DG Allen, SB Wild, DL Wood. Organometallics 5:1009, 1986.
36. PK Weiner, PA Kollman. J Comput Chem 2:287, 1981.
37. WC Still, F Mohamadi, NGJ Richards, WC Guida, M Lipton, R Liskamp, G Chang,
     T Hendrickson, FD Gunst, W Hasel. "Macromodel V3.0." Department of Chemis-
     try, Columbia University, New York.
38. O Schwalm, J Weber, B Minder, A Baiker. Int J Quant Chem 52:191–197, 1994.
39. "CAChe WorkSystem V3.7." CAChe Scientific Inc., Beaverton, OR, 1995.
40. F Agbossou, J-F Carpentier, A Mortreux, G Surpateanu, AJ Welch. New J Chem
     20:1047–1060, 1996.
41. "Cerius²." Molecular Simulations, Inc., San Diego, CA, 1999.
42. M Diéguez, A Ruiz, C Claver, MM Pereira, AMdAR Gonsalves. J Chem Soc Dal-
     ton Trans:3517–3522, 1998.
43. JP Collman, LS Hegedus, JR Norton, RG Finke. Principles and Applications of
     Organotransition Metal Chemistry. 2nd ed. Mill Valley, CA: University Science
     Books, 1987.
44. TW Swaddle. Inorganic Chemistry: An Industrial Environmental Perspective. San
     Diego, CA: Academic Press, 1997.
45. YV Kissin, RI Mink, T Nolin, AJ Brandolini. Top Catal 7:69–88, 1999.
46. EJ Arlman, P Cossee. J Catal 3:99, 1964.
47. KJ Ivin, JJ Rooney, CD Stewart, MLH Green, R Mahtab. J Chem Soc Chem Com-
     mun:604–606, 1978.
48. HW Turner, RR Schrock. J Am Chem Soc 104:2331–2333, 1982.
49. J Soto, ML Steigerwald, RH Grubbs. J Am Chem Soc 106:4479–4480, 1982.
50. L Clawson, J Soto, SL Buchwald, ML Steigerwald, RH Grubbs. J Am Chem Soc
     107:3377–3378, 1985.
51. H Kraulendat, H-H Britzinger. Ang Chem Int Ed Engl 29:1412, 1990.
52. WE Piers, JE Bercaw. J Am Chem Soc 112:9406–9407, 1990.
53. SL Mayo, BD Olafson, WA Goddard. J Phys Chem 94:8897, 1990.
54. LA Castonguay, AK Rappé. J Am Chem Soc 114:5832–5842, 1992.

55. H Kawamura-Kuribayashi, N Koga, K Morokuma. J Am Chem Soc 114:8687–8694, 1992.

56. G Guerra, L Cavallo, V Venditto, M Vacatello, P Corradini. Makromol Chem Macromol Symp 69:237–246, 1993.

57. G Guerra, L Cavallo, G Moscardi, M Vacatello, P Corradini. J Am Chem Soc 116:2988–2995, 1994.

58. G Guerra, P Corradini, L Cavallo, M Vacatello. Macromol Symp 89:307–319, 1995.

59. G Guerra, P Longo, L Cavallo, P Corradini, L Resconi. J Am Chem Soc 119:4394–4403, 1997.

60. G Guerra, L Cavallo, P Corradini, R Fusco. Macromolecules 30:677–684, 1997.

61. L Cavallo, G Guerra, P Corradini. Gaz Chim Ital 126:463–467, 1996.

62. L Cavallo, G Guerra. Macromolecules 29:2729–2737, 1996.

63. P Corradini, V Busico, L Cavallo, G Guerra, M Vacatello, V Venditto. J Mol Catal 74:433–442, 1992.

64. M Toto, L Cavallo, P Corradini, G Moscardi, L Resconi, G Guerra. Macromolecules 31:3431–3438, 1998.

65. DY Yoon, PR Sundararajan, PJ Flory. Macromolecules 8:776, 1975.

66. UW Suter, PJ Flory. Macromolecules 8:765, 1975.

67. PR Sundararajan, PJ Flory. J Am Chem Soc 96:5025, 1974.

68. A Schafer, E Karl, L Zsolnai, G Huttner, HH Brintzinger. J Organomet Chem 328:87, 1987.

69. T Ooi, RA Scott, K Vandekooi, HA Scheraga. J Phys Chem 46:4410, 1967.

70. RD Hancock. Prog Inorg Chem 37:187, 1989.

71. D Cozak, M Melnik. Coord Chem Rev 74:53, 1986.

72. P Ammendola, G Guerra, V Villani. Makromol Chem 185:2599, 1984.

73. LS Hegedus. Transition Metals in the Synthesis of Complex Organic Molecules. 2nd ed. Sausalito, CA: University Science Books, 1999.

74. H Klein, H Fuess. In: J Weitkamp, HG Karge, H Pfeifer, W Holderich, eds. Studies in Surface Science and Catalysis. Amsterdam: Elsevier Science, 1994, pp 2067–2074.

75. H Klein, C Kirschhock, H Fuess. J Phys Chem 98:12345–12360, 1994.

76. BF Mentzen, M Sacerdote-Peronnet. Mater Res Bull 29:1341–1348, 1994.

77. JA Horsley, JD Fellmann, EG Derouane, CM Freeman. Catalysis 147:231–240, 1994.

78. TA Hagler, S Lifson, P Dauber. J Am Chem Soc 101:5122, 1979.

79. E Jaramillo, SM Auerbach. J Phys Chem B 103:9589–9594, 1999.

80. R Millini, S Rossini. In: H Chon, S-K Ihm, YS Uh. ed. Studies in Surface Science and Catalysis. Amsterdam: Elsevier Science, 1997, pp 1389–1396.

81. K Teraishi. J Mol Catal A Chem 132: 73–85, 1998.

82. BM Bhawal, R Vetrivel, TI Reddy, ARAS Deshmukh, S Rajappa. J Phys Org Chem 7:377–384, 1994.

83. H Klein, H Fuess, S Ernst, J Weitkamp. Microporous Materials 3:291–304, 1994.

84. ER Freitas, CR Gum. Chemical Engineering Progress. 1979, pp 73–76.

85. MA Pietsch, AK Rappé. J Am Chem Soc 118:10908, 1996.

86. AK Rappé, MA Pietsch, DC Wiser, JR Hart, LM Bormann-Rochotte, WM Skiff. Mol Engin 7:385–400, 1997.

272 White and Douglass

87. DS Brown, NO Gonzales, WM Skiff, JH Worstell. Applying molecular simulation to homogeneous catalysis. American Institute of Chemical Engineers, 1998 Annual Meeting, Miami Beach, FL, 1998.
88. JT Golab. Industrial chemistry modeling: in the new millennium. 216th American Chemical Society National Meeting, Boston, MA, 1998.
89. E Hauptman, RM Waymouth, JW Ziller. J Am Chem Soc 117:11586, 1995.
90. BP Amoco Chemicals. New Business Development Business Group. Private document.
91. P-O Norrby, B Akermark, F Haeffner, S Hansson, M Blomberg. J Am Chem Soc 115:4859, 1993.
92. PG Anderson, A Harden, D Tanner, P-O Norrby. Chem Eur J 1:12, 1995.
93. E Pena-Cabrera, P-O Norrby, M Sjogren, A Vitagliano, VD Felice, J Oslob, S Ishii, D O'Neill, B Akermark, P Helquist. J Am Chem Soc 118:4299–4313, 1996.
94. PS Pregosin, H Ruegger, R Salzmann, A Albinati, F Lianza, RW Kunz. Organometallics 13:83–90, 1994.
95. JD Oslob, B Akermark, P Helquist, P-O Norrby. Organometallics 16:3015–3021, 1997.
96. NL Allinger, X Zhou, J Bergsma. Theochem 118:69–83, 1994.
97. P-O Norrby, HC Kolb, KB Sharpless. J Am Chem Soc 116:8470–8478, 1994.
98. EJ Larson, VL Pecoraro. J Am Chem Soc 113:3810, 1991.
99. P-O Norrby, C Linde, B Akermark. J Am Chem Soc 117:11035–11036, 1995.
100. TR Cundari, L Saunders, LL Sisterhen. J Phys Chem A 102:997–1004, 1998.
101. KR Adam, M Antolovich, LG Brigden, LF Lindoy. J Am Chem Soc 113:3346–3351, 1991.
102. MM Gugelchuk, KN Houk. J Am Chem Soc 116:330–339, 1994.
103. TR Cundari, LL Sisterhen, C Stylianopoulos. Inorg Chem 36:4029–4034, 1997.
104. JW Lauher. J Am Chem Soc 108:1521–1531, 1986.
105. L Bencze, R Szilagyi. J Organomet Chem 465:211–219, 1994.
106. VJ Burton, RJ Deeth, CM Kemp, PJ Gilbert. J Am Chem Soc 117:8407, 1995.
107. RJ Deeth, VJ Paget. J Chem Soc Dalton Trans, 537–546, 1997.
108. G Alberti, A Grassi, GM Lombardo, GC Pappalardo, R Vivani. Inorg Chem 38:4249–4255, 1999.
109. RW Harrison. J Comp Chem 14:1112–1122, 1993.
110. AL Swain, M Miller, J Green, DH Rich, J Schneider, SBH Kent, A Wlodawer. Proc Natl Acad Sci USA 87:8805–8809, 1990.
111. IT Weber, RW Harrison. Protein Engineering 9:679–690, 1996.
112. FC Bernstein, TF Koetzle, GJB Williams, EF Meyer, MD Brice, JR Rodgers, O Kennard, M Shimanouchi, M Tasumi. J Mol Biol 112:535–542, 1977.
113. CE Sansom, J Wu, IT Webber. Protein Engineering 5:659–667, 1992.
114. W Goertz, W Keim, D Vogt, U Englert, MDK Boele, LAvd Veen, PCJ Kamer, PWNMv Leeuwen. J Chem Soc Dalton Trans:2981–2988, 1998.
115. P Faye, E Payen, D Bougeard. J Chem Soc Faraday Trans 92:2437–2443, 1996.
116. P Faye, E Payen, D Bougeard. In: GE Froment, B Delmon, P Grange, eds. Hydrotreatment and Hydrocracking of Oil Fractions. Amsterdam: Elsevier Science, 1997, pp 281–292.
117. D Gleich, R Schmid, WA Herrmann. Organometallics 17:2141–2143, 1998.

118. LA Castonguay, AK Rappé, CJ Casewit. J Am Chem Soc 113:7177–7183, 1991.
119. JP Bowen, NL Allinger. J Org Chem 52:2937, 1987.
120. CP Casey, LM Petrovich. J Am Chem Soc 117:6007–6014, 1995.
121. CP Casey, GT Whiteker. Isr J Chem 30:299–304, 1990.
122. M Kranenberg, YEMvd Burgt, PCJ Kramer, PWNMv Leeuwen. Organometallics 14:3081–3089, 1995.
123. F Torrens. J Mol Cat A: Chem 119:393–403, 1997.
124. L Fan, D Harrison, TK Woo, T Ziegler. Organometallics 14:2018–2026, 1995.
125. Z Lin, TJ Marks. J Am Chem Soc 112:5515–5525, 1990.
126. MP Doyle, WR Winchester, JAA Hoorn, V Lynch, SH Simonsen, R Ghosh. J Am Chem Soc 115:9968–9978, 1993.
127. MP Doyle, LJ Westrum, WNE Wolthuis, MM See, WP Boone, V Bagheri, MM Pearson. J Am Chem Soc 115:958–964, 1993.
128. AA Varnek, AS Glebov, OM Petrakhin, RP Ozerov. Koord Khim 15:740, 1989.
129. HR Allcock, DC Ngo, M Parvez, RR Whittle, WJ Birdsall. J Am Chem Soc 113: 2628, 1991.
130. AA Varnek, A Maia, D Landini, A Gamba, G Morosi, G Podda. J Phys Org Chem 6:113–121, 1993.
131. RH Boyd, SM Breitling, M Mansfield. Am Inst Chem Eng 19:1016, 1973.
132. RD Hancock, SM Dobson, A Evers, PW Wade, MP Ngwenya, JCA Boeyens, KP Wainright. J Am Chem Soc 110:277, 1988.
133. RD Hancock, MP Ngwenya, PW Wade, JCA Boeyens, SM Dobson. Inorg Chim Acta 164:73–84, 1989.
134. E Brunet, AM Poveda, D Rabasco, E Oreja, LM Font, MS Batra, JC Rodriguez-Ubis. Tetrahedron: Asymmetry 5:935–948, 1994.
135. F Verpoort, AR Bossuyt, L Verdonck, B Coussens. J of Mol Catal A Chem 115: 207–218, 1997.
136. L Bencze, RK Szilagyi. In: Y Imamoglu, ed. Metathesis Polymerization of Olefins and Polymerization of Alkynes. Dordrecht, The Netherlands: Kluwer Academic, 1998, pp 411–443.
137. B Altava, MI Burguete, JM Fraile, JI Garcia, SV Luis, JA Mayoral, AJ Royo, MJ Vicent. Tetrahedron: Asymmetry 8:2561–2570, 1997.
138. J Castro, A Moyano, MA Pericas, A Riera. Tetrahedron 51:6541–6556, 1995.
139. PD Beer, EL Tite, MGB Drew, A Ibbotson. J Chem Soc Dalton Trans 2543–2550, 1990.
140. Y-D Wu, KN Houk, JS Valentine, W Nam. Inorg Chem 31:718–720, 1992.
141. W Kaminsky, O Rabe, A-M Schauwienold, GU Schupfner, J Hanss, J Kopf. J Organomet Chem 497:181–193, 1995.
142. N Balcioglu, I Uraz (Unalan), C Bozkurt, F Serin. Polyhedron 16:327–334, 1997.
143. L Cavallo, G Guerra, G Moscardi, P Corradini. Polym Mater Sci Eng 73:465–466, 1995.
144. L Bencze, R Szilagyi. J Mol Cat 76:145–156, 1992.
145. M Clarke, RD Cramer, N Vanopden-Bosch. J Comput Chem 10:982, 1989.
146. RD Hancock, AE Martell, D Chen, RJ Motekaitis, D McManus. Can J Chem 75: 591–600, 1997.
147. AF Wells. Structural Inorganic Chemistry. Clarence, UK: Oxford, 1986, pp. 289.

148. JCA Boeyens, FA Cotton, S Han. Inorg Chem 24:1750, 1985.
149. NL Allinger, MI Quinn, K Chen, B Thompson, MR Frierson. J Mol Struct 194:1, 1989.
150. B Didillon, C Houtman, T Shay, JP Candy, JM Basset. J Am Chem Soc 115:9380–9388, 1993.
151. F Humblot, D Didillon, F Lepeltier, JP Candy, J Corker, O Clause, F Bayard, JM Basset. J Am Chem Soc 120:137–146, 1998.
152. MGB Drew, PCH Mitchaell, S Kasztelan. J Chem Soc Faraday Trans 86:697, 1990.
153. TM Brunier, MGB Drew, PCH Mitchell. J Chem Soc Faraday Trans 88:3225–3232, 1992.
154. S Ramdas, JM Thomas, AK Cheetham, PW Betteridge, EK Davies. Angew Chem Int Ed Engl 23:671–679, 1984.
155. PA Wright, JM Thomas, AK Cheetham, AK Nowak. Nature 318:611–614, 1985.
156. P Demontis, S Yashonath, ML Klein. J Phys Chem 93:5016–5019, 1989.
157. RL June, AT Bell, DN Theodorou. J Phys Chem 94:8232–8240, 1990.
158. S Yashonath. J Phys Chem 95:5877–5881, 1991.
159. G Schrimpf, M Schlenkrich, J Brickmann, P Bopp. J Phys Chem 96:7404–7410, 1992.
160. F Bosselet, M Sacerdote, J Bouix, BF Mentzen. Mat Res Bull 25:443–450, 1990.
161. M Sacerdote, F Bosselet, BF Mentzen. Mat Res Bull 25:593–599, 1990.
162. DE Williams. Acta Cryst A36:715–723, 1980.
163. DE Williams, DJ Houpt. Acta Cryst B42:286–295, 1986.
164. AL Kieselev, AA Lopatkin, AA Shulga. Zeolites 5:261, 1985.
165. MA Pietsch, AK Rappé. J Am Chem Soc 118:10908–10909, 1996.
166. JR Hart, AK Rappé. J Am Chem Soc 115:6159–6164, 1993.
167. DC Wiser, AK Rappé. Polym Mater Sci Eng 74:423–424, 1996.
168. TR Cundari, J Deng, W Fu, TR Klinckman, A Yoshikawa. J Chem Inf Comput Sci 38:941–948, 1998.
169. AK Cheetham, JD Gale, AK Nowak, BK Peterson, SD Pickett, JM Thomas. Faraday Discuss Chem Soc 87:79–90, 1989.

11

Titanium Chemistry

Mark S. Gordon and Brett M. Bode
Iowa State University, Ames, Iowa

Simon P. Webb
Pennsylvania State University, University Park, Pennsylvania

Takako Kudo
Gunma University, Kiryu, Japan

Jerzy Moc
Wroclaw University of Technology, Wroclaw, Poland

Dmitri G. Fedorov
University of Tokyo, Tokyo, Japan

Gyusung Chung
Konyang University, Chungnam, Korea

1. INTRODUCTION

The chemistry of titanium is of considerable importance, primarily because of its roles as a catalyst in various chemical reactions (e.g., silane polymerization (1), hydrosilation (2), and Ziegler–Natta (3) polymerization), as materials and materials precursors, and as the basis for electronic and magnetic devices. In the past several years, the interest in titanium chemistry in this group has focused on its fundamental molecular and electronic structure in a variety of chemical environments, on its function as a catalyst in the hydrosilation and bis-silylation reactions, and on the nature of the structure, bonding, and mechanism of formation of metallocarbohedrenes.

Because it is electron deficient in most of its molecular environments, electronic structure calculations on compounds that contain titanium are generally more complicated than are analogous calculations on species that contain only lighter main group elements. One reason for this is that its electron deficiency results in the formation of unusual structural arrangements that are difficult to describe using simple computational methods. Likewise, since it is frequently impossible to draw one simple Lewis structure for Ti-containing compounds, the usual methods that are based on single configuration wavefunctions are often inappropriate.

This chapter reviews a range of recent calculations on several different problems involving titanium chemistry, performed primarily by this group. We begin, in Section 2, by considering the theoretical and computational methods that have been used. This is followed, in Section 3, by a discussion of unusual structures and associated potential energy surfaces that occur in titanium chemistry due in large part to the electron-deficient nature of this element. In Section 4, the potential use of divalent Ti as a catalyst is discussed. A summary and discussion of future topics is presented in Section 5.

2. THEORETICAL METHODS

Since titanium is a moderately heavy element, it can be beneficial to make use of effective core potentials (ECPs), in which the inner-shell electrons are replaced with a model potential (4). The advantage of this approach is that the computational effort is significantly reduced, since only the valence electrons are explicitly considered. Considerable effort has been expended in the development of efficient methods for obtaining analytic first and second energy derivatives, gradients, and hessians, in order to make geometry optimizations and frequency calculations more feasible. The primary disadvantage is that the most common ECPs, those developed by Hay and Wadt (5), Stevens and coworkers (6), and Christiansen et al. (7), use relatively small basis sets, since their initial developments occurred before the use of systematically large basis sets became commonplace. So one can expect at most semiquantitative accuracy using ECPs. It should be emphasized, however, that ECPs should be thought of as alternative basis sets, so that all of the common methods for recovering electron correlation can be used with them.

Fortunately, Ti is still small enough (22 electrons) that one can frequently perform all-electron calculations, at least to obtain the final energetics. Although far more effort has been expended in developing extended basis sets for main group elements, there are valence triple zeta (TZV) basis sets available for the first-row transition metals, and these are frequently used in this laboratory, augmented by polarization functions (8). This means one uses p functions on hydrogens, d functions on main group elements, and f functions on transition metals. Because Ti has four valence electrons (s^2d^2), one expects that the chemistry of this element may bear some resemblance to that of the Group IVA elements carbon and silicon (s^2p^2). On the other hand, a simple picture for the ground electronic valence state of C or Si would have all valence orbitals singly occupied, with none empty, whereas an analogous picture for Ti would leave two empty d orbitals. In this sense, titanium is more similar to the electron-deficient elements boron and aluminum. From this perspective, one might expect titanium compounds to have unusual structures, both molecular and electronic. An important consequence of this observation is that it will often be impossible to write a single, simple Lewis structure for compounds that contain Ti. This generally means that it is difficult to find a single electronic configuration that adequately describes what the electrons are doing in such species. Then a single configuration wavefunction, such as that employed by the Hartree–Fock molecular orbital method or by density functional theory, is unlikely to be appropriate, even as a starting point for subsequent correlated calculations. In such cases a multiconfigurational (MC) wavefunction must be considered. The most common approach is the complete active space (CAS) self-consistent field (SCF) or fully optimized reaction space (FORS) method. Both of these approaches are specific examples of the more general MCSCF method (9).

The key in carrying out MCSCF calculations is the determination of a reasonable "active space," that set of orbitals and electrons that are directly involved in the chemistry to be described. The active space in turn defines the set of electronic configurations that determine the MCSCF wavefunction. The choice of active spaces is described in a recent review (9). While the MCSCF wavefunction provides a qualitatively correct description of a system, it does not account for the bulk of the electron correlation, usually referred to as "dynamic" correlation.

For systems that are adequately described by single-configuration wavefunctions, dynamic correlation is most commonly accounted for by second-order perturbation theory, referred to as many-body perturbation theory (MBPT2 (10)) or Moeller–Plesset perturbation theory (MP2 (11)). While higher orders of perturbation theory are frequently used, the reliability of these higher-order methods and the convergence of the perturbation expansion has been increasingly called into question (12,13). The more reliable endpoint for accurate energies is coupled cluster (CC) theory. These are most commonly implemented at the single- and double-excitation level, with triple excitations included perturbatively, CCSD(T) (14). When MCSCF wavefunctions are used as the reference, the most commonly used methods for recovering the electron correlation are multireference configuration interaction (MRCI) (15) and multireference perturbation theory (MRPT) (16). The former is usually implemented at the single- and double-excitation level, MR(SD)CI, but it is so computationally demanding that this level of theory is still limited to small active spaces. Second-order MRPT is more efficient, as indeed is its single-reference analog.

Geometry optimizations are generally carried out at the SCF, MCSCF, or

MP2 level of theory and are followed by hessian calculations to determine whether a particular structure is a minimum or a saddle point on the potential energy surface. If one is interested in determining a transition state for a chemical reaction, it is advisable to determine the minimum-energy path (MEP) that connects the transition state with the corresponding reactants and products. Especially for complex species, it is not always obvious what these connecting minima are without mapping out the MEP. The most common MEP algorithm is the second-order method developed by Gonzalez and Schlegel (17).

Because atomic titanium has low-lying electronic excited states, potential energy surfaces of Ti compounds, especially unsaturated compounds, often cross. When such crossings occur, nonradiative transitions can occur via spin-orbit coupling. In such cases, spin-orbit coupling probabilities must be evaluated. Such calculations usually are performed with MCSCF-based wavefunctions. In addition, relativistic effects can have a significant effect on chemical properties, even for compounds containing elements in the first transition series. Fortunately, new models have been developed to treat relativistic effects for all-electron basis sets (18).

One of the goals of computational inorganic chemistry is to perform electronic-structure calculations on the specific system of interest, not on some simplified prototype. Improved hardware and software both contribute to achieving this goal. A particularly important development in this regard is scalable computing—the use of multiple computer nodes to perform complex calculations. This capability has recently been enhanced by the development of a distributed data interface (DDI) (19) that permits a distributed parallel computer to look like a shared memory (SMP) computer to take maximum use of the aggregate memory and disk space. This method, which facilitates the prediction of structures at high levels of theory, can run on computers ranging from expensive ''supercomputers'' to networks of PCs.

All of the methods just outlined have been used for some of the chemistry discussed next, generally using the electronic structure program GAMESS (20) (General Atomic and Molecular Electronic Structure System). Coupled cluster calculations were performed using either ACES II (21) or Gaussian94 (22). Multireference CI calculations are commonly performed using the MOLPRO (23) code.

3. STRUCTURES AND POTENTIAL ENERGY SURFACES

3.1. TiX$_2$

In the Group IVA analogs, CH_2 has a triplet ground state, while SiH_2 and the heavier congeners have singlet ground states, although such terminology is not particularly relevant to Pb. The low-lying states of TiH_2 have been analyzed with state-averaged MCSCF wavefunctions and all-electron extended basis sets, followed by multireference configuration interaction (24). At this level of theory, several triplet states are found to lie within a 5-kcal/mol range, with the $^3B_1$ state predicted to be the ground state. However, the $^3A_1$ state is found to be only 1 kcal/mol higher in energy, so the ground state prediction may not be definitive. It is clear that the ground state is a triplet, since the lowest-energy singlet state is predicted to be nearly 1 eV above the ground state. The lowest quintet states are even higher in energy.

For all of the low-lying states, the H–Ti–H bend potential is very flat. All of these states are predicted to be mildly bent in C_{2v} symmetry. These states are also susceptible to distortion along the asymmetric stretching mode to C_s symmetry with little energy expense. Although multireference wavefunctions are clearly necessary for a proper description of TiH_2, it is interesting that the $^1A_1$ structure is well reproduced by second-order perturbation theory, MP2. The latter level of theory also predicts $TiCl_2$ to have a slightly bent triplet ground state (25).

An important divalent titanium species is titanocene, the Ti analog of ferrocene. While there have been a few experimental (26) or theoretical (27) studies of this species, none are definitive. On the experimental side, it is not clear whether the ground state is singlet or triplet, or indeed even whether the ground state species studied was the monomer. The only previous theoretical study of $TiCp_2$ (Cp = cyclopentadienyl) was an early semiempirical calculation that could not distinguish spin states. Since this is such an important compound, a series of MCSCF calculations have been initiated on $TiCp_2$, in which all of the Ti valence electrons and orbitals are included in the active space. Preliminary results suggest that, like the simpler TiH_2 and $TiCl_2$ analogs, the ground state is a triplet (28).

3.2. Ti_2H_6

Superficially, Ti_2H_6 is a simple titanium analog of ethane and disilane. Indeed, early calculations on this species assumed that this is the case and therefore used single configuration, restricted Hartree-Fock (RHF)–based methods to predict its molecular and electronic structure (29). However, this assumption ignores the fact, discussed in Section 1, that Ti is an electron-deficient atom with more unoccupied valence d orbitals than the empty valence p orbitals in carbon or silicon. A careful examination of the electrons and orbitals in Ti_2H_6 reveals that one must treat this molecule as a potential diradical. This means the lowest state could either be an open-shell singlet or a triplet, depending on the extent of Ti–Ti bonding, but most assuredly not as a closed-shell singlet.

The calculations on Ti_2H_6 were performed (30) using a basis set of triple zeta plus polarization quality, including f functions on Ti. For the singlet states, both restricted open-shell Hartree–Fock (ROHF) and FORS MCSCF reference wavefunctions were used, while ROHF reference wavefunctions were used for the triplet states. In order to account for dynamic electron correlation and obtain reliable energetics, these reference wavefunctions were augmented by second-order perturbation theory. For ROHF- and MCSCF-based wavefunctions, this means RMP2[10] and MCQDPT2[16], respectively.

Five minima were found on both the singlet and triplet potential energy surfaces. All of these minima have hydrogens bridging the two titaniums, more reminiscent of diborane than ethane or disilane. The $(\mu\text{-}H)_3$ staggered and eclipsed structures that had previously been described by closed-shell RHF reference wavefunctions actually require *at least* a 2-electron, 3-orbital FORS MCSCF wavefunction for a qualitatively correct description. The analogous triplets require an ROHF wavefunction in which two degenerate states are averaged. All minima are predicted to be lower in energy than two separated TiH_3 radicals, with binding energies in excess of 50 kcal/mol for the lowest-energy structures. The global minimum for Ti_2H_6 is found to be the triply bridged C_s triplet ($^3A''$), but this species is only 0.3 kcal/mol lower in energy than the analogous $^1A''$ state. Doubly bridged D_{2h} and quadruply bridged D_{4h} $(\mu\text{-}H)_4$ species are only about 5 kcal/mol higher in energy than these C_s structures. For these more symmetric structures, the singlets are lower than the corresponding triplets by about 1.4 kcal/mol. The $(\mu\text{-}H)_3$ staggered and eclipsed singlets are more than 20 kcal/mol higher. The corresponding staggered and eclipsed triplets are much more stable than the singlets.

No Ti–Ti bonding is possible in the triplet minima, since the two unpaired electrons have the same spin. Somewhat more surprising is the fact that little Ti–Ti bonding is found in the singlet minima as well. In the $(\mu\text{-}H)_3$ C_{3v} minima, both unpaired electrons reside on the less saturated Ti. The $(\mu\text{-}H)_3$ singlet and triplet C_s minima are purely diradical. Natural orbital analyses of the $(\mu\text{-}H)_2$ D_{2h} and $(\mu\text{-}H)_4$ D_{4h} singlet wavefunctions show a large amount of diradical character, although a slight bonding interaction is predicted in both of these singlets. This is reflected in the observation that these singlets are slightly lower in energy than the corresponding triplets. The lack of Ti–Ti bonding may be understood using the localized charge distribution (LCD) analysis (31). This analysis suggests that the lack of Ti–Ti bonding in the D_{2h} isomer, where bonding should be most likely, is due to steric crowding, that is, unfavorable interactions of the bond with the surrounding molecule. Comparison of the calculated frequencies of Ti_2H_6 isomers with the experimental spectra of Chertihin and Andrews (32) suggests the presence of Ti_2H_6 in the matrix.

The D_{2h} $H_2Ti(\mu\text{-}H)_2TiH_2$ structure is an excellent prototype for the many homodinuclear titanium(III) compounds known experimentally (33), such as titanocene dimer. Experimental evidence suggests either a Ti–Ti bond or a large singlet-triplet energy gap in this compound. Since the study of the Ti_2H_6 prototype predicts little Ti–Ti bonding and a small singlet-triplet splitting, it may be that the presence of the cyclopentadienyl rings and/or distortion of the bridge out of the molecular plane must significantly modify the electronic structure. This is the subject of ongoing calculations.

In dinuclear complexes comprising two metal centers, each with an unpaired electron, the interaction is said to be antiferromagnetic if singlet coupling of the electrons is favored over triplet coupling. If the reverse is true, the interaction is ferromagnetic (34). According to these definitions, D_{2h} (μ-H)$_2$ Ti$_2$H$_6$ is antiferromagnetic, with the isotropic exchange interaction $J = -250$ cm^{-1}, where

$$J = 0.5 \, [E(\text{singlet}) - E(\text{triplet})].$$

In general, the magnitude and sign of the singlet-triplet energy gap is a measure of the isotropic exchange interaction, in the absence of spin-orbit coupling and magnetic dipole–dipole interactions. When the singlet-triplet splitting is very small, as is the case for Ti$_2$H$_6$, the normally subtle spin-orbit and magnetic effects can become important. Since EPR experiments are able to detect such subtle effects, and since these subtle effects have rarely been addressed using ab initio quantum mechanics methods, the excited-state energies and spin-orbit coupling matrix elements connecting the various states of Ti$_2$H$_6$ were calculated (35) using configuration interaction (CI) wavefunctions and the one-electron Z_{eff} method (36). An overall ferromagnetic effect of 0.660 cm^{-1} is predicted on the ground-state singlet–first excited triplet energy gap. Future calculations will expand these studies to more complex bridged compounds.

3.3. Ti₂X₈ Compounds

TiH$_4$ has a tetrahedral structure that is reminiscent of methane and silane. It is well represented by a single-configuration wavefunction and therefore a single, simple Lewis structure. However, in view of the foregoing discussion, one might (correctly) suspect that this is not the entire story of this "simple" molecule. Indeed, TiH$_4$ is predicted (37) to dimerize with no energy barrier to Ti$_2$H$_8$. Since multireference wavefunctions are not necessary for this species, the calculations employed MP2 geometries, followed by single points at the coupled cluster, CCSD(T) level of theory. The exothermicity of the reaction 2TiH$_4$ → Ti$_2$H$_8$ is estimated to be ~40 kcal/mol. Both doubly and triply bridged dimers have been found, and the interconversion among the various isomers requires little energy.

Because bridged compounds such as Ti$_2$H$_6$ and Ti$_2$H$_8$ are important prototypes for more complex analogs, a systematic study of Ti$_2$X$_8$ species (X = F, Cl, Br) was carried out (38). Geometry optimizations were performed at the MP2 level of theory, using effective core potentials, with a double zeta basis set augmented by polarization functions. The importance of higher-level dynamic correlation was explored using CCSD(T). Basis set effects were examined by single-point calculations that employed all-electron extended basis sets. Ti$_2$F$_8$

is predicted to be a bound C_{2h} dimer with two bridging bonds, lower in energy than the separated monomers by 10.5 kcal/mol. In contrast, Ti_2Cl_8 and Ti_2Br_8 are predicted to be weakly bound dimers with D_{3d} symmetry, whose structures resemble associated monomers. This is consistent with experimental data, which suggests that solid-state TiF_4 is a bridged polymer chain, whereas solid-state $TiCl_4$ and $TiBr_4$ remain molecular in structure. Ti_2Cl_8 is predicted to be bound by 4.9 kcal/mol at the CCSD(T)//MP2 level of theory. Coupled cluster calculations were not possible for the Br analog, but one can speculate that this species has a similar binding energy. Constrained optimizations along the dimerization pathway show that the formation of the F and Cl dimers occur with no energy barrier. Transition states with symmetrically equivalent bridging halides represent possible routes to halide exchange between monomers.

Dynamic correlation, at the MP2 level of theory, is very important for the prediction of accurate energetics for these species. While MP2 was not very important for the prediction of geometries for the fluorine dimer, it was essential for the geometry predictions for the heavier congeners. The effective core potential basis set overestimates the dimerization energy of TiF_4 by about 4 kcal/mol, but it was quite reliable for the heavier species.

The LCD analysis was used to support the notion that the dimerization process is governed by a continuous competition between unfavorable endothermic monomer distortions and favorable exothermic monomer interactions. It was shown that the internal monomer energies indeed increase as the dimerization proceeds at the same time that the interaction between monomers drives the energy down. The net energy change along the dimerization path is a competition between these two. In the fluorine system, the favorable monomer interaction energy dominates, leading to a strongly bound dimer. In the chlorine and bromine systems, the two effects roughly cancel, so the net effect is a very weak binding.

3.4. $H_2Ti{=}EH_2$

There has been a long-standing interest in the nature of multiple bonding between transition metals and main group elements (39). To quantitatively assess the nature of Ti=C vs. Ti=Si π bonding, the internal rotation process was studied for the molecules $H_2Ti{=}CH_2$ and $H_2Ti{=}SiH_2$ (40). Geometries were determined using MCSCF wavefunctions, optimizing the geometry of both the lowest singlet and lowest triplet states as a function of the angle between the TiH_2 and EH_2 (E = C, Si) planes. Both effective core potentials and the triple zeta + polarization (TZVP) basis set were used. In general, ECPs provided a good representation of the all-electron results. Final energetics were determined with the MC-QDPT2 method and the TZVP basis set. At the best level of theory, MC-QDPT2 with the TZVP basis set, the rotation barrier on the singlet surface is 15.8 kcal/mol for the carbon compound and 8.6 kcal/mol for the silicon species. To some degree this reflects the greater ability of C than Si to form π bonds, but note that due to the participation of the Ti d orbitals, there is significant π bonding at all rotation angles. Note also that the triplets for both species are minima at the twisted structure. Unlike the singlet states, the energy decreases for the triplets are very similar for C and Si.

Whereas the singlet remains below the triplet at all rotation angles for the carbon compound, the two curves cross in the silicon species, so the triplet is the ground state at the twisted structure. It is also interesting that the singlet TiC distance actually *decreases* upon rotation, whereas the TiSi distance increases, as expected. The origin of this behavior may be understood from the natural orbital occupation numbers (NOONs) for the π bonding orbitals. Normally, one expects this NOON to decrease from roughly 2.0 to roughly 1.0 as the π bond is broken. This behavior is just what is found for Si. In contrast, the NOON actually increases in the carbon case. This very likely reflects the greater ability of carbon to interact with the empty Ti d orbitals as the rotation occurs. So, while the rotation barrier cannot be equated with a π bond energy, the relative rotation barriers for C vs. Si do reflect the stronger π bonds formed by C. The determination of the total Ti$=$Si and Ti$=$C bond energies is in progress.

3.5. TiH₃OH Decomposition

TiH_3OH is the simplest prototype for a Ti–O bond. Therefore the thermal decomposition processes of this species is the simplest prototype for the chemical vapor deposition of titanium oxide. Titanol can dissociate by the following unimolecular routes (41):

$$TiH_3OH \rightarrow TiH_2{=}O + H_2 \tag{1}$$

$$\rightarrow HTiOH + H_2 \tag{2}$$

$$\rightarrow TiH_2 + H_2O \tag{3}$$

$$\rightarrow TiH_3 + OH \tag{4}$$

$$\rightarrow TiH_4 + O \tag{5}$$

In addition, one can have isomerization between the first two products. The ground electronic state of TiH_3OH is a singlet, so the analysis of its decomposition pathways was performed on the singlet surface. All geometries for this reaction were determined using MCSCF wavefunctions with a triple zeta + polarization (TZVP) basis set. Final energies were obtained using the CASPT2 method with the same basis set. The structure of the parent molecule is found to be C_{3v}, with a linear Ti–O–H angle. Reactions (4) and (5) are predicted to proceed monotonically uphill, with no intervening barrier. At the MCSCF level of theory, the endothermicities for these reactions are 114.3 and 203.2 kcal/mol, respectively. Since these are very high-energy processes, they were not pursued further. Reaction (2) is predicted to occur without barrier. The CASPT2 endothermicity, including zero-point vibrational corrections, is 30.9 kcal/mol. The endothermicity for the production of titanone, $H_2Ti=O$, is only 2.5 kcal/mol at the same level of theory, so this isomer is much lower in energy than HTiOH. However, there is an intervening barrier of 18.6 kcal/mol between TiH_3OH and $H_2Ti=O$. [Note that the ground electronic state of HTiOH is a triplet, as is the case for TiH_2. On the triplet surface, the relative stabilities of these two isomers is reversed.] For the remaining reaction (3), the production of water and TiH_2, there is an intermediate complex in which the two products are weakly bound at about 55 kcal/mol above the reactant. The products are an additional 12.5 kcal/mol higher in energy. So, the lowest-energy decomposition path is clearly the one that leads to $H_2Ti=O$. Note also that the isomerization barrier from $H_2Ti=O$ to HTiOH is 48 kcal/mol, so the lowest-energy route to the higher-energy isomer is directly from TiH_3OH.

It is interesting to compare the electronic structure of $H_2Ti=O$ with its main group analogs $H_2C=O$ and $H_2Si=O$. The percentage decrease in bond length on going from the singly bound to the doubly bound species is about 7.5% for both the Ti and Si compound. In contrast, the percentage decrease for C is 14.6%, nearly twice as large. This reflects the much greater propensity for C to form double bonds. The $Ti=O$ and $Si=O$ bonds are also much more polar than $C=O$: the Mulliken charges on the metal are 0.102 (C), 0.825 (Si), 0.713 (Ti).

3.6. $Ti^+ + C_2H_6$

Experimental studies of gas-phase reactions of first-row transition metal cations with simple alkanes provide valuable insight into the mechanism and energetics of C–H and C–C bond activation. Insertion of the metal into C–H or C–C bonds is common and leads eventually to elimination of H_2 or small alkanes. Because C_2H_6 is the simplest alkane in which there is a competition between hydrogen and smaller alkane (i.e., methane) elimination, and since there have been several experimental gas-phase studies of Ti^+ reactions with ethane (42), a comprehensive theoretical study of this reaction was initiated (43). The theoretical methods used include the B3LYP version (44) of density functional theory (DFT) with a valence triple zeta plus polarization basis set (TZVP), CCSD(T) with an extended basis set, and CASSCF + MCQDPT2 using the TZVP basis set. The use of MCSCF-based wavefunctions proved to be essential, since parts of the potential energy surfaces are inherently multireference in nature. Geometries were determined using either B3LYP or CASSCF. Since Ti^+ has low-lying doublet and quartet states, the potential exists for surface crossings, in which case spin-orbit coupling can become important. The spin-orbit coupling matrix elements were

calculated with the CASSCF/TZVP basis set, using the full Breit–Pauli Hamiltonian (45).

The main questions addressed in this work were:

1. What is the nature of the adduct ions observed in the flow tube experiments?
2. Why are only the H_2 elimination products, not CH_4, observed at low
   energies?
3. Does the H_2 elimination proceed by a 1,1- or a 1,2-mechanism?
4. How important are spin-orbit effects on the $Ti^+ + C_2H_6$ reaction?

It has been determined that both the H_2 and CH_4 reactions are initiated by
the formation of an η^3 coordinated Ti^+—C_2H_6 ion-induced dipole complex in
the quartet state. Because the ground state of Ti^+ is a quartet, whereas the C–H
and C–C insertion products have ground-state doublets, doublet-quartet surface
crossings occur, and the two reactions subsequently proceed on the doublet surface. So the nontrivial spin-orbit coupling predicted by these calculations plays
an important role in the mechanism.

The net H_2 elimination process is predicted to be nearly thermoneutral. The
highest-energy transition state on this path has an energy requirement of 5–9
kcal/mol, depending on the level of theory employed. This may be contrasted
with recent DFT results on Co^+ and Fe^+ mediated H_2 elimination from C_2H_6.
These latter calculations find a lower C–H insertion transition state, perhaps due
to deeper wells corresponding to the initial M^+-ethane complexes. The calculations also predict that the 1,2-elimination is favored over the 1,1-elimination
mechanism. The net CH_4 elimination is predicted to be rather endothermic at all
levels of theory, with a high C–C insertion barrier. This is consistent with the
lack of experimental observation of CH_4.

4. TITANIUM AS A CATALYST

Divalent titanium has been implicated as a catalyst in several reactions, including
silane polymerization (1) and hydrosilation (2). Considerable theoretical effort
has therefore been expended recently in analyzing the role of divalent titanium
as a catalyst in the hydrosilation and bis-silylation reactions.

The hydrosilation reaction is the addition of a silane, R_3Si–H to a C=C
double bond, to form new carbon–silicon bonds. In the absence of a catalyst,
this reaction occurs only with a very large (50–60 kcal/mol) barrier (46).

In order to assess several levels of theory, the first study was performed
on the prototypical system $SiH_4 + C_2H_4$, using TiH_2 as catalyst (46). These calculations were performed using the TZVP basis set, at the Hartree–Fock, MP2, and
CCSD(T) levels of theory. It is very clear from a detailed analysis of the reaction
path that correlation corrections are essential to obtain a reliable prediction of

the energetics for this reaction. On the other hand, MP2 parallels the CCSD(T) calculations very well. This is an important result, since the MP2 calculations are affordable for larger species, while CCSD(T) would be out of the question. The key bottom line for this reaction is that the initial steps in the reaction, the formation of complexes between the catalyst and the two substrates, is so exothermic, nearly 70 kcal/mol at the highest level of theory, that all subsequent steps have barrier heights that are more than 30 kcal/mol *below* the starting reactants. The overall exothermicity for this reaction is predicted to be 28 kcal/mol.

With the foregoing results in hand, several additional reactions were considered with increasingly complex reactants and catalyst. The three reactions studied were:

$$TiH_2 + SiCl_3H + C_2H_4 \tag{6}$$

$$TiCl_2 + SiCl_3H + C_2H_4 \tag{7}$$

$$TiCp_2 + SiCl_3H + C_2H_4 \tag{8}$$

where Cp is cyclopentadienyl. The calculations on these reactions (48) were made possible by two recent advances in theory and code development: the development of very effective parallel computer codes for MP2 gradients (19) and the development of much more efficient gradient and hessians for effective core potentials (49). Due to the size of these systems, using the TZVP basis set and CCSD(T) calculations would be prohibitively expensive.

Fortunately the unsubstituted work demonstrated that the MP2 level of theory is adequate for this reaction. To reduce the size of the basis set, ECPs were used for all elements except H and C. Effective core potentials are a very effective means for studying heavier elements, since they replace inner shells with more easily evaluated one-electron potentials leading to greater computational efficiency. To test the accuracy of ECPs they were used for Ti and Si in the unsubstituted reaction and checked against the previous all-electron results. The RMS difference in energy was shown to be 0.5 kcal/mol, which was deemed acceptable. Thus, ECPs were used in the substituted systems for all atoms except H and C, which remained all-electron.

The results for reactions (6)–(8) differ in detail, but not in substance, from those for the parent reaction just discussed. Initial formation of complexes formed between the catalyst and the substrates are consistently very exothermic, so all subsequent barriers are well below the starting reactants in energy. Thus there are no net barriers for any of these reactions. The overall exothermicities for these three reactions are 37.2 kcal/mol.

Bis-silylation is a more complex reaction in which a disilane adds across a $C{=}C$ double bond, to form two new Si–C bonds:

$$X_3Si-SiX_3 + R_2C{=\!=}CR_2 \rightarrow X_3Si-CR_2-CR_2-SiX_3 \qquad (9)$$

A preliminary study (50) demonstrated that in the absence of a catalyst, this reaction has such a high barrier that it could not be a viable process. The most commonly used catalysts for bis-silylation are Pt and Ni phosphines. However, in view of the computational results, summarized earlier, on the hydrosilation reaction, the bis-silylation reaction has been studied, again using divalent Ti as a prototypical catalyst (51). Since this work is still in progress, only a very brief summary will be presented here. The reaction under consideration has R = H and X = Cl, with the $TiCl_2$ catalyst. Geometries have been determined using both Hartree–Fock and MP2 methods. Two very important points have already emerged from this study. One is that, as for the hydrosilation reaction, correlation is essential for even a qualitatively sensible picture of the reaction. There are very large MP2 corrections relative to the Hartree–Fock results. The second important point is that effective core potentials faithfully reproduce the all-electron results at both the Hartree–Fock and MP2 levels of theory. This is important, since it means that ECPs are a viable, efficient alternative to the more expensive all-electron wavefunctions. It is advisable, however, to use the all-electron basis sets for first-row atoms such as carbon.

5. SUMMARY AND POTENTIAL FOR MATERIALS CHEMISTRY

Several titanium species are of interest due to their potential as materials with desirable properties or as materials precursors. One species that has generated considerable attention is the metallocarbohedrenes (met-cars), M_8C_{12}, first synthesized by the Castleman group (52). Although by far the most attention has been paid to M = Ti, other met-cars have been synthesized. Although the experimental structure for these interesting species has yet to be determined, the overwhelming computational evidence favors the T_d structure (53). Much more interesting than the met-car structure in the long run is the mechanism by which these species form and why the growth of these cage species stops at the Ti_8C_{12} structure. The experiments are initiated by laser ablation of Ti metal in the presence of small hydrocarbons. Therefore, systematic calculations have been initiated in this laboratory to explore the building-up mechanisms that lead to the Ti_8C_{12} structure (54,55).

It has been known for some time that SiO cage compounds, polyhedral oligomeric silsesquioxanes (POSS), are important three-dimensional materials for both lubricants and coatings (56). Very recently, there has been increasing interest in the titanium analogs of the POSS, both fully and partially substituted POSS. Important issues include the structure of the Ti cage compounds, their mechanism of formation, and the effect of substituents, solvent, and catalysts on the mechanism. All of these issues are currently being addressed.

Finally, we would note that the state of the art in computational chemistry has improved dramatically during the last decade. This is especially important for transition metals, for which quantitative experimental data (for example, thermodynamic quantities) are frequently unavailable. There is now an opportunity to use high-quality calculations to systematically study the molecular and electronic structures, the bond energies, and the nature of chemical bonding of transition metals in a very broad range of chemical environments. Such studies will provide quantitative tests and analyses of the very successful qualitative models that we have come to rely on.

ACKNOWLEDGMENTS

The work described here has been supported by grants from the National Science Foundation, the Air Force Office of Scientific Research, and the U.S. Department of Energy through the Ames Laboratory.

REFERENCES

1. JF Harrod, T Ziegler, V Tschinke. Organometallics 9:897–902, 1990.
2. TJ Barton, P Boudjouk. Organosilicon chemistry—a brief overview. In: T Ziegler, FWG Fearon, eds. Advances in Silicon-Based Polymer Science. Advances in Chemistry Series No. 224. Washington, DC: American Chemical Society, 1990, pp 3–46.
3. N Koga, K Morokuma. J Am Chem Soc 110:108–112, 1988.
4. LR Kahn, P Baybutt, DG Truhlar. J Chem Phys 65:3826–3853, 1976.
5. PJ Hay, WR Wadt. J Chem Phys 82:270–310, 1985.
6. (a) WJ Stevens, H Basch, M Krauss. J Chem Phys 81:6026–6033, 1984; (b) WJ Stevens, H Basch, P Jasien. Can J Chem 70:612–630, 1992; (c) TR Cundari, WJ Stevens. J Chem Phys 98:5555–5565, 1993.
7. SA Wildman, GA DiLabio, PA Christiansen. J Chem Phys 107:9975–9979, 1997, and references cited therein.
8. VA Glezakou, MS Gordon. J Phys Chem 101:8714–8719, 1997.
9. MW Schmidt, MS Gordon. Ann Rev Phys Chem 49:233–266, 1998.
10. WJ Lauderdale, JF Stanton, J Gauss, JD Watts, RJ Bartlett. Chem Phys Lett 187:21–28, 1991.
11. JA Pople, JS Binkley, R Seeger. Int J Quantum Chem S10–S19, 1, 1976.
12. J Jensen, O Christiansen, H Koch, P Jorgensen. J Chem Phys 105:5082–5090, 1996.
13. T Dunning Jr. In preparation.
14. (a) J Paldus. In: GL Malli, ed. Relativistic and Electron Correlation Effects in Molecules and Solids. New York: Plenum, 1994, pp 207–282; (b) RJ Bartlett. In: DR Yarkony, ed. Modern Electronic Structure Theory. I. Singapore: World Scientific, 1995, pp 1047–1131.

15. HJ Werner, PJ Knowles. J Chem Phys 89:5803–5814, 1988.

16. H Nakano. J Chem Phys 99:7983–7992, 1993.

17. C Gonzales, HB Schlegel. J Chem Phys 90:2154–2161, 1989.

18. See, for example: (a) KJ Dyall. J Chem Phys 100:2118–2127, 1994; (b) K Dyall. J Chem Phys 106:9618–9626, 1997; (c) T Nakajima, K Hirao. Chem Phys Lett 302: 383–391, 1999; (d) R Samzow, BS Hess, G Jansen. J Chem Phys 96:1227–1231, 1992.

19. GD Fletcher, MW Schmidt, MS Gordon. Adv Chem Physics 110:267–294, 1999.

20. MW Schmidt, KK Baldridge, JA Boatz, ST Elbert, MS Gordon, JH Jensen, S Koseki, N Matsunaga, KA Nguyen, S Su, TL Windus, M Dupuis, JA Montgomery Jr. J Comp Chem 14:1347–1363, 1993.

21. ACES II is a program product of the Quantum Theory Project, University of Florida. Authors: JF Stanton, J Gauss, JD Watts, M Nooijen, N Oliphant, SA Perera, PG Szalay, WJ Lauderdale, SR Gwaltney, S Beck, A Balkova, DE Bernholdt, KK Baeck, P Rozyczko, H Sekino, C Huber, RJ Bartlett. Integral packages: VMOL (J Almlof, PR Taylor); VPROPS (PR Taylor); ABACUS (T Helgaker, HJAa Jensen, P Jorgensen, J Olsen, PR Taylor).

22. MJ Frisch, GW Trucks, M Head-Gordon, PMW Gill, MW Wong, JB Foresman, BG Johnson, HB Schlegel, MA Robb, ES Replogle, R Gomperts, JL Andres, K Raghavachari, JS Binkley, C Gonzalez, RL Martin, DJ Fox, DJ Defrees, J Baker, JJP Stewart, JA Pople. Gaussian 92, Revision A. Gaussian, Inc., Pittsburgh PA, 1992.

23. (a) HJ Werner, PJ Knowles, University of Sussex, 1991. (b) HJ Werner, PJ Knowles. J Chem Phys 82:5053–5063, 1985; (c) PJ Knowles, HJ Werner. Chem Phys Lett 115:259–267, 1985.

24. T Kudo, MS Gordon. J Chem Phys 102:6806–6811, 1995.

25. MS Gordon. Unpublished results.

26. AK Fischer, G Wilkinson. J Inorg Nuc Chem 2:149–152, 1956.

27. JW Lauher, R Hoffmann. J Am Chem Soc 98:1729–1742, 1976.

28. M Freitag, MS Gordon. In preparation.

29. A Garcia, JM Ugalde. J Phys Chem 100:12277–12279, 1996.

30. SP Webb, MS Gordon. J Am Chem Soc 120:3846–3857, 1998.

31. JH Jensen, MS Gordon. J Phys Chem 99:8091–8107, 1995.

32. GV Chertihin, L Andrews. J Am Chem Soc 116:8322–8327, 1994.

33. See, for example: (a) S Xin, JF Harrod, E Samuel. J Am Chem Soc 116:11562–11563, 1994; (b) SI Troyanov, H Antropiusová, K Mach. J Organomet Chem 427: 49–55, 1994; (c) A Davidson, SS Wreford. J Am Chem Soc 96:3017–3018, 1974; (d) HH Brintzinger, JE Bercaw. J Am Chem Soc 92:6182–6185, 1970.

34. (a) O Kahn. Angew. Chem Int Eng Ed 24:834–850, 1985; (b) O Kahn. Molecular Magnetism. New York: VCH, 1993.

35. SP Webb, MS Gordon. J Chem Phys 109:919–927, 1998.

36. (a) S Koseki, MW Schmidt, MS Gordon, T Takada. J Phys Chem 96:10768–10772, 1992; (b) S Koseki, N Matsunaga, MW Schmidt, MS Gordon. J Phys Chem 99: 12764–12772, 1995; (c) S Koseki, MW Schmidt, MS Gordon. J Phys Chem 102: 10430–10435, 1998.

37. S Webb, MS Gordon. J Am Chem Soc 117:7195–7201, 1995.

38. SP Webb, MS Gordon. J Am Chem Soc 121:2552–2560, 1999.
39. (a) TR Cundari, MS Gordon. J Am Chem Soc 113:5231–5243, 1991; (b) TR Cundari, MS Gordon. J Phys Chem 96:631–636, 1992.
40. G Chung, MS Gordon. In preparation.
41. T Kudo, MS Gordon. J Phys Chem A 102:6967–6972, 1998.
42. (a) LS Sunderlin, PB Armentrout. Int J Mass Spectrom. Ion Processes 94:149–177, 1989; (b) RS MacTaylor, WD Vann, AW Castleman Jr. J Phys Chem 100:5329–5333, 1996.
43. J Moc, DG Fedorov, MS Gordon. J Chem Phys 112:10247, 2000.
44. (a) AD Becke. Phys Rev A 38:3098–3100, 1988; (b) AD Becke. J Chem Phys 98:1372–1377, 1993; (c) AD Becke. J Chem Phys 98:5648–5652, 1993.
45. HA Bethe, EE Salpeter. Quantum Mechanics of the One-, Two-Electron Atoms. New York: Plenum, 1977.
46. PN Day, MS Gordon. Theor Chim Acta 91:83–90, 1995.
47. BM Bode, PN Day, MS Gordon. J Am Chem Soc 120:1552–1555, 1998.
48. BM Bode, MS Gordon. Theoretical Chemistry Accounts 102:366–376, 1999.
49. BM Bode, MS Gordon. J Chem Phys in press.
50. F Raaii, MS Gordon. J Phys Chem A 102:4666–4668, 1998.
51. Y Alexeev, MS Gordon. In preparation.
52. BC Gua, KP Kerns, AW Castleman Jr. Science 255:1411–1412, 1992.
53. (a) Z Lin, MB Hall. J Am Chem Soc 114:10054–10055, 1992; (b) RW Grimes, JD Gale. JCS Chem Comm 1222–1224, 1992; (c) MM Rohmer, P de Vaal, M Benard. J Am Chem Soc 114:9696–9697, 1992; (d) L Lou, T Guo, P Nordlander, RE Smalley. J Chem Phys 99:5301–5305, 1993; (e) M Methfessel, M van Schilfgaarde, M Scheffler. Phys Rev Lett 70:29–32, 1993.
54. VA Glezakou, MS Gordon. In preparation.
55. S Varganov, MS Gordon. In preparation.
56. T Kudo, MS Gordon. J Am Chem Soc 120:11432–11438, 1998.

12

Spin-Forbidden Reactions in Transition Metal Chemistry

Jeremy Noel Harvey
University of Bristol, Bristol, England

1. INTRODUCTION

The material in this book, and indeed the very fact that the book exists, provides ample evidence for the growing maturity of computational organometallic chemistry. Computational methods, especially those based on ab initio quantum chemistry, are increasingly providing useful support and guidance to the experimental investigation of the properties of organometallic compounds. In this chapter, I want to address a very common type of reactivity of these compounds for which standard computational methods have not yet proved as successful and to outline the strategies that have been and will be used to tackle such systems.

After describing why there is a problem, I will briefly summarize the theoretical description of spin-forbidden reactions. It will be useful at this point to draw parallels with other types of nonadiabatic chemistry, in particular, electron transfer. Then I will review some of the typical contexts in which spin-forbidden behavior occurs in transition metal systems, to try to illustrate how widespread it is. This will be followed by a presentation of strategies used for characterizing and understanding spin-forbidden reactions, based on the use of energies and gradients derived from ab initio electronic structure calculations. Finally, I will discuss the appropriateness of the different electronic structure methods for such work, before a concluding section, which will provide some guidelines for how best to set about the computational investigation of a spin-forbidden reaction.

Because spin changes are so widespread in transition metal chemistry, many studies have touched upon the subject in more or less detail. The focus of this chapter will be on presenting strategies for the investigation of this chemistry, with some examples, mostly drawn from work I have myself been involved in, for illustration. I have therefore not attempted a comprehensive review of all the work in the area, and there are certainly many important studies of which I am aware, and probably many more of which I am not, that are not cited here, for which I would of course like to apologize.

2. THE PROBLEM

The partially filled shell of d electrons on the metallic center dominates the chemistry of transition metal compounds. Bare transition metal atoms have unpaired electrons in this shell, leading to high-spin configurations. In the presence of ligands, the degeneracy of the d orbitals is split to a greater or lesser extent, which may lead to low-spin configurations, if the gain in energy from occupying a lower-lying orbital is large enough to compensate for the pairing energy. In many compounds, however, the d-orbital splitting, caused by the ligand field, is intermediate in magnitude, so high- and low-spin configurations lie relatively close in energy. During a chemical reaction, if the ligand field changes, the ground state of the products may have a different spin state than that of the reactants. Such a reaction is referred to as *spin-forbidden*, because in the absence of spin-orbit coupling, passing from a potential energy surface of one spin to another of different spin is "forbidden" and does not happen.

There has been considerable debate in the literature concerning the effect of spin-state change on the rates of chemical reactions. Thus, one recent paper (1) raised the question "Can spin-state change slow organometallic reactions?" in the context of a study of CO ligand addition/elimination in a cobalt complex (Tp' is a modified tris(pyrazolyl)borate ligand):

$$Tp'Co(CO)_2 \leftrightarrow Tp'Co(CO) + CO$$

The left-hand dicarbonyl complex is a singlet, whereas the monocarbonyl derivative is a triplet, so the process is clearly spin-forbidden. Despite this, the authors observed the equilibrium between the two to be rapid, and they concluded: "It seems thus implausible that changes in spin state should slow organometallic transformations as a general rule."

As against this, there have been many observations of spin-forbidden reactions that *are* in fact slow. One example is provided by comparing the following two reactions (also CO association processes!):

$$Fe(CO)_3 + CO \rightarrow Fe(CO)_4$$

$$Fe(CO)_4 + CO \rightarrow Fe(CO)_5$$

The first of these two gas-phase reactions is observed to proceed essentially at the collision rate, whereas the second is 500 times slower (2). This difference in speed cannot be blamed on the existence of a barrier in the second case, because the rate is not appreciably changed by temperature. Instead, one can note that $Fe(CO)_3$ and $Fe(CO)_4$ have triplet ground states, whereas $Fe(CO)_5$ is a singlet, so the first reaction is spin-allowed, whereas the second is spin-forbidden. Clearly, spin *does* play a role in this and many other cases.

In fact, such apparently contradictory observations are reconciled within the theoretical description of spin-forbidden reactions. As discussed at greater length later, such reactions can be roughly considered to require (1) that the system reach a configuration where the potential energy of the two spin states are nearly equal, and (2) that the spin-orbit coupling between spin states should then cause the system to change spin. In cases where spin-orbit coupling is strong, requirement (2) is easily fulfilled, so the corresponding reactions will not necessarily be slower than comparable spin-allowed ones. However, requirement (1) may lead to very slow reactions, because the potential energy surfaces (PESs) of the two states may be so different that they cross only at rather high energy, leading to a substantial activation barrier. The speed of the reaction will thus depend on the strength of the spin-orbit coupling and on where the relevant PESs approach each other.

The problem referred to in the title of this section is that locating the regions where PESs of the reactant and product spin states approach one another is difficult within computational ab initio quantum chemistry. Also, the way in which the two factors just mentioned combine to determine the rates of spin-forbidden reactions has sometimes been ill appreciated. Taken together, these facts have meant that computational studies have not contributed to understanding the *mechanistic role* of spin changes in quite the same way that they have helped to clarify the mechanisms of many other reactions of transition metal systems.

Solving the electronic Schrödinger equation for anything but the simplest systems is an extremely complex numerical problem, which relies heavily on the use of approximations. One of the most important of these is already present in the choice of the Hamiltonian, which in almost all computational studies is taken to include only the kinetic energy and Coulombic charge interaction terms, because these are usually quantitatively the most important (3):

$$H_{elec} = \sum_i^{n_{el}} \frac{-\nabla_i^2}{2} + \sum_i^{n_{el}} \sum_{j>i}^{n_{el}} \frac{1}{r_{ij}} + \sum_a^{n_{nucl}} \sum_i^{n_{el}} \frac{Z_a}{r_{ia}}$$

Inclusion of Pauli's exclusion principle leads to the standard methods of ab initio computational chemistry. Within these methods, molecular systems containing the same nuclei and the same number of electrons, but having a different total electronic spin, can roughly speaking be said to be *different systems*. Thus, matrix elements of the Hamiltonian between Slater determinants corresponding to different spin states will all be zero, and they will not interact or mix at all. The wavefunctions obtained will be pure spin states.*

This will be true even in the vicinity of a crossing between PESs corresponding to different spin states. This should not in fact be so, because small (or not-so-small!) terms have been omitted from the Hamiltonian (4), and these terms lead to coupling and therefore mixing of different spin states in the wavefunction. The most important of these terms, for our purposes, is the spin-orbit coupling. Though it is possible to include this term in ab initio computations (5,6), this leads to significant computational difficulties, so it is not realistic to do so in general computations of PESs. For example, analytical gradients are not available for spin-orbit–coupled methods. These gradients play a very important role in characterizing the mechanisms of adiabatic reactions, by facilitating the location of saddle points or transition states on the corresponding adiabatic PESs. The unavailability of gradients makes it nearly impossible at present to locate stationary points on spin-mixed PESs. Because of the computational difficulties, advances in this area do not appear to be forthcoming, so for the foreseeable future, ab initio calculations, at least for large systems, will be restricted to single-spin-state PESs. Because the regions where one PES crosses another have no special features if one considers each PES individually, locating these regions is not straightforward. This is the fundamental difficulty that impedes the exploration of spin-forbidden reaction mechanisms in transition metal systems.

3. OVERVIEW OF THEORY

Spin-forbidden reactions are a subset of the broader class of *electronically nonadiabatic* processes, which involve more than one PES. The fundamental theory of how such processes occur is well understood (7–9), and a very large amount of research is being performed with the aim of elucidating more details in all the areas of nonadiabatic chemistry. It is not possible to present this work here, so I will instead provide an outline of the most important theoretical insights in the

---

Unrestricted wavefunction methods (see Ref. 3) lead to mixed spin states, but this is of no help in the present context.

context of spin-forbidden reactions, as well as outlining the similarities with other nonadiabatic processes occurring in transition metal systems.

In simple terms, one may describe chemistry using a semiclassical model involving motion of the system through phase space on several coupled potential energy surfaces (PESs). The potential energy is the electronic energy of the system at a given nuclear configuration. The following discussion refers to the schematic PESs shown in Figure 1.

The PESs (solid lines) are shown here as *crossing* each other. If the potential energy is obtained by diagonalizing the complete electronic Hamiltonian of the system, including the spin-orbit-coupling terms, PESs tend *not* to cross, although they may do so in regions called *conical intersections*, which do not directly concern us here(8). As already mentioned, though, most ab initio calculations do not use the full electronic Hamiltonian, using only the Coulomb and kinetic terms, and within this simplified Hamiltonian, PESs corresponding to different spin states do not mix at all and may therefore cross each other without restriction, as shown. In the language of nonadiabatic theory, such PESs are *diabatic*. Upon including spin-orbit coupling in the Hamiltonian, the different spin

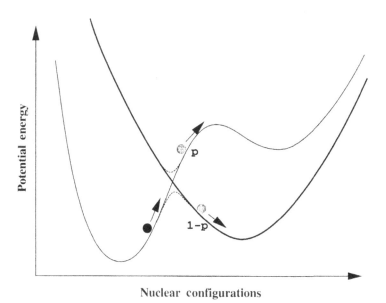

FIGURE 1   Schematic PESs, with the diabatic curves, corresponding to well-defined spin states (solid lines), and the adiabatic curves (dashed lines). The hopping event of a classical particle as it moves through the crossing region is schematically illustrated.

states can mix, leading to *adiabatic* PESs that do not cross (neglecting conical intersections), shown as dashed lines in Figure 1. Because of the mixing, the same adiabatic surface may correspond to different spin states in different regions of phase space. Where the diabatic surfaces cross, the corresponding adiabatic surfaces will show an *avoided crossing*, which may lead to a barrier on the lowest adiabatic PES. It is important to note that for spin-forbidden processes, the PESs for individual spin states are *diabatic* and do not correspond to the eigenstates of the full Hamiltonian.

To return now to the semiclassical model of nonadiabatic behavior, one can describe reactions on the spin-state (diabatic) PESs as follows: The system will move throughout phase space on the reactant PES until it reaches a point where the product PES has the same energy as the reactant one. At that point, it may either remain on the reactant PES or hop over onto the product one. The Landau–Zener formula for curve crossing in one-dimensional systems has often been used in a multidimensional context (10) as a useful approximation for the probability p with which this hop occurs, leaving $(1 - p)$ of the trajectories to continue on the initial PES (Fig. 1):

$$p_{LZ} = 1 - \exp \left(\frac{-4\pi^2 \, H_{rp}^2}{hv \, \Delta F} \right)$$

(H_{rp} is the magnitude of the coupling between the reactant and product surfaces at that point, v is the speed of the system upon passing the point where the PESs cross, and ΔF is the magnitude of the difference between the slopes on the two surfaces.) It can be seen that as the coupling becomes large, the exponential becomes small, so the hopping probability approaches 1. This constitutes the *adiabatic* limit,* meaning that the system will remain on the same adiabatic PES throughout (as discussed earlier, this will involve a change of spin!). In the opposite limit, where H_{rp} goes to zero, the hopping probability is zero too, leading to the *diabatic* limit: The system keeps the same spin throughout, even though this involves changing from one adiabatic PES to another. Spin-orbit coupling in transition metal atoms can be very large, especially for elements of the second and third transition rows, and this often applies to compounds containing the metals as well, so that in many cases spin-forbidden processes may in fact occur in the adiabatic limit. In other cases, the electronic structure may lead to the

---

*The use of the term *adiabatic* can be understood here by considering the effect of the v (speed) term. As the speed goes to zero, the exponential goes to zero, whatever the strength of the coupling. Thus, a system moving infinitely slowly upon approaching the crossing between the diabatic surfaces will be guaranteed to behave adiabatically, that is, to remain on the same adiabatic PES. This is analogous to thermodynamic systems, which behave adiabatically when changes to parameters such as pressure and volume are introduced infinitely slowly.

coupling between spin states being much weaker than suggested by the atomic coupling constants, leading instead to *nonadiabatic* behavior, corresponding to a hopping probability somewhere between 0 and 1.

This model leads to a very natural distinction between two factors influencing reactivity in spin-forbidden systems: the electronic and nuclear factors. The first of these is the strength of the coupling between the spin states. If it is very weak, a spin-forbidden process can be very slow, however favorable the nuclear factor. When spin-forbidden processes are discussed with the focus on this first factor alone, the conclusion is often that the large spin-orbit coupling typical of transition metal system will rule out reactions involving spin changes being slow. This is in fact not the case, mainly because the nuclear factor is often more important than the strength of the coupling in determining the speed of spin-forbidden reactions. This second factor refers to the amount of energy needed to move the nuclei from the reactant minimum to the region where the two spin-state PESs cross, and it is analogous to the activation energy for adiabatic reactions. If the geometries of the different spin states are sufficiently different, this energy can be quite large, and the reaction leading from one spin state to the other can be very slow, even if the electronic coupling is so strong that one is effectively at the adiabatic limit. So whatever the electronic factor, it is usually the nuclear factor that has the largest effect on the speed of spin-forbidden reactions, and computational work therefore revolves around locating the regions where the relevant PESs cross.*

It should be pointed out that though the semiclassical model is qualitatively correct, one needs also to include the possibility that the system will *tunnel* from one diabatic surface to the other. This is a manifestation of the quantum nature of the nuclei, which cannot be said to be precisely localized in phase space but are instead smeared out in the vicinity of the classical "position." This means that the system does not quite need to reach the crossing region to hop from one spin-state PES to the other, but may instead tunnel across. To use a classically based analogy, this involves changing from one electronic state to the other while simultaneously and instantly changing the nuclear configuration also. Because nuclei are heavy, the smearing of their positions is limited. In practice, this therefore means that tunneling can occur only in the vicinity of the region where the PESs cross, so the semiclassical model whereby the spin change occurs at the crossing point is not significantly wrong.

---

*When spin-orbit coupling is *very* strong, individual spin-state PESs are quite different from the adiabatic PESs and may not provide even a qualitatively useful 0th order representation of the system. This is, however, unusual, although one can expect that in systems with strong coupling, the avoided crossing between spin-state PESs may be quite strong, so the highest point on the lower state is substantially lower in energy than the corresponding diabatic state at that point.

The preceding discussion has been very general, and much of it applies to other types of nonadiabatic process. By far the most important of these is electron transfer (ET). This involves a transition from one diabatic state to another, with each state corresponding to a different charge distribution:

$$D—A \leftrightarrow D^+—A^-$$

where D and A are donor and acceptor molecules (or functional groups within a same molecule). In some cases, the two (diabatic) states can be very strongly coupled (e.g., in inner-sphere ET), in which case the lowest adiabatic PES will be directly accessible by ab initio calculations and will have a transition state (TS) as a signature of the avoided crossing of the two diabatic states. This TS may then be localized using standard ab initio methods. In many other cases, however, the coupling will be weak, so even if both diabatic states have the same spin and space symmetry, they will mix only weakly in the crossing region. Similar models to those used for spin-forbidden processes can describe the behavior of such systems, a point that has been made in several theoretical contributions (9,11).

The parallel can be extended to the methods used for ab initio calculations on electron transfer. Strictly speaking, the lowest ab initio adiabatic PES for an electron-transfer system has a TS where the diabatic surfaces cross (unless the two diabatic states have different spin or symmetry). However, the weak mixing of the diabatic states in the crossing region will lead to numerical instabilities, so it is often impossible to locate the TS. It is then easier to study the crossing region using ab initio methods that "decouple" the diabatic surfaces. One way to do this relies on the fact that D interacts only weakly with A and involves calculating PESs for D, D^+, A, and A^- separately. The "reactant" diabatic PES is then given as the sum of the D and A PESs, the "product" PES as the sum of the D^+ and A^- PESs. These decoupled PESs no longer interact at all, like spin-state PESs in the study of spin-forbidden processes, and the crossing region can therefore be studied using the same techniques. Although this approach has not yet been applied in studies of ET in transition metal systems, it has yielded very useful results for a broad range of processes, such as self-exchange in O_2^-/ O_2 (12) and NH_2OH^+/NH_2OH (13), and for models of electron transfer within the photosynthetic reaction center (14).

As suggested by the foregoing discussion of ET with strong coupling, even processes that occur adiabatically on a single ab initio PES can often be discussed usefully in terms of the crossing of two diabatic surfaces. This is the insight at the heart of the curve-crossing model (15) of organic reactivity. The approach can also be of practical use in computational work. Thus, most molecular mechanics force fields are unable to describe chemical reactions because the reactants and products have different bonding patterns, corresponding to different expressions for the PESs, which do not smoothly merge in the TS region. This suggests that the two expressions be treated as diabatic surfaces and that one then search for the crossing region between these PESs, which should provide a good approximation to the TS. This approach has been shown to be reasonably successful by comparison to ab initio calculations (16).

In conclusion, the theory of nonadiabatic processes suggests that the key to predicting spin-forbidden reactivity in transition metal system is the ability to predict the properties of the region where the corresponding spin-state PESs cross. This does not mean that one assumes the occurrence of nonadiabatic behavior (although this is possible), but simply reflects the fact that the system often needs a considerable amount of energy to reach the crossing region. As discussed shortly, the PESs actually cross at many different nuclear configurations; the chemically significant ones will be those of lowest energy, and it is these that computational chemists need to locate.

4. SPIN-FORBIDDEN CHEMISTRY IN TRANSITION METAL SYSTEMS

In this section, I will try to illustrate the role of spin-forbidden reactions in the chemistry of transition metals. The section is intended to give some idea of the *breadth* of this chemistry; doing justice to the depth would be impossible.

Surely the longest-established type of spin-forbidden process in transition metal chemistry is the phenomenon known as *spin crossover* (17,18). As mentioned in Section 1, unligated metal atoms and ions have high-spin states, with ligands tending to stabilize lower-spin states through the ligand field. It is well known that the strength of the latter can be modulated by careful choice of the ligands, so it is easy to imagine that for the proper choice of metal and ligands, the high-spin and low-spin states may be formally isoenergetic. In practice, the different states will have different optimal geometries, so systems having such a finely balanced ligand field are able to have distinct minima corresponding to the spin states, and these can be separately observed. Because the difference in geometry can be rather small, the corresponding PESs can cross at low energies, leading to facile interconversion of the spin states (19). In some cases, it has also been proven that the reactivity is dominated by tunneling behavior at low temperature (20).

An important development in this field has been the discovery for some spin-crossover systems that the more stable of the two spin states can be converted to the higher state by irradiation with visible light, a phenomenon referred to as light-induced excited spin-state trapping (LIESST) (21). The reverse transformation can be induced with light of a different color, which has led to intense interest in the possibility of using such systems as light-addressable information storage devices. Obviously, this depends on the *thermal* rate of interconversion being low, and the design of systems having this property is dependent on having a full understanding of the nuclear factor of the corresponding spin-forbidden process. Spin crossover has also been observed for many bioinorganic systems (22).

Much of the theoretical work in this field has focused on the use of empirical potentials for the two minima, and good results have been obtained (23,24). This can be expected, because the minima corresponding to the different spin states are generally rather close, and they are well approximated by harmonic expansion of the surfaces around the minima. However, ab initio computations can be expected to contribute positively to this field if they are able to characterize the crossing of the PESs in a less ad hoc manner.

Ligand association and dissociation are more typical "reactions," in that they involve a change in the coordination sphere of the metal atom. This leads to changes in the ligand field, and so may also be accompanied by a change in spin state. The CO dissociation reactions mentioned in Sec. 1 are good examples of such processes. Many other dissociation/association reactions have been suggested or observed to occur with a change in spin. Thus, phosphine exchange in singlet $CpCo(PPh_3)_2$ was suggested to occur via a triplet intermediate, $CpCo(PPh_3)$ (25). This has been investigated in a recent computational study, which has shown that the related $CpCo(PH_3)$ complex does indeed have a triplet ground state, with the singlet lying considerably higher in energy (26).

For this reaction, spin-forbidden ligand dissociation should be *faster* in this system than the spin-allowed process, providing the crossing region is energetically accessible and that spin-orbit coupling is not too small. For systems having similar low-spin to *low*-spin dissociation energies, reactions such as this one that have a lower-energy *high*-spin dissociation channel will be faster than reactions where the intermediate has a low-spin ground state. This idea, whereby spin-forbidden reactions can turn out to be *faster* than related spin-allowed ones, has received considerable experimental and computational attention (27). This was prompted by the observation of facile ligand exchange in the doublet $CpMoCl_2(PR_3)_2$ complexes, which was suggested to occur via quartet $CpMoCl_2(PR_3)$ intermediates (28). Although these have been shown indeed to have quartet ground states (27), the details of the crossing behavior have only recently (29) been established using one of the methods presented here, and these results will be discussed in a later section. Given the basic role of ligand dissociation and association reactions in inorganic and organometallic chemistry, computational contributions to understanding the role of spin-state changes are especially desirable.

The ligand association reactions of CO and O_2 with haem in its quintet state to form singlet haem-CO or haem-O_2 (Fig. 2) are bioinorganic spin-forbidden processes of tremendous importance. The relevance of the spin-forbidden nature of these reactions became significant very early on, and extensive theoretical as well as experimental work has been performed on these reactions (30–

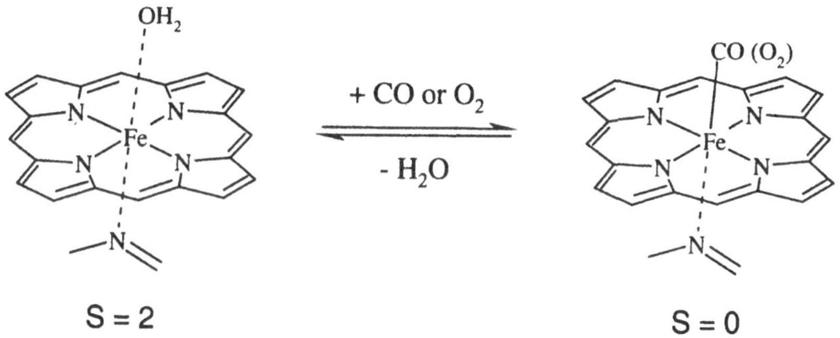

Figure 2 Spin-forbidden addition of singlet CO or triplet O_2 to quintet haem to form singlet carbonyl-haem or oxy-haem.

32). It would be of great importance to explore the relevant PESs and ascertain whether surface crossings lead to effective barriers for these reactions.

Oxidative addition, really only a more complicated form of ligand association, also leads to large changes in the coordination properties of the metallic center that is involved. It is therefore not surprising that this process, as well as its reverse, reductive elimination, also involves changes in spin in many cases. Such addition processes, when they involve C–H bonds, are key steps in the catalytic functionalization of alkanes and have received considerable experimental attention due to their potential industrial importance (33). CpIrCO and CpRhCO have been found to insert efficiently into the C–H bond of methane, whereas the first transition metal row congener, CpCoCO, does not (33,34). Computational studies have shown that the iridium and rhodium compounds have singlet ground states or very low excitation energies to singlet states, whereas the cobalt compound is a triplet, meaning that a crossing must occur en route to the inserted CpCoH(CO)(CH$_3$) species (34). In the same paper, it was estimated that this crossing would occur at relatively high energies, and the difficulty of locating the crossing region more accurately was discussed at length.

Ultrafast kinetic studies have been performed on the insertion of photolytically generated singlet and triplet CpM(CO)$_2$ (M = Re, Mn) intermediates into the Si–H bond of the (C$_2$H$_5$)$_3$SiH molecule. The rhenium compound is a singlet, whereas the manganese derivative is primarily a triplet under the experimental conditions, and this has been shown to affect the kinetics of the two processes (35).

Oxidation chemistry is a general term covering a range of processes, many of which include metal oxides. Because of the small spatial extent of metal-centered d orbitals, spin pairing in multiply bonded oxo species is rather weak, leading to similar energies for the low spin $M{=}O$ and high spin $\cdot M{-}O\cdot$ configurations, which have completely different reactivities. Given this fact, it is not surprising that spin-forbidden processes are particularly common in this field, as has been evidenced by both experimental and computational studies. Organic synthesis requires new selective oxidation methods, so lots of examples of spin-forbidden reactions come from this field, such as the epoxidation of alkenes by manganese-based catalysts (36) and the oxidation chemistry of chromyl chloride CrO_2Cl_2 (37,38). Studies of bioinorganic oxidation chemistry has also highlighted the occurrence of many spin-forbidden reactions, e.g., in the manganese-based photosynthetic oxygen evolution in Photosystem II (39) or the oxidation chemistry of cytochrome P-450 via iron-oxo species (40).

Gas-phase chemistry has extensively explored the effect of spin on reactivity in transition metal chemistry, largely thanks to the unrivaled possibility to perform state-specific chemistry. Gas-phase chemistry of *neutrals* has by and large mostly addressed the chemistry of small molecular species such as $Fe(CO)_5$ (41) and derivatives thereof. The spin-forbidden recombination of CO with the product of $Fe(CO)_5$ photodissociation, $Fe(CO)_4$, has already been mentioned (2). Many other similar reactions have been studied, e.g., the association reactions of $Fe(CO)_4$ with other ligands, such as H_2, C_2H_4, N_2 (42,43), and the reactions of substituted derivatives such as $Fe(CO)_3(C_2H_4)$, which are also assumed to be triplets (44). While most of these reactions are observed to be slower than the gas collisional rate, not all of them are, and the substitution effects have been discussed in some detail (42). Such discussion would undoubtedly be clarified by an accurate characterization of the PESs and of their crossings using ab initio methods. In turn, this is likely to provide valuable insight that can be transposed to mechanisms in condensed-phase systems.

Considerable work has already been carried out using ab initio calculations to predict the photodissociation dynamics of gas-phase metal carbonyls (45). This is a fertile area for computational work, given the extensive experimental results available, which include the use of ultrafast methods to characterize the short time behavior in photoexcited states. There is considerable evidence that surface crossings, especially of a spin-forbidden nature, play a considerable part in the dynamics. Much of the theoretical work so far has focused on reduced-dimensionality models of the PESs, which have been used in quantum mechanical studies of the nonadiabatic nuclear dynamics, in which spin-forbidden transitions are frequently observed (45). Here, too, the potential benefits to be derived from a proper understanding of the spin-state chemistry are considerable, due to the importance of light-induced processes in organometallic and bioinorganic systems.

Mass spectrometric investigation of "bare" metal ions and of their reactions with small molecules has also contributed enormously to the understanding of transition metal chemistry. It might seem that bare metal ions, as well as ions bearing a few ligands only, are not representative of typical catalysts due to their extensive coordinative unsaturation. However, it has been observed many times that they do in fact display remarkably similar reactivity patterns to the condensed-phase analogs (46), and they therefore provide excellent models for study, especially in terms of the level of experimental detail that is accessible. It should also be stated that the limited molecular size of such models lends itself to achieving much higher levels of computational accuracy than is currently possible with larger molecular arrays.

This field of investigation has accordingly grown considerably in recent years, and only a few highlights can be mentioned here. State-selected reaction rates have been measured for reactions with various hydrocarbons, clearly demonstrating the higher reactivity of low-spin states (47). This work has also been extended recently to reactions of neutral metal atoms (48,49). Trends have been established by studying *identical* reactions for the whole 10 elements of a given transition metal period, which is completely impossible for condensed-phase compounds (50,51). In many cases, these periodic comparisons have shown that spin has an important effect on thermochemistry and kinetics of transition metal compounds. The intricate electronic structure of metal-oxo species has been examined and related to the extreme complexity of the reactivity of such compounds (52). The chemistry of the iron oxide cation, FeO^+, has received considerable attention, with a particular focus on the spin-forbidden steps that occur during oxidation by this species (53,54). This work on the gas-phase FeO^+ ion has led to valuable insight into the role of spin in the bioinorganic oxidation chemistry of cytochrome P-450, already mentioned earlier (40).

Spin-forbidden processes are extremely common in transition metal chemistry. Both experimentalists and computational chemists are increasingly confronted by this fact, creating a need for useful models of how spin changes influence reactivity. Computational chemistry should play a key role in developing such models, using the methods discussed in the next section.

5. COMPUTATIONAL STUDY OF POTENTIAL ENERGY SURFACE CROSSING

As already indicated in Section 3, the computational chemist needs to locate the regions of configurational space where PESs cross so as to be able to understand, rationalize, and provide useful predictions for all of the fascinating chemistry discussed in the previous section. More precisely, he or she must locate the regions of lowest energy where crossing occurs.

5.1. Construction of Global Potential Energy Surfaces

As in computational studies of adiabatic processes, for systems having more than three or four atoms, it is not at all realistic to compute the PESs completely and systematically. However, one such study has been performed for the $Sc^+ + H_2$ system, within the side-on C_{2v} approach mode, and 2D PESs have been constructed as a function of the H–H and Sc–H_2 midpoint distances, using calculations at the MR-CISD level of theory (55). This seminal study is important because it explicitly considered the implication of the PESs for the kinetics and dynamics of the spin-forbidden insertion reaction:

$$Sc^+ \, (^3D) + H_2 \rightarrow ScH_2^+ \, (^1A_1)$$

Thus, all the triplet surfaces were found to be mainly repulsive, because triplet scandium (^3D, $[Ar]3d^14s^1$), with its two parallel valence electrons, cannot form bonds to both hydrogen atoms. However, the $^1A_1$ state is found to cross one of the triplet states near the singlet transition state for dissociation into Sc^+ (^1D). The authors point out that this crossing region is accessible for moderately energetic Sc^+ (^3D) and H_2 reagents, allowing the spin-forbidden reaction to occur by surface hopping (55).

Although similar studies are impossible for more realistic systems, the general topology of the PESs involved in spin-forbidden ligand recombination and oxidative addition reactions is often rather similar to that obtained for ScH_2^+. High-spin PESs tend to be repulsive, due to the impossibility of forming bonds with the incoming ligands, whereas at least one of the low-spin PESs is attractive, at least in parts, and the two tend to cross in the entrance channel. The details of the PESs and of their crossing region determine where the crossing occurs and, consequently, the kinetics of the reaction. Because of the generality of this picture, the paper by Rappé and Upton (55) has been of great use to experimental and computational chemists investigating similar spin-forbidden processes, especially those involving small gas-phase ions.

It should be noted that it is of course possible to construct complete PESs even for polyatomic systems, if one treats only a limited number of degrees of freedom. One can thus perform ab initio calculations on a grid of points generated by varying only a few structural parameters. This method has been used in the context of quantum dynamical studies of photodissociation in metal carbonyl derivatives (45). The study of $HCo(CO)_4$ dissociation, for example, was performed on PESs that considered only the Co–H and Co–C_{axial} bond lengths, with the other parameters being frozen (56). This approach is certainly useful. But like other approaches, discussed later, by completely neglecting relaxation of the frozen geometrical parameters, it will not always be very accurate.

5.2. Qualitative Evaluation of Crossing Profiles

Qualitative information on the crossing behavior of spin-state PESs is frequently able to give a good enough idea of the likely importance of the nuclear factor on the kinetics of a spin-forbidden process. For example, if the minima of two spin states are rather similar in geometry, as well as lying close in energy, one can fairly reliably predict that they will cross at fairly low energies. Additional information can be derived from considering the force constants for motion toward the other minimum.

In fact, such approaches can be used for assessing *adiabatic* reactivity also: Two structurally similar minima are likely to be connected by a relatively low-lying transition state, especially if the force constants corresponding to motion toward the other minimum are small. In practice, with the development of efficient gradient-based methods for optimizing transition states (57,58) and their general availability, such qualitative reasoning is rarely employed, except for trivial transition states, e.g., for rotation around single bonds.

For spin-forbidden reactions, because the better methodologies are less available at present, many studies have in fact relied on the inspection of surfaces. It is to be stressed that this may often be quite satisfactory when qualitative considerations alone are of interest. To take an example, Fe^+ has a ^6D ground state, which is unreactive with alkanes in the gas phase. The low-lying excited ^4F state is able to insert into C–H bonds. Computational studies of the gas-phase reactions of Fe^+ with methane (59) and ethane (60) have shown that the quartet and sextet complexes between the metal ion and the alkane lie close in energy. The sextet energy rises well above the quartet as the system then moves toward the relatively high-lying quartet C–H or C–C insertion TS. In such cases, it is reasonable to suppose that the system is able to reach the PES crossing region at relatively low energies, so interconversion of the spin states will occur rapidly compared to the insertion. According to the Curtin–Hammett principle (61), this means that the spin-forbiddenness of the insertion will not affect its kinetics.

Computing single-point energies on one PES at the minima, TSs, or other important points on the other PES leads to a refined understanding of the nature of the surface crossing. This is due to the fact that PESs can be assumed to be relatively smooth, which narrows down the energy range where crossing can be expected to occur. This is illustrated in Figure 3, where the position of the PES corresponding to the spin state of the products (bold line) is shown at several critical points on the PES of the reactant spin state (solid line). It can be seen that a reasonable estimate of the crossing position can be obtained, providing the product PES does not have the unexpected shape shown by the dashed line.

An even more refined approach involves computing the energy on one PES at a whole range of geometries on the other PES, e.g., along an optimized IRC. This approach has been used to characterize PES crossings in the $FeO^+ + CH_4$ system (62), as well as in the model system for chromyl chloride oxidation of alkenes, $CrO_2Cl_2 + C_2H_4$ (37). This procedure gives some insight as to where the crossing between PESs occurs along a reaction coordinate.

Overall, such qualitative approaches for characterizing the crossing region between two PESs can be extremely useful, especially in cases where the exact

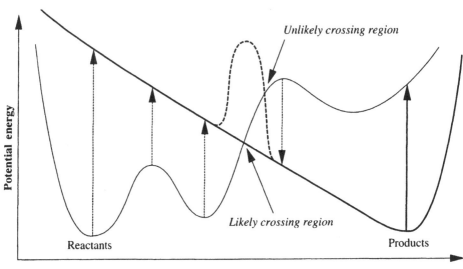

FIGURE 3 Schematic representation of a spin-forbidden reaction, showing how knowledge of the energy on one PES at selected points on the other PES can lead to reasonable assumptions as to the location of a surface crossing. For these assumptions to be wrong, unexpected behavior of one or both of the PESs, as shown by the bold dashed line, is necessary.

details of where the crossing occurs are not expected to affect the kinetics of a reaction. Quantitative accuracy, however, is unfortunately impossible. This should be obvious for the simpler approaches discussed earlier. However, upon performing ab initio studies on multiple PESs of different spin, there is a tendency to plot the results together on a single diagram, giving potential energy as a function of a single "reaction coordinate." This leads to something resembling Figure 3. However, at most points, the vertical difference in energy between the lines representing different PESs on such a figure will be meaningless. At best, when single points have been computed on one PES at optimized points on the other PES, one knows the energy at a few points on both surfaces simultaneously. In the worst case, which is also the more common one, one knows the energy at one set of geometries on one PES and at another set of *different* geometries on the other PES. Even when one has computed the "same" intermediate on both surfaces, the corresponding geometries will have been optimized and will frequently differ significantly. It is therefore misleading to plot such points above one another, to draw lines between them, and, especially, to attribute any precise significance to where the lines cross. This completely neglects the role of differential *relaxation* on the two PESs of the multiple geometric parameters that cannot possibly be represented on the one-dimensional plot.

To conclude, it should be stressed that the objections in the previous paragraph concerning the temptation to interpret one-dimensional pictures of the crossing of PESs in a *quantitative* way are in no way aimed at discrediting qualitative approaches for characterizing surface crossings. It should simply be realized that these approaches cannot give exact energies and structures for the lowest point where two PESs cross. In cases where the crossing appears to lie just above or just below some critical energy for the system being considered, more quantitative methods are thus needed.

5.3. The Method of Partial Optimization

One of these more systematic approaches, which has been called the method of partial optimization (29), relies on identifying a single geometrical coordinate that has very different optimal values in the two spin states. This coordinate can typically be a bond length, a bond angle, or a combination of several such coordinates. Next, a set of partial geometry optimizations are performed on each surface while holding the value of the unique coordinate fixed. Repeating this operation for several values of the unique coordinate, and on both PESs, leads to two one-dimensional sections through the PESs, which, providing the coordinate has not been poorly chosen, will cross at some point. This leads to one-dimensional graphs similar to Figures 2 and 3, and in principle allows a much more precise evaluation of the energy and geometry of the crossing region.

This approach has been used, e.g., in studies of the recombination of CO with $Fe(CO)_4$ (63). As mentioned earlier, $Fe(CO)_4$ has a triplet ground state, whereas the recombined product $Fe(CO)_5$ is a singlet. The triplet PES is essentially repulsive for all distances as the incoming ligand approaches, whereas the singlet PES is attractive throughout, so there is a crossing, as shown schematically in Figure 4. The most natural unique coordinate to use for partial optimization in this system is the Fe–C distance of the dissociating CO ligand, and this is indeed the choice made in Ref. 63.

The calculations in that study were performed using the BLYP density functional, with an ECP on the iron atom, together with large triple- to quadruple-zeta polarized basis sets (63). At this level, the $^3B_2$ state of $Fe(CO)_4$ lies 3.8 kcal $\cdot$ mol^{-1} below the $^1A_1$ state. The authors did not give precise details concerning the location of the crossing point, but on their graph showing the partial optimization curves, the crossing can be seen to occur for $r(Fe–C)$ between 3.5 and 4.0 Å and at an energy that is almost the same as that of singlet $Fe(CO)_4$ + CO. This suggests that the nuclear factor is quite substantial in this reaction, with a considerable barrier to recombination being formed. It appears that this barrier occurs because the singlet PES is not yet significantly attractive at the Fe–C

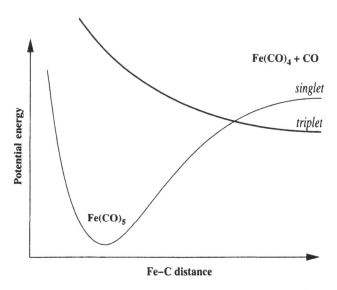

FIGURE 4 Schematic representation of the singlet and triplet PESs in the $Fe(CO)_4$ + CO system.

distances where the triplet PES is already rather repulsive. The crossing behavior in this system will be discussed at greater length in the next section.

An advantage of the partial optimization method is that it leads to the location of a refined "crossing" point compared to the mere qualitative consideration of the PESs. It also provides appealing one-dimensional plots of the potential, apparently giving considerable insight into the mechanism, e.g., as to the "early" or "late" nature of the crossing with respect to the two minima. However, the approach does not lead to quantitative results, for much the same reasons as discussed earlier. The crossing "point" that is found is in fact *two* points, lying on the projections of the two PESs onto two *different* one-dimensional sections of configurational space. This is because the curves describe the energies of the two states at *separately optimized* configurations on the two PESs, with only the unique coordinate guaranteed to be equal. If the other coordinates have very different equilibrium values on the two PESs along the curves, as can be expected given the differences in electronic structure, one will need to climb in energy from both sides to reach configurations that are the same for each spin state. The energy of the "crossing point" is only a rough lower estimate of the energy of the lowest point where the multidimensional PESs really cross. Of course, careful selection of the unique coordinate may somewhat alleviate this problem, but it cannot be completely eliminated.

The main disadvantage of the partial optimization method is that it is computationally rather expensive, since it requires multiple geometry optimizations on both PESs. The method discussed next is not only more accurate, but also requires only the equivalent of one geometry optimization on each surface, so it should really be preferred in most cases when the qualitative approaches of the previous section are not suitable.

5.4. Explicit Optimization of the Minimum on the Crossing Seam

The foregoing methods for characterizing the regions where different PESs cross are intrinsically approximate, even though they may often yield information of sufficient accuracy for the individual case being studied. As noted, some of them can also turn out to be of considerable computational expense. When an accurate result is desired, it is actually possible to locate directly and exactly the point of lowest energy where two PESs of different spin cross. Locating the minimum-energy crossing point (MECP), as it is often called, can be done in the same way that one finds minima or transition states on single PESs, by geometry optimization using energy gradients. Methods for finding MECPs have been described by several groups (64–71), with all relying in essence on minimizing a generalized gradient derived from the energies and gradients on the two PESs. The version used in much of the work described here is based on the following gradient, g (65):

$$g = \alpha(V_1 - V_2)(g_1 - g_2) + g_1 - h(g_1 \cdot h)$$

where α is an arbitrary constant, V_1 and V_2 are the potentials on the two PESs, g_1 and g_2 the corresponding gradients, and h is a unit vector parallel to $(g_1 - g_2)$. The minimization is performed with the BFGS approximate second-order method (72), and the whole optimization is driven by a shell script that calls the ab initio program to generate V_i and g_i, extracts them from the output, and uses them to update the geometry until convergence is reached.* The advantage of this script method, which has been used before (68), is that it can very easily be adapted for use with a variety of ab initio program packages. The version used here can be obtained upon request from the author.

Several features of the method are worthy of note. First, the (local) minimum on the crossing seam is fully optimized. As well as providing the energy, this provides a considerable amount of *structural* information, including of course the most favorable geometry for spin change. This can be of considerable use in understanding reaction mechanisms, by examining the deformations the system undergoes with respect to the minimum. Thus, bond stretching or shortening can

---

*With a reasonable starting geometry, speed of convergence is similar to that obtained for minima.

be related to bond-breaking or formation or to changes in bond order, changes in angles can be related to changes in hybridization, etc. Also, the gradients on the two PESs are parallel at the MECP, and they thus jointly define one unique direction, which is analogous to the transition vector of adiabatic reactions, providing a clue to the orientation of the reaction coordinate at the MECP. A more detailed reaction coordinate can be obtained by optimizing two "reaction paths," steepest descent paths from the MECP on the two PESs. Also, it is possible to check that the MECP is indeed a minimum within the seam of crossing between the PESs and not a saddle point, by computing and diagonalizing a generalized hessian (70) in the same way as one computes frequencies at an adiabatic stationary point.

Second, there exist simple methods, analogous to transition-state theory, for predicting rates of spin-forbidden reactions, based on the properties of the MECP (71,73,74). Although these have not yet been applied to systems containing transition metals, they could of course be useful for exploring the extent to which the *electronic* factor plays a role in such cases.

Third, the method involves a *single* geometry optimization, using information at each point from the two surfaces, and should thus be roughly as computationally demanding as one geometry optimization on each PES. In many cases, this should be much less than for the partial optimization method—as well as leading to greater accuracy.

Finally, it is possible to use a hybrid low-/high-level approach (70) to locate MECPs, which can be useful when analytical gradients are unavailable at the level of theory needed to describe the two PESs and their relative energy correctly. Of course, it is essentially meaningless to perform single-point calculations at a high level at the geometry of an MECP optimized at a lower level, because there is no guarantee that the PESs will have the same energy at the higher level. However, one can use the lower level to optimize the geometry, under the constraint that the *higher-level* energies are equal (70). This relies on the fact that the lower level should reproduce the features of the PES *orthogonal* to the reaction path reasonably accurately, even if it completely fails to describe the relative energies of the two PESs. This principle is also the basis of the IRCMax method (75) for accurately determining adiabatic barrier heights using high-level energies and low-level gradients.

The MECP optimization technique has, as yet, been used for only a handful of transition metal systems. I will discuss a few of these studies so as to provide insight into how the method compares with the approaches discussed earlier. The gas-phase reactions of Sc^+ with small molecules such as H_2O (76) and CH_4 (77) are relatively simple examples of spin-forbidden reactions that will serve as a good introduction. As already mentioned, Sc^+ has a ^3D ($[Ar]3d^14s^1$) ground state with a low-lying singlet excited state. The triplet is unable to insert into the X–

H bonds of H_2O or CH_4, because it cannot form two new bonds, whereas the singlet can, leading to relatively stable closed-shell intermediates $H-Sc^+-XH_n$. The Sc^+/H_2O system has been the object of a number of computational studies (78–80), which have all located the key stationary points on the singlet and triplet PESs. These include the singlet and triplet Sc^+-OH_2 ion–molecule complexes, the insertion transition state leading to $H-Sc^+-OH$, the H_2 elimination TS, and the resulting ScO^+-H_2 ion–molecule complex, with the latter three points all having singlet ground states. The most recent paper (80) uses B3LYP density functional calculations, with DZVP basis sets, together with CCSD(T) single-point calculations with TZV basis sets and multiple polarization functions. The authors loosely discuss the effect of spin on this reaction and claim that the singlet PES crosses below the triplet one somewhere between the reactant ion–molecule complexes and the inserted $HScOH^+$, at an energy below that of the separated reagents, Sc^+ and H_2O. This is based on the fact that the triplet reactant complex is more stable than the singlet, whereas the inserted product has a singlet ground state, with the corresponding triplet lying substantially higher in energy. The inserted species are very different in nature, since $^1HScOH^+$ is a well-characterized molecule with two metal–element bonds, whereas "inserted" triplet $HScOH^+$ is in fact better described as a very loose complex between $ScOH^+$ ($^2\Delta$) and an H atom, and has a very long Sc–H "bond."

The singlet and triplet energies along the reaction profile are thus derived from *separately* optimized and significantly different geometries, so qualitative reasoning cannot be used to determine the crossing energy accurately, as discussed in previous sections. This is especially important because the crossing is estimated to occur close to the TS for O–H bond insertion on the singlet PES and because this TS lies only slightly lower in energy than the reactants, Sc^+ + H_2O (80). Under thermal conditions, the reaction would be unlikely to occur if the crossing occurs somewhat higher than the reactants' energy, and this cannot really be excluded on the basis of qualitative estimates of the crossing energy.

To clarify this issue, new B3LYP calculations, using the Gaussian 98 program suite (81) together with our script for optimization of MECPs (71), have been performed to investigate the crossing behavior more closely (76). A similar basis set to that of reference (80), i.e. the SVP basis set (82), augmented by an f polarization function on Sc, diffuse s functions on H, and diffuse s and p functions on O, has been used. As expected, results similar to those described in previous studies are obtained for the stationary points (Fig. 5).

The new result is the optimization of the MECP between the $^1A'$ and $^3A'$ PESs. This point is indeed found to lie lower in energy than the Sc^+ + H_2O reactants, and very close, both in geometry and in energy, to the TS. As can be seen in Figure 5, the triplet state still lies somewhat lower in energy than the singlet at the TS itself, however, indicating that the crossing occurs "after" the

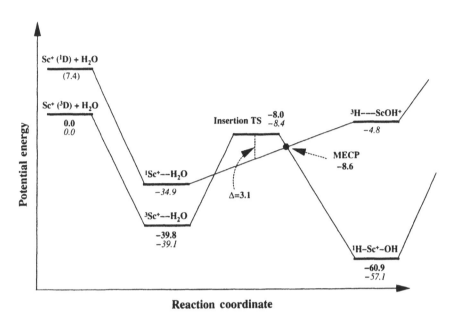

FIGURE 5   PESs in the $Sc^+ + H_2O$ system. Energies are in $kcal \cdot mol^{-1}$ relative to reactants, with the numbers in bold taken from Ref. 83 and the numbers in italics from Ref. 80. The excitation energy of Sc^+ is the experimental value. (From Refs. 80 and 83.)

TS. This is further confirmed by the fact that following the gradient downhill from the MECP on the singlet surface leads directly to the inserted species $HScOH^+$. It is noteworthy that the energy difference at the TS is sufficiently small to make it relatively obvious that the MECP occurs at low energies, using the qualitative reasoning discussed earlier. However, the explicit optimization of the MECP provides completely unambiguous proof that it does indeed lie at an energy that is thermally accessible, and thus the reaction should be fairly fast, unless spin-orbit coupling is very small.

The corresponding reaction with methane has PESs that are rather similar to those described here, so it will not be discussed in detail. It should, however, be pointed out that the description of the MECP in this system appears to have been one of the first such reports for a transition metal system in the literature (77). The study used CASSCF calculations with four electrons in an eight-orbital active space, combined with polarized double-zeta basis sets (77). Dynamic correlation is not very important in this system, partly because Sc^+ has only one d electron, so this level of theory produces rather good agreement with experiment for quantities such as the $^1D \leftarrow {}^3D$ excitation energy, the complexation energy of CH_4 and Sc^+, and the insertion energy to form singlet $H-Sc^+-CH_3$. It can thus be expected that the region of the surface crossing is well described too. The authors find an MECP lying slightly lower in energy than the TS for C–H bond insertion on the singlet PES, and crossing is predicted to be facile in this system too, in agreement with experiment.

The second example, the association reaction of CO with triplet $Fe(CO)_4$ to form singlet $Fe(CO)_5$ (2), has been discussed in previous sections. A partial optimization study of the crossing behavior in these systems has been performed (63), leading to the conclusion that crossing occurs at long Fe–C distances, where the singlet PES is not yet attractive, but is not repulsive either, and the triplet state is repulsive. A substantial barrier to association was predicted, in poor agreement with experiment. To examine the accuracy of the partial optimization method, new calculations (83) have been performed to locate the MECP explicitly, using the BLYP and B3LYP density functionals as implemented in Gaussian 98 (81), together with the SVP basis set (82). A schematic representation of the optimized structure of the C_{2v}-symmetric MECP is shown in Figure 6, and geometric parameters of the MECP and various minima together with relative energies are presented in Table 1.

The results of this study are quite different from those obtained using the partial optimization method, which shows the importance of structure relaxation. Thus, the MECP has a much more compact structure, in terms of the dissociating Fe–C bond distance, than it does in the previous study, where r was between 3.5 and 4.0 Å. At the lower values of r predicted here, the singlet surface is already significantly bonded, so in the case of the B3LYP method, the MECP is actually significantly *lower* in energy than the dissociated products on the singlet PES. A more striking difference from the behavior expected from partial optimization occurs for the data derived using the BLYP functional. As with other ''pure'' density functionals, the energy difference between the singlet and triplet states of $Fe(CO)_4$ is predicted to be much smaller than that obtained using the B3LYP functional, which partially includes HF ''exact'' exchange (this appears to be a general trend for pure and hybrid functionals (84,85)). Because BLYP

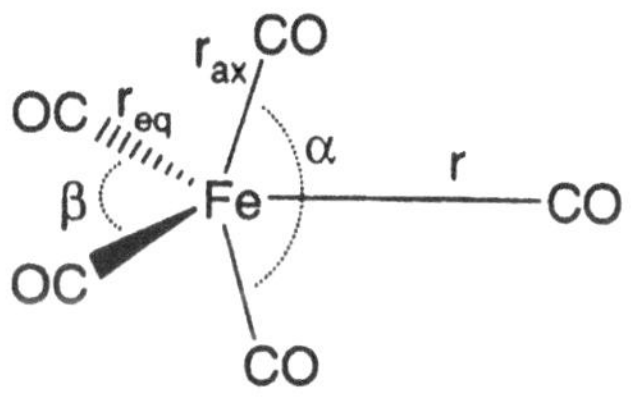

FIGURE 6   Structure of the MECP between the lowest singlet and triplet PESs of $Fe(CO)_5$, with a definition of the structural parameters referred to in Table 1.

TABLE 1 Geometric Parameters (Å, degrees) and Relative Energies (kcal · mol^{-1}) in the Fe(CO)$_5$ System

		r	r_{ax}	r_{eq}	α	β	E_{rel}
B3LYP	Fe(CO)$_5$ ($^1A_1'$)	1.813	1.822	1.813	180.0	120.0	−33.3
	Fe(CO)$_4$ ($^3B_2$)	/	1.875	1.847	146.9	98.9	0.0
	Fe(CO)$_4$ ($^1A_1$)	/	1.821	1.784	154.5	128.7	9.1
	MECP	2.557	1.901	1.843	165.8	102.3	5.2
BLYP	Fe(CO)$_5$ ($^1A_1'$)	1.825	1.827	1.825	180.0	120.0	−42.1
	Fe(CO)$_4$ ($^3B_2$)	/	1.868	1.831	147.3	98.8	0.0
	Fe(CO)$_4$ ($^1A_1$)	/	1.806	1.806	141.0	141.0	0.5
	MECP (BLYP)	2.933	1.867	1.823	159.2	103.1	3.0

The molecules belong to the following point groups: Fe(CO)$_5$: D$_{3h}$; singlet and triplet Fe(CO)$_4$ and MECP: C$_{2v}$; except for singlet Fe(CO)$_4$ at the BLYP level: D$_{2d}$.
Source: Ref. 83.

predicts the triplet to lie only 0.5 kcal · mol^{-1} below the singlet for $r = \infty$, the partial optimization method would predict the PESs to cross at very long values of r, since the triplet surface needs to rise only very slightly, or the singlet surface to become slightly attractive, for the energy difference to be annihilated. However, this neglects the fact that the structures of singlet and triplet Fe(CO)$_4$ are significantly different, so both will need to be distorted so as to reach the MECP. This explains why the present calculations find the MECP to lie significantly higher than both the singlet and triplet dissociation products, and at a fairly small value of r. The optimized r is, however, slightly higher than that obtained at the B3LYP level, indicating that the crossing does occur somewhat "earlier" along the association pathway than in that case.

These results show that the partial optimization method can yield misleading results. In terms of the chemistry, definitive conclusions are not forthcoming, because the BLYP and B3LYP functionals do not agree as concerns the singlet–triplet energy difference of Fe(CO)$_4$, and it is not known what the exact result is. However, the disagreement concerning the barrier to recombination is less severe, with both levels predicting a barrier of 3–5 kcal · mol^{-1} with respect to triplet Fe(CO)$_4$. This will need to be tested by more extensive calculations using other density functionals or methods (83).

A final example concerns the dissociation reactions of the doublet complexes CpMoCl$_2$(PR$_3$)$_2$, which have already been mentioned. So as to definitively prove that this system is directly able to reach the quartet PES en route to dissociation into CpMoCl$_2$(PR$_3$) and phosphine, we have undertaken a study of the crossing behavior in the CpMoCl$_2$(PH$_3$)$_2$ model system (29). Because previous work

TABLE 2 Structure of the MECP in the $CpMoCl_2(PH_3)_2$ System, Together with Relative Energies (in kcal · mol^{-1})

	r	E_{rel}
$^2CpMoCl_2(PH_3)_2$		−8.7
$^2CpMoCl_2(PH_3) + PH_3$		9.2
$^4CpMoCl_2(PH_3) + PH_3$		0.0
Partial Optimization Crossing Point	3.43 Å	2.3
MECP	3.20 Å	4.1

Source: Adapted from Ref. 29.

on related systems (86) had used the partial optimization method, we compared the results derived from the method with the fully optimized MECP. The calculations (29) were performed with the B3LYP density functional, together with the standard LanL2DZ basis set, which includes effective core potentials (ECPs) to treat the innermost core electrons on Mo, Cl, and P and using the Gaussian 98 program (81). The results are summarized in Table 2.

In this case, the agreement between the two methods is reasonable, with both predicting the crossing to occur very slightly above the energy of the dissociated intermediates, in the quartet state. Because this energy region is well below the *doublet* dissociation energy, both methods predict that $CpMoCl_2(PR_3)_2$ complexes should be able to dissociate faster by a spin-forbidden route to quartet $CpMoCl_2(PR_3)$ than by the spin-allowed route, in agreement with the original prediction (28). As shown in Table 2, there is also reasonable agreement concerning the geometry of the crossing, although here again the more accurate method also predicts a more compact structure for the crossing. However, it is important to note that the partial optimization calculations in this case were considerably more expensive than the MECP optimization. The latter required about 35 energy + gradient computations on each PES (= 70 in total), compared to a total of 282 energy + gradient computations for the five partial optimizations on both surfaces.

6. AB INITIO METHODS AND CROSSING POINTS

As for all computational studies, spin-forbidden processes require a very careful selection of the electronic structure method to be applied. In the main, the choice is made in the same way. Thus, one tries to find a level of theory that is both affordable and accurate. Accuracy can be guaranteed by choosing the highest levels of ab initio theory, multireference CI with a large number of reference configurations, and a very large basis set. As always, and especially in transition metal chemistry, this is very rarely a realistic approach, and one seeks instead to calibrate a lower level of theory to the problem one is treating. This can involve simply comparing the predictions of the level chosen with available experimental data for the system. A better calibration can often be obtained by comparing a *set* of increasingly accurate computational methods with experiment. Even if the methods chosen are all cheaper than the level intended to be used, this can give valuable information as to how well the predicted properties are converged with respect to the level of theory. Comparison with *more* accurate methods often requires performing the calibration calculations on a smaller model system or on parts of the PESs to be studied only.

This sort of reasoning is of course available in the general literature, so there is no need to extend the discussion further. One point that does need to be raised, however, is that the method chosen to locate an MECP needs to reproduce the energy difference between the spin states as well as possible. It frequently happens that a computational method leads to the PES of one spin state of a system being substantially higher than it should be, e.g., because that state has more correlation energy, and the method obtains only a small part of the correlation energy in each state. In such cases, the energy "offset" will lead sometimes to a quite serious error on the energy and geometry of the MECP. An example of this can be seen in the previous calculations on the MECP in the $Fe(CO)_4$ + CO system, where the BLYP and B3LYP levels lead to different predictions for the energy difference between singlet and triplet states. This leads to different MECP geometries and significantly different energies as well. Of course, in this case it is not known which of the methods is correct (or least incorrect!), as mentioned earlier. In cases where the level of theory needed to treat the two PESs in a balanced way does not yield analytical gradients, one can use the hybrid method of Ref. 65, as already discussed.

A final point of some importance concerns the use of single-reference methods. Density functional theory is often the method of choice in computational transition metal chemistry, simply because it is often the only affordable method that yields a qualitatively accurate description of the PESs. However, it is in some respects a "single-reference," method in that it constructs a "wavefunction" from a single Slater determinant and cannot therefore describe systems that inherently require a representation in terms of multiple determinants, such as low-spin open-shell systems. This makes it inappropriate for treating some regions on PESs, although experience shows that DFT is somewhat less sensitive to this effect than other single-determinant systems.

One could argue that this makes DFT unsuitable for the location of *all* MECPs and the description of spin-state crossing regions in general. This is because the *adiabatic* wavefunctions at the MECP are inherently multireferential in nature, since they are mixtures of diabatic states of different spin. However,

MECP location *never* treats the adiabatic wavefunction—only the diabatic states. This is why finding crossing regions is such a problem for ab initio methods. Were spin-state–mixed ab initio calculations generally available, it would probably be the case that single-determinant approaches would be unsuitable. However, in the present state of affairs, this is not necessarily so. Although some PES crossings occur in regions where one or both of the diabatic states is inherently multireferential (think of the H_2 molecule with $r = \infty$, where singlet and triplet states "cross"), this is far from always being the case. In fact, most of the systems described here do not have this problem, and single-reference approaches are perfectly acceptable. It is therefore *not* automatically necessary to use multireference methods such as MCSCF to treat MECPs between spin states (although it is again stressed that *some* systems *will* need such a treatment). In fact, MCSCF may frequently fail the test of giving a balanced treatment of different spin states, given the fact that it often does not recover a large proportion of the dynamic correlation in a system. In such cases, MCSCF will actually be a very bad electronic structure method to use for examining MECPs!

7. CONCLUSIONS

In this chapter, I have tried to give a feeling for the importance of spin-forbidden processes in transition metal chemistry. This is certainly relevant to much of inorganic and organometallic chemistry, to catalysis, and to bioinorganic chemistry. I have also described how understanding such processes is a challenge for computational chemists. This is because it requires the simultaneous consideration of multiple PESs, even in cases where spin-orbit coupling is so strong that the reaction can be considered to occur adiabatically on a single PES that changes smoothly from one spin state to the other. This is because the Hamiltonian used in ab initio calculations does not include spin-orbit coupling, and so can only calculate *diabatic* PESs having a definite spin.

I have also shown that a number of rather simple methods can be used to characterize the crossing behavior in such systems. Which is the most appropriate of these methods depends on what one wishes to find out about the system. However, for cases where the most accurate information possible is desired about the crossing behavior, the explicit optimization of the MECP is recommended in preference to the partial optimization, due to its less ambiguous conclusions and lower computational expense. Overall, the flowchart in Figure 7 can be used to design a computational investigation of a spin-forbidden process. In this chart, the starting point is the identification of any possible spin-forbidden behavior, based on experimental results or on empirical models of likely spin-pairing in the reagents, products, and hypothetical intermediates. In some cases, this may already allow one to exclude the significant occurrence of spin changes during the reaction to be investigated. It also enables the design of an appropriate model

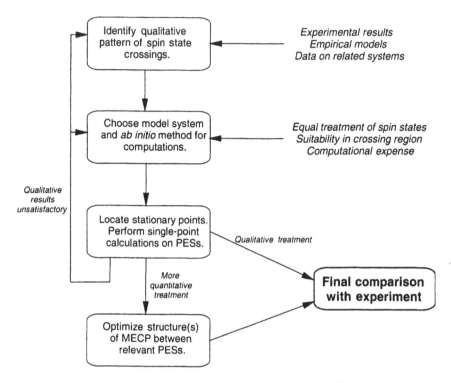

FIGURE 7   Flowchart for planning a computational study of a spin-forbidden reaction.

for the ab initio investigation. Great care should be taken at this stage to ensure that the chosen model and computational method will give a reasonably accurate description of the relative energetics of different spin states. After locating the minima and transition states that are present in the initial reaction scheme, one can compute single-point energies on other PESs at these optimized geometries. In some cases, this may lead to sufficiently useful qualitative understanding of the spin-forbidden behavior, and no further work is then needed.

However, in other cases, the initial calculations may lead to a reevaluation of the whole reaction scheme, and possibly to the consideration of supplementary spin states. Less drastically, it may lead to a change in which atoms of the real system are included in the model or to the use of a different computational method should the initial one prove too inaccurate or too computationally demanding.

As a final step, if quantitative information is required about the crossing behavior, the structure of MECPs relevant to the mechanism should be optimized. When analytical gradients are unavailable at the level of theory required to provide a reliable description of the relative energies of the different PESs, the hybrid method can be used to optimize the MECP.

Overall, one can certainly expect that the coming years will see a growing number of contributions addressing the role of spin in transition metal chemistry. It is hoped that the present overview will provide useful insights into the challenges implicit in ab initio computational studies of spin-forbidden processes, and will to some degree assist in attaining a higher degree of predictive power in such work.

REFERENCES

1. JL Detrich, OM Reinaud, AL Rheingold, KH Theopold. J Am Chem Soc 117: 11745–11748, 1995.
2. TA Seder, AJ Ouderkirk, E Weitz. J Chem Phys 85:1977–1987, 1986.
3. A Szabo, NS Ostlund. Modern Quantum Chemistry: Introduction to Advanced Electronic Structure Theory. Mineola, NY: Dover, 1996, p 43.
4. For a description of other terms entering into the electronic Hamiltonian, see any textbook on quantum mechanics, e.g., C Cohen-Tannoudji, B Diu, F Laloë. Quantum Mechanics. New York: Wiley, 1977, Chap XII.
5. BA Hess, CM Marian, SD Peyerimhoff. In: DR Yarkony, ed. Modern Electronic Structure Theory. Singapore: World Scientific, 1995, pp 152–278 and work cited therein.
6. S Koseki, MW Schmidt, MS Gordon. J Phys Chem 96:10768–10772, 1992.
7. DS Sholl, JC Tully. J Chem Phys 109:7702–7710, 1998.
8. DR Yarkony. Acc Chem Res 31:511–518, 1998.
9. See, e.g., M Bixon, J Jortner. In: Adv Chem Phys 106:35–202, 1999.
10. For a typical application, see JC Tully, RK Preston. J Chem Phys 55:562–572, 1971.
11. N Sutin. Acc Chem Res 15:275–282, 1982.
12. K Ohta, K Morokuma. J Phys Chem 91:401–406, 1987.
13. ML Hung, ML McKee, DM Stanbury. Inorg Chem 33:5108–5112, 1994.
14. O Kitao, H Ushiyama, N Miura. J Chem Phys 110:2936–2946, 1999.
15. S Shaik, A Shurki. Angew Chem Int Ed 38:586–625, 1999.
16. F Jensen. J Am Chem Soc 114:1596–1603, 1992.
17. E König, G Ritter, SK Kulshreshtha. Chem Rev 85:219–234, 1985.
18. P Gütlich, HA Goodwin, DN Hendrickson. Angew Chem Int Ed Engl 33:425–427, 1994.
19. JK Beattie. Adv Inorg Chem 32:1–53, 1988.
20. C-L Xie, DN Hendrickson. J Am Chem Soc 109:6981–6988, 1987.
21. CC Wu, J Jung, PK Gantzel, P Gütlich, DN Hendrickson. Inorg Chem 36:5339–5347, 1997, and work cited therein.
22. KM Barkigia, NY Nelson, MW Renner, KM Smith, J Fajer. J Phys Chem B 103: 8643–8646, 1999.
23. E Buhks, G Navon, M Bixon, J Jortner. J Am Chem Soc 102:2918–2923, 1980.
24. BS Brunschwig, N Sutin. Coord Chem Rev 187:233–254, 1999.
25. AH Janowicz, HE Bryndza, RG Bergman. J Am Chem Soc 103:1516–1518, 1981.

26. R Poli, KM Smith. Eur J Inorg Chem 877–880, 1999.
27. R Poli. Acc Chem Res 20:494–501, 1997.
28. AA Cole, JC Fettinger, DW Keogh, R Poli. Inorg Chim Acta 240:355–366, 1995.
29. KM Smith, R Poli, JN Harvey. New J Chem 24:77–80, 2000.
30. E Buhks, J Jortner. J Chem Phys 83:4456–4462, 1985.
31. BS Gerstman, N Sungar. J Chem Phys 96:387–398, 1992.
32. AP Shreve, S Franzen, MC Simpson, RB Dyer. J Phys Chem B 103:7969–7975, 1999, and work cited therein.
33. BA Arndtsen, RG Bergman, TA Mobley, TH Peterson. Acc Chem Res 28:154–162, 1995.
34. PEM Siegbahn. J Am Chem Soc 118:1487–1496, 1996.
35. H Yang, KT Kotz, MC Asplund, MJ Wilkens, CB Harris. Acc Chem Res 32:551–560, 1999.
36. C Linde, B Åkermark, PO Norrby, M Svensson. J Am Chem Soc 121:5083–5084, 1999.
37. M Torrent, L Deng, T Ziegler. Inorg Chem 37:1307–1314, 1998.
38. C Limberg, R Köppe. Inorg Chem 38:2106–2116, 1999.
39. PEM Siegbahn, RH Crabtree. J Am Chem Soc 121:117–127, 1999.
40. S Shaik, M Filatov, D Schröder, H Schwar. Chem Eur J 4:193–199, 1998.
41. N Leadbeater. Coord Chem Rev 188:35–70, 1999.
42. E Weitz. J Phys Chem 98:11256–11264, 1994.
43. PG House, E Weitz. Chem Phys Lett 266:239–245, 1997.
44. PG House, E Weitz. J Phys Chem A 101:2988–2995, 1997.
45. MC Heitz, K Finger, C Daniel. Coord Chem Rev 159:171–193, 1997.
46. K Eller, H Schwarz. Chem Rev 91:1121–1177, 1991.
47. JC Weisshaar. Acc Chem Res 26:213–219, 1993.
48. Y Wen, A Yethiraj, JC Weisshaar. J Chem Phys 106:5509–5525, 1997.
49. PA Willis, HU Stauffler, RZ Hinrichs, HF Davis. J Chem Phys 108:2665–2668, 1998.
50. PB Armentrout. Acc Chem Res 28:430–436, 1995.
51. D Walter, PB Armentrout. J Am Chem Soc 120:3176–3187, 1998.
52. D Schröder, H Schwarz. Angew Chem, Int Ed Engl 34:1973–1995, 1995.
53. A Fiedler, D Schröder, S Shaik, H Schwarz. J Am Chem Soc 116:10734–10741, 1994.
54. N Harris, S Shaik, D Schröder, H Schwarz. Helv Chim Acta 82:1784–1797, 1999.
55. AK Rappé, TH Upton. J Chem Phys 85:4400–4410, 1986.
56. MC Heitz, C Ribbing, C Daniel. J Chem Phys 106:1421–1428, 1997.
57. HB Schlegel. In: DR Yarkony, ed. Modern Electronic Structure Theory. Singapore: World Scientific, 1995, pp 459–500.
58. PY Ayala, HB Schlegel. J Chem Phys 107:375–384, 1997.
59. DG Musaev, K Morokuma. J Chem Phys 101:10697–10707, 1994.
60. MC Holthausen, A Fiedler, H Schwarz, W Koch. J Phys Chem 100:6236–6242, 1996.
61. FA Carey, RJ Sundberg. Advanced Organic Chemistry. 2nd ed. Part A: Structures and Mechanisms. New York: Plenum, 1984, p 219.
62. K Yoshizawa, Y Shiota, T Yamabe. J Chem Phys 111:538–545, 1999.

63. SA Decker, M Klobukowski. J Am Chem Soc 120:9342–9355, 1998.
64. N Koga, K Morokuma. Chem Phys Lett 119:371–374, 1985.
65. A Farazdel, M Dupuis. J Comput Chem 12:276–282, 1991.
66. DR Yarkony. J Phys Chem 97:4407–4412, 1993.
67. MJ Bearpark, MA Robb, HB Schlegel. Chem Phys Lett 223:269–274, 1994.
68. KM Dunn, K Morokuma. J Phys Chem 100:123–129, 1996.
69. JM Anglada, JM Bofill. J Comput Chem 18:992–1003, 1997.
70. JN Harvey, M Aschi, H Schwarz, W Koch. Theor Chem Accts 99:95–99, 1998.
71. JN Harvey, M Aschi. Phys Chem Chem Phys 1:5555–5563, 1999.
72. WH Press, SA Teukolsky, WT Vetterling, BP Flannery. Numerical Recipes in Fortran 77. Cambridge, UK: Cambridge University Press, 1996.
73. JC Lorquet, B Leyh-Nihant. J Phys Chem 92:4778–4783, 1988.
74. Q Cui, K Morokuma, JM Bowman, SJ Klippenstein. J Chem Phys 110:9469–9482, 1999.
75. DK Malick, GA Petersson, JA Montgomery Jr. J Chem Phys 108:5704–5713, 1998.
76. JN Harvey. To be published.
77. DG Musaev, K Morokuma. J Phys Chem 100:11600–11609, 1996.
78. JL Tilson, JF Harrison. J Phys Chem 95:5097–5103, 1991.
79. S Ye. J Mol Struct (Theochem) 417:157–162, 1997.
80. A Irigoras, JE Fowler, JM Ugalde. J Am Chem Soc 121:574–580, 1999.
81. Gaussian 98, Revision A. 6. MJ Frisch, GW Trucks, HB Schlegel, GE Scuseria, MA Robb, JR Cheeseman, VG Zakrzewski, JA Montgomery Jr, RE Stratmann, JC Burant, S Dapprich, JM Millam, AD Daniels, KN Kudin, MC Strain, O Farkas, J Tomasi, V Barone, M Cossi, R Cammi, B Mennucci, C Pomelli, C Adamo, S Clifford, J Ochterski, GA Petersson, PY Ayala, Q Cui, K Morokuma, DK Malick, AD Rabuck, K Raghavachari, JB Foresman, J Cioslowski, JV Ortiz, BB Stefanov, G Liu, A Liashenko, P Piskorz, I Komaromi, R Gomperts, RL Martin, DJ Fox, T Keith, MA Al-Laham, CY Peng, A Nanayakkara, C Gonzalez, M Challacombe, PMW Gill, B Johnson, W Chen, MW Wong, JL Andres, C Gonzalez, M Head-Gordon, ES Replogle, JA Pople, Gaussian, Inc., Pittsburgh, PA, 1998.
82. A Schäfer, H Horn, R Ahlrichs. J Chem Phys 97:2571–2577, 1992.
83. JN Harvey. To be published.
84. W Wang, E Weitz. J Phys Chem A 101:2358–2363, 1997.
85. O Gonzalez-Blanco, V Branchadell. J Chem Phys 110:778–183, 1999.
86. DW Keogh, R Poli. J Am Chem Soc 119:2516–2523, 1997.

13

Oxidative Addition of Dihydrogen to M(PH$_3$)$_2$Cl, M = Rh and Ir: A Computational Study Using DFT and MO Methods

Margaret Czerw,
Takeyce K. Whittingham, and
Karsten Krogh-Jespersen
Rutgers, The State University of New Jersey, New Brunswick, New Jersey

1. INTRODUCTION

Modern electronic structure methodology offers a highly powerful approach to the detailed understanding of many aspects concerning the structure and reactivity of organometallic systems. Given the dearth of high-quality thermodynamic data available for organometallic reactions, the ability to extract energy parameters from electronic structure calculations may arguably be their greatest asset. Thus, reliable and accurate prediction of reaction and activation energies can provide potentially valuable guidance in determining the factors that control the rates and thermodynamics of organometallic reaction mechanisms, including those relevant to catalysis (1,2). Rational catalyst design and optimization on the basis of electronic structure calculations is within reach (3). High-level computational methods, which are widely applicable to a large variety of problem situations, are in strong demand, and there is a continuous need to evaluate new procedures and extend old ones. Organometallic chemistry provides a particularly diverse and fruitful field for such endeavors. Procedures such as effective core potentials, now well established in chemistry, have found some of their most impressive applications here (4), and many of the recent advances in density functional theory have had particular impact in organometallic chemistry (5).

Complexes of the Group 9 metals containing the moiety ML_2X (M = Rh, Ir; L = tertiary phosphine; X = a formally anionic ligand) form a group of important and widely used catalysts. For example, $Rh(PPh_3)_3Cl$ (Wilkinson's catalyst) is perhaps the best-known catalyst for olefin hydrogenation (6). Related reactions catalyzed by $Rh(PR_3)_2Cl$-containing complexes, including the photo- and transfer-dehydrogenation of alkanes, have been reported (7,8). More recently, $H_2Ir(PCP)$ [PCP = η^3-1,3-$C_6H_3(P'Bu_2)_2$] was found to catalyze the thermochemical dehydrogenation of alkanes to give alkenes and dihydrogen (9). Efficient methods for alkane functionalizations, such as oxidation and dehydrogenation, have tremendous value from industrial and environmental perspectives. These catalyses undoubtedly involve formal oxidative addition reactions to three-coordinate, $14e^-$ M(I) complexes to give five-coordinate, $16e^-$ M(III) complexes. Whether the actual operation of a catalyst such as $H_2Ir(PCP)$ proceeds via an oxidative addition/reductive elimination mechanism, formally including both Ir(III) and Ir(V) complexes, or by a series of concerted displacements is a topic of current research (10,11).

In this chapter, we will focus on oxidative addition of one or two molecules of dihydrogen (H_2) to coordinatively unsaturated $M(PH_3)_2Cl$, M = Rh and Ir. We will examine the performance of first-principles computational methods based on the traditional molecular orbital approach and on density functional theory, with a focus on thermodynamic and kinetic parameters.

2. COMPUTATIONAL METHODS

It is well appreciated that thermodynamic and kinetic parameters are difficult to compute for organometallic molecular systems (see, e.g., Refs 12–14 and Chap. 4 by Frenking in the present volume). In particular, such quantities cannot be predicted within an independent-particle, single-determinant Hartree–Fock type of approach; electron correlation must be included in the computational methods applied to achieve reliable and accurate results. In this work, we examine the performance of three first-principles methods, generally acknowledged by the abbreviations BLYP, B3LYP, and MP2. The first two are methods based on density functional theory (DFT) (15); the latter is an ab initio, molecular orbital (MO)–based method (16).

If we write the expression for the total energy (17) as the sum of the one-electron kinetic energy, E_T, the electron–nuclear attraction and nuclear–nuclear repulsion energies, E_V, the electron–electron repulsion energy, E_J, and the electron–electron exchange and correlation energies, $E_{XC} = E_X + E_C$,

$$E = E_T + E_V + E_J + E_{XC} \qquad (1)$$

we may further identify the principal computational methods used here as follows. The MP2 method is defined by setting E_X equal to the full Hartree–Fock exchange and evaluating E_C from second-order perturbation theory with the Hartree–Fock Hamiltonian as the reference, zeroth order Hamiltonian (18). Some of our calculations incorporated the electron correlation (E_C) from Møller–Plesset perturbation theory applied fully through fourth-order (MP4(SDTQ)) (19) or through the coupled-cluster, single and double excitation method (with triple excitations treated noniteratively), CCSD(T) (20,21). The last method is generally considered state of the art at present. In the BLYP method, E_X is obtained from Becke's 1988 nonlocal exchange functional (22) and E_C is produced by the nonlocal correlation functional of Lee et al. (23). Finally, in the method denoted B3LYP, the three-parameter exchange functional proposed by Becke in 1993 (24), which incorporates some exact Hartree–Fock exchange, replaces his 1988 exchange functional.

We make use of an effective core potential (ECP) on the metal atom and basis sets of valence double-zeta or better quality. The Hay–Wadt relativistic, small-core ECPs and corresponding basis sets (split valence double-zeta) were used for Rh and Ir (LANL2DZ model) (25). These ECPs release the penultimate electrons ($4s$, $4p$ for Rh; $5s$, $5p$ for Ir) for explicit basis function coverage along with the valence electrons. We used Dunning/Huzinaga all-electron, full double-zeta plus polarization function basis sets for the third-row elements (P, Cl) (26). Hydrogen atoms in H_2, which formally become hydrides in the product complexes, were described by the 311G(p) basis set (27); hydrogen atoms in phosphine groups carried a 21G basis set (28). In selected cases, we replaced the hydrogen atoms on the phosphines with methyl groups in which the C atoms were described by the Dunning/Huzinaga double-zeta plus polarization function basis set (D95d, 26), and H atoms were described by the STO-3G basis set (29).

Reactant, transition-state, and product geometries were fully optimized using gradient methods (30); symmetry constraints were imposed, when appropriate. The stationary points were further characterized by normal-mode analysis. The (unscaled) vibrational frequencies formed the basis for the calculation of vibrational zero-point energy corrections and, together with thermodynamic corrections for finite temperature, provided the data needed to convert from internal energies to reaction or activation enthalpies (ΔH; $T = 298$ K, $P = 1$ atm) (31). The higher-level MP4/CCSD calculations always used MP2 optimized geometries, and energy–enthalpy conversions were made based on the data derived at the MP2 level. All the electronic structure methods used here are implemented in the GAUSSIAN 98 series of computer programs (32).

3. MOLECULAR STRUCTURES AND SPIN STATES OF M(PH₃)₂Cl, M = Rh and Ir

The two limiting Jahn–Teller structures available to singlet ML_3 fragments with d^8 metal electronic configuration may be characterized as T and Y, respectively, both of molecular C_{2v} symmetry (33). When the composition of the fragment is ML_2L', additional intermediate distorted structures become possible. Early computational studies of $Rh(PH_3)_2Cl$ using ab initio methods (34–38) considered only the T structure, with Cl at the base of the T (*trans*-1a, T_{Cl}) (39). In some cases a low-lying triplet state was identified as the ground state for *trans*-1a, although it was recognized that this result could be due to insufficient or unbalanced treatment of electron correlation (37,38). Margl et al. (40) found during a detailed DFT study of C–H bond activation by $Rh(PH_3)_2Cl$, that a second T-type structure with a phosphine at the base of the T (*cis*-1a, T_{PH3}) was distinctly the ground state, 16.5 kcal/mol below T_{Cl}. These authors included relativistic energy corrections in their calculations and computed the energies of the triplet (as well as open-shell singlet) states to be well above the energies of the (closed-shell) singlet states for 1a. However, the recent B3LYP study of Su and Chu (41) reported a triplet ground state for *trans*-1a and did not consider any cis structures.

All computational methods applied here agree that both *trans*-1a (T_{Cl}) and *cis*-1a (T_{PH3}) structures exist as discrete minima in singlet states (Fig. 1), and that there are no additional minima (Y_{Cl}, etc.) of low energy on the singlet potential energy surface for $Rh(PH_3)_2Cl$. The singlet cis–trans enthalpy difference is 10–12 kcal/mol from the DFT methods, smaller than that obtained from ab initio perturbation theory (14–16 kcal/mol); with the most accurate MO-based model used (CCSD(T)), the cis–trans enthalpy difference is 8.0 kcal/mol (Table 1). The *trans*-1a structures formally attain the electronic configuration $d_{xy}(2)d_{xz}(2)d_{yz}(2)$

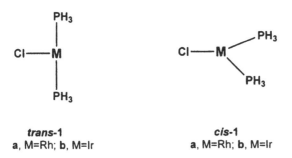

trans-1
a, M=Rh; b, M=Ir

cis-1
a, M=Rh; b, M=Ir

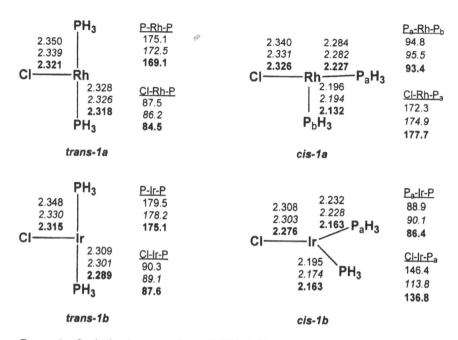

FIGURE 1   Optimized geometries of $M(PH_3)_2Cl$ isomers, M = Rh and Ir (singlet *trans*-1, singlet *cis*-1). Bond lengths in Å, angles in degrees. BLYP: regular font; B3LYP: *italics* font; MP2: **bold** font.

$d_{z^2}(2)d_{x^2-y^2}(0)$ (Rh, P, and Cl form the xy plane), and the low-lying triplet state formally has one electron promoted from the d_{z^2} orbital to $d_{x^2-y^2}$. Triplet *trans*-1a maintains C_{2v} symmetry with the two DFT methods, and the energy is more than 10 kcal/mol above the singlet state. However, the MP2 method breaks the molecular symmetry for triplet *trans*-1a and collapses the structure toward a cis conformation. The *cis*-1a isomer formally has an electronic configuration identical to that of *trans*-1a. All three computational methods locate minima for triplet *cis*-1a, but the BLYP method is unique in predicting a slightly pyramidal structure; the two other methods predict planar structures for triplet *cis*-1a. The energies of the *cis*-1a triplet states are computed well above the *cis*-1a singlet states in the DFT methods (11–13 kcal/mol), but the difference is smaller from MP2/MP4 calculations (8–9 kcal/mol) and even less, 5.6 kcal/mol, with CCSD(T).

All methods thus predict that singlet *cis*-1a (T_{PH_3}) represents the global minimum for $Rh(PH_3)_2Cl$ (Table 1). The overall energetic ordering of the isomers based on the two DFT methods is *cis*-1a (singlet) < *trans*-1a (singlet) < *cis*-1a (triplet) < *trans*-1a (triplet), confirming the results of Margl et al. (40). From the MO-based methods, the ordering is *cis*-1a (singlet) < *cis*-1a (triplet) < *trans*-

TABLE 1 Relative Enthalpies (ΔH, kcal/mol) of M(PH$_3$)$_2$Cl (1) Species

Species	Isomer	State	BLYP	B3LYP	MP2	MP4(SDTQ)	CCSD(T)
$M = Rh$							
1a	cis	Singlet	0.0	0.0	0.0	0.0	0.0
1a	cis	Triplet	13.3	10.7	8.4	9.3	5.6
1a	trans	Singlet	11.8	9.9	14.3	15.9	8.0
1a	trans	Triplet	19.4	12.6	a		
$M = Ir$							
1b	cis	Singlet	0.0	0.0	0.0	0.0	0.0
1b	cis	Triplet	20.0	18.5	27.6	23.2	21.0
1b	trans	Singlet	18.3	17.4	26.0	25.4	18.4
1b	trans	Triplet	a	23.5	a		

a The structure collapses to cis; see text.

1a (singlet). There is only limited structural or energetic data available on these elusive 14e species. $Rh(PPh_3)_3^+$ is known to be diamagnetic, showing a T-type structure with a P–Rh–P angle equal to 159° (42). Our B3LYP optimization on $Rh(PH_3)_3^+$ produces a T with a P–Rh–P angle of 175° (43).

As expected, the general features of the potential energy surfaces for $Ir(PH_3)_2Cl$ strongly resemble those found for $Rh(PH_3)_2Cl$. The *trans*-**1b** (T_{Cl}) structure for $Ir(PH_3)_2Cl$ has been considered by several research groups (41,44–46); when investigated, a triplet state was calculated lower in energy than the singlet (41). Previous work by us also considered only the T_{Cl} structure, with the cautionary note (45) that although a structure with *cis*-phosphines (Y_{Cl} or T_{PH3}) was conceivable, a trans conformation would, for steric reasons, presumably be the minimum-energy structure with the bulky phosphines favored by experimentalists ('Bu, ^{I}Pr, Ph, etc.). Our present set of calculations show that singlet *trans*-**1b** and *cis*-**1b** structures do exist with all three methods (Fig. 1), but with none of the methods does singlet *trans*-**1b** represent the global minimum; singlet, distorted T_{PH3}-type *cis*-**1b** structures are always considerably lower in energy. The singlet cis–trans enthalpy difference is 17–18 kcal/mol from the DFT methods, increases to 25–26 kcal/mol at the MP2/MP4 levels, and is 18.4 kcal/mol at the CCSD(T) level (Table 1). The P–Ir–P angle is very close to 90° in these optimized *cis*-**1b** structures, but the remainder of the "T" is not as well formed as in the case of $Rh(PH_3)_2Cl$. The distortion toward a Y_{Cl} type of geometry, judged by the extent of equalization of the two Cl–Ir–P angles, is largest for BLYP, smaller for MP2, and smallest for B3LYP (Fig. 1). Only with the B3LYP method does a triplet *trans*-**1b** exist as a minimum (6 kcal/mol above singlet *trans*-**1b**); triplet *trans*-**1b** collapses to triplet *cis*-**1b** with both the BLYP and MP2 methods. The singlet–triplet *cis*-**1b** enthalpy separation is substantial: nearly 20 kcal/mol at the DFT level and considerably higher with MP2/MP4 (23–27 kcal/mol); the CCSD(T) result is 21.0 kcal/mol. The larger singlet–triplet separation in **1b** relative to **1a** may at least partially be traced (47) to the different electronic configurations of the atomic ground states (Ir: $5d^76s^2$, quartet; Rh: $4d^85s^1$, quartet) and the magnitudes of the excitation energies to the lowest doublet states (large in Ir, small in Rh). The larger ligand field splitting generally encountered in the third vis-à-vis second transition metal series should be a contributing factor as well. The ordering of states in **1b** is similar to what we found earlier for **1a**, i.e., *cis*-**1b** (singlet) < *trans*-**1b** (singlet) < *cis*-**1b** (triplet), but the energetic separation of the global minimum (singlet *cis*-**1b**) from the competing local minima is considerably larger in **1b** than in **1a**.

The increased stability of cis (T_{PH3} type) over trans (T_{Cl}-type) structures was, in the case of $Rh(PH_3)_2Cl$, ascribed by Margl et al. (40) to the larger (static) trans influence exerted by PH_3 relative to Cl (48). Covalent metal-ligand interactions are significant contributors of trans influence, and we would expect such interactions to be more pronounced with a metal capable of stronger covalent bonding (Ir). Indeed, the singlet cis–trans enthalpy difference is 8 kcal/mol and 18 kcal/mol for M = Rh and Ir (CCSD(T), Table 1), respectively. The speculation (40,45) that bulky phosphines may reverse the cis–trans energetic order receives some support from additional calculations with alkylated phosphines (at the B3LYP level only). A methyl group provides a simple model for the heavily substituted phosphines inevitably used as protecting groups in experimental studies employing these transition metal complexes. Substitution of CH_3 for H in PH_3 renders the phosphine a stronger σ-electron donor (more basic) (49), but does not impart significant bulk to the phosphine. From B3LYP calculations, we find the singlet cis–trans energy difference in $Rh(PMe_3)_2Cl$ is 13.3 kcal/mol and that of $Ir(PMe_3)_2Cl$ is 21.7 kcal/mol, 3–4 kcal/mol larger than the energy differences in the parent complexes (10.2 kcal/mol and 17.6 kcal/mol, respectively). However, the energetic order is reversed when methyl is replaced by tert-butyl ($PMe_3 \rightarrow P^tBu_3$). For $Ir(P(^tBu)_3)_2Cl$, the cis–trans energy difference is 9.7 kcal/mol in favor of the trans T_{Cl}-type structure. The T_{PH3} structure for $Ir(P(^tBu)_3)_2Cl$ is highly distorted but remains cis (P–Ir–P = 123.2°, P–Ir–Cl = 98.9° and 137.0°).

Major experimental problems in the application of tricoordinate Rh- or Ir-halogen complexes in catalytic processes include thermal (phosphine) degradation and dimer complex formation. The dimerization reaction

$$2\ M(PH_3)_2Cl \rightarrow (PH_3)_2M(Cl)(Cl)M(PH_3)_2 \qquad (2)$$

is highly exothermic (Table 2). When M = Rh, we find $\Delta H = -39.1$ kcal/mol (BLYP), -43.8 kcal/mol (B3LYP), and -64.1 kcal/mol (MP2). A lower limit of -17.4 kcal/mol has been provided for the dimerization enthalpy of

TABLE 2   Reaction Enthalpies (ΔH, kcal/mol) for Dimerization of $M(PH_3)_2Cl$ (Reaction 2) and for H_2 Addition to $M(PH_3)_2Cl$ Species (Reactions 3–5)

Reaction	BLYP	B3LYP	MP2	MP4(SDTQ)	CCSD(T)
M = Rh					
2	−39.1	−43.8	−64.1	a	a
3	−7.9	−7.9	−21.5	−16.6	−17.9
4	−27.0	−26.2	−46.7	−42.7	−35.0
5	−15.2	−16.3	−32.5	−26.7	−27.0
M = Ir					
2	−40.5	−45.9	−66.7	a	a
3	−23.6	−25.7	−35.8	−32.6	−36.6
4	−49.0	−51.2	−71.9	−66.9	−64.0
5	−30.7	−33.8	−45.9	−41.5	−45.6

a Calculation not attempted.

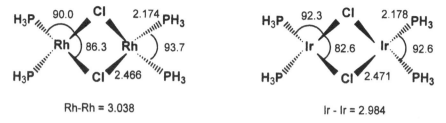

Rh-Rh = 3.038   Ir - Ir = 2.984

FIGURE 2   B3LYP-optimized geometries of the $(PH_3)_2M(Cl)(Cl)M(PH_3)_2$ dimer. Bond lengths in Å, angles in degrees.

$Rh(P^iPr_3)_2Cl$ (50). On the basis of this result, a lower limit of 32.5 kcal/mol was derived for the exothermicity of the hydrogenation reaction $Rh(P^iPr_3)_2Cl + H_2$. Later (45), the hydrogenation enthalpy value was revised up to 39 kcal/mol (i.e., $\Delta H(H_2$ addition) < -39 kcal/mol), which implies that the dimerization enthalpy for $Rh(P^iPr_3)_2Cl$ must exceed 23.9 kcal/mol (i.e., ΔH(dimerization) < -23.9 kcal/mol). Although, strictly speaking, there is no disagreement between the predicted theoretical and experimental values for the dimerization enthalpy, the substantial difference in magnitude between the predictions [-24 kcal/mol (exp.) vs. -39 kcal/mol to -64 kcal/mol (theory)] is striking. It appears that the presence of bulky phosphines or solvation significantly influences the value of the dimerization enthalpy. When M = Ir, the dimerization enthalpy is predicted to be 1–2 kcal/mol larger than when M = Rh: -40.5 kcal/mol (BLYP), -45.9 kcal/mol (B3LYP), and -66.7 kcal/mol (MP2). The dimeric product (C_2 symmetry) has Cl atoms bridging symmetrically between the two metal centers (40,51) as shown in Figure 2.

The degradation and aggregation problems may be circumvented through the use of rigid "pincer" ligands such as tridentate 1,5-bis(dialkylphosphinomethyl)phenyl (PCP). (PCP)-based catalysts are thermally stable above 200 °C, cannot undergo dimerization, and efficiently dehydrogenate alkanes to form the corresponding alkene plus dihydrogen (9–11,52).

3.1.   Oxidative Addition of H_2 to $M(PH_3)_2Cl$, M = Rh and Ir: Reaction Products and Transition States

Two singlet isomers for the three-coordinate M(I) complexes (*cis*-1, *trans*-1) have been identified in the previous section. Addition to either complex of a dihydrogen molecule with the formation of two M–H bonds formally oxidizes the metal atom to M(III). Assuming a least-motion pathway, the *cis*-1 reactant will lead to the square pyramidal (SQP) product *cis*-2, reaction (3); similarly, addition of H_2 to *trans*-1 produces the trigonal-bipyramidal (TBP) product *trans*-2, reaction (4).

$$
\text{Cl—M}\!\!\underset{PH_3}{\overset{PH_3}{<}} \quad + \quad \text{H-H} \quad \longrightarrow \quad H_3P\text{—M—H} \atop Cl\;H \qquad (3)
$$

cis-1
a, M=Rh; b, M=Ir

cis-2
a, M=Rh; b, M=Ir

$$
\text{Cl—M} \atop PH_3 \;(PH_3) \quad + \quad \text{H-H} \quad \longrightarrow \quad \text{Cl—M} \overset{H}{\underset{H}{}} \atop PH_3 \qquad (4)
$$

trans-1
a, M=Rh; b, M=Ir

trans -2
a, M=Rh; b, M=Ir

When M = Rh, H_2 addition to *cis*-1a is exothermic by about 8 kcal/mol at the DFT level and by more than twice that in the MPn/CCSD calculations (Table 2). The product *cis*-2a has a distorted SQP shape with one hydride apical (Fig. 3). H_2 addition to *trans*-1a is exothermic by 26–27 kcal/mol at the DFT level and by more than 40 kcal/mol at the MPn levels; the CCSD(T) value is 35.0 kcal/mol. The TBP product, *trans*-2a, has a distinct Y_{Cl} shape characterized by a very narrow H–Rh–H angle (~62°). *Trans*-2a is located approximately 10 kcal/mol below *cis*-2a in enthalpy (39) according to all the computational methods employed (Table 3).

Dihydrogen addition reaction (3) is, on a comparable basis, about 15–18 kcal/mol more exothermic when M = Ir than when M = Rh (Table 2). Thus, reaction (3) is exothermic by 26–27 kcal/mol at the DFT level and by more than 40 kcal/mol at the MPn levels; the CCSD(T) value is 36.6 kcal/mol, essentially twice the value for reaction (3) when M = Rh. For reaction (4), the differential exothermicity increase (M = Ir vs. Rh) is larger by approximately 25 kcal/mol. We obtain exothermicities near 50 kcal/mol at the DFT level and well above 60 kcal/mol with the MO-based correlation methods. The calculation at the CCSD(T) level predicts an exothermicity of 64.0 kcal/mol, 29 kcal/mol larger than when M = Rh. The SQP–TBP enthalpy difference is again on the order of 10

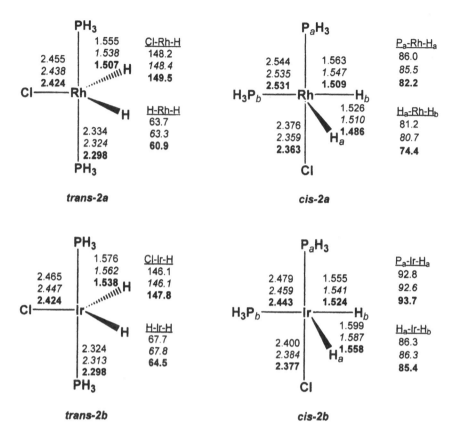

FIGURE 3 Optimized geometries of $H_2M(PH_3)_2Cl$ isomers, M = Rh and Ir (*trans*-2, *cis*-2). Bond lengths in Å, angles in degrees. BLYP: regular font; B3LYP: *italics* font; MP2: **bold** font.

TABLE 3 Relative Enthalpies (ΔH, kcal/mol) of $H_2M(PH_3)_2Cl$ (2) Species

Species	Isomer	BLYP	B3LYP	MP2	MP4(SDTQ)	CCSD(T)
M = Rh						
2a	trans	0.0	0.0	0.0	0.0	0.0
2a	cis	7.3	8.4	10.9	10.2	9.1
M = Ir						
2b	trans	0.0	0.0	0.0	0.0	0.0
2b	cis	7.1	8.1	10.0	9.8	8.7

kcal/mol in favor of the TBP structure, ***trans*-2b**, and the difference is essentially independent of the computational method used. ***Trans*-2b** features the characteristic highly acute H–M–H angle ($\sim$65°, Y_{Cl} shape; Fig. 3).

The transformation from a $14e^-$ three-coordinate fragment to a $16e^-$ five-coordinate complex is fully allowed by orbital symmetry (53); see Figure 4. The LUMO of the three-coordinate fragment is a hybrid orbital composed from the $(n)d_{x^2-y^2}$, $(n + 1)s$, and $(n + 1)p$ orbitals, extending into the space of the vacant, in-plane coordination site. This orbital has perfect symmetry and orientation to interact with the two electrons in the H_2 σ-bond orbital. Conversely, one of the doubly occupied, in-plane d orbitals (d_{xy}) has the proper local π-type symmetry and extension to interact with the antibonding LUMO of H_2 (σ*). The formation of the two M–H bonds progresses smoothly as the H–H bond dissociates. The spherical symmetry of the H $1s$-orbitals makes it possible not only to achieve M-H orbital overlap early, and possibly even form a $M(PH_3)_2Cl$-H_2 "precursor" complex, but also to maintain strong overlap throughout the concerted addition process.

The considerable exothermicities associated with reactions (3) and (4) reflect the unsaturated nature of **1**. In accordance with the Hammond principle (54), we can expect low or nonexistent activation energy barriers. For M = Ir, no transition state or "precursor" complex can be located for reaction (4) with any of the computational methods applied here. For reaction (3) it is possible to locate a strongly bound "precursor" complex with the B3LYP method ($\Delta H \sim -15.5$ kcal/mol relative to isolated reactants, H–H = 1.00 Å), but the transition state for H–H bond cleavage is only 0.3 kcal/mol above this complex (H–H = 1.01 Å). On the BLYP and MP2 surfaces for reaction (3), M = Ir, H_2 addition proceeds smoothly along the least-motion path without the appearance of any intermediate stationary points. Thus, there does not appear to be an activation energy barrier for dihydrogen addition to **1b** under ambient conditions.

The picture may at first appear slightly more complex when M = Rh. For reaction (4), there is neither a transition state nor a "precursor" complex with the BLYP or MP2 methods; at the B3LYP level, a "precursor" complex (Rh–H

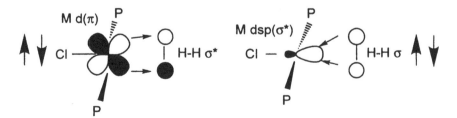

FIGURE 4   Favorable orbital interactions between **1** and H_2.

= 1.63 Å, H–H = 0.96 Å) can be located with a binding energy $\Delta E = -27.6$ kcal/mol relative to the reactants. The transition state for H–H dissociation lies within 0.2 kcal/mol of the "precursor" energy (Rh–H = 1.58 Å, H–H = 1.15 Å), and the product (Rh–H = 1.54 Å, H–H = 1.61 Å) is only 0.8 kcal/mol below the energy of the "precursor". For the less exothermic reaction (3), M = Rh, all computational methods predict "precursor" complexes and/or transition states. For example, at the B3LYP level a complex (Rh–H = 1.75 Å, H–H = 0.83 Å) can be located with a binding energy $\Delta E = -11.8$ kcal/mol relative to the reactants. A transition state lies 5.7 kcal/mol uphill from this "precursor" (Rh–H = 1.56 Å and 1.54 Å, H–H = 1.32 Å), and the *cis-2a* product (Rh–H = 1.51 Å and 1.55 Å, H–H = 1.98 Å) is 2.0 kcal/mol above the energy of the "precursor" complex. No "precursor" complex could be found at the MP2 level, but a transition state was found 16.8 kcal/mol below the separated reactants. The products of reaction (3), *cis-2a*, and the "precursor" complexes are always very close in energy and well below the isolated reactants. The electronic structure calculations refer to an idealized gas-phase path and neglect dynamics, and these "transition states" and "precursor" complexes would not appear to carry any significance on the potential energy surface at room temperature. We do not think there inherently is an activation barrier for H_2 addition to **1** (*cis* or *trans*) at ambient conditions.

Since *cis-1* is considerably more stable than *trans-1* (Table 1), a solution of **1** would contain almost exclusively *cis-1* and hence presumably overwhelmingly form *cis-2* upon oxidative addition of H_2. However, *trans-2* is more stable than *cis-2* by ≈ 10 kcal/mol (Table 3), and it is thus possible that the dihydrogen addition bypasses *cis-2* altogether. Interconversion of TBP and SQP complexes often proceeds with low activation energies (33,48), and we have located the transition state for conversion of *cis-2* to the thermodynamically favored product *trans-2*. When M = Rh (*cis-2a* → *trans-2a*), the transition state is about 10 kcal/mol above *cis-2a*. The transition state for the SQP → TBP interconversion when M = Ir (*cis-2b* → *trans-2b*) is also about 10 kcal/mol above the SQP conformer (*cis-2b*), so with both metals the rearrangement should be facile at ambient temperatures. The activation energy for *cis-2* → *trans-2* interconversion should be even less with sterically bulky phosphines that selectively destabilize **cis-2**.

From kinetics studies of H_2 adding to $M(P^iPr_3)_2Cl$, M = Rh and Ir (45,50), a lower limit of 39 kcal/mol was derived for the reaction exothermicity in the case of Rh; an absolute value of 48 kcal/mol was obtained when M = Ir. The proper reaction for comparison purposes would appear to involve the lowest-energy reactant and product structures, i.e., *cis-1* and *trans-2*, reaction (5), and the computed data are included in Table 2. For the parent species, M = Rh, we find reaction exothermicities of 15–16 kcal/mol with the DFT methods and 27–32 kcal/mol with the MPn methods; CCSD(T) predicts $\Delta H = -27.0$ kcal/mol. These values are well below the experimental estimate. When M = Ir, we find exothermicities that are more than 15 kcal/mol larger: 31–34 kcal/mol with the DFT methods and 42–46 kcal/mol with the Mpn methods; CCSD(T) predicts $\Delta H = -45.6$ kcal/mol. If we use methylated phosphine species in our comparison (B3LYP level only) to better include the electronic effects of alkylated phosphines, we find a decrease in exothermicity of ~3–4 kcal/mol. As in the case of the dimerization reaction (2), there are significant differences between the enthalpies predicted by the DFT- and MO-based methods. For reaction (5), the DFT-based methods appear to produce far too low exothermicities. With the reasonable assumption that the few kcal/mol increase in exothermicity ($PH_3 \rightarrow PMe_3$) predicted by B3LYP would carry over to the MO-based calculations, our best computed value for reaction (5) is near 30 kcal/mol when M = Rh, still approximately 10 kcal/mol too low. It is possible that bulky phosphines (such as ^iPr) will preferentially destabilize the three-coordinate reactant (pushing it toward the T_{Cl} structure) and thus increase the exothermicity of H_2 addition. On the other hand, the computed exothermicities when M = Ir are close to the experimentally derived (absolute) value, in particular, the presumably most accurate method (CCSD(T)) supplies a value very close to the experimental one.

The addition reactions involve cleavage of the H–H bond and formation of two (equivalent in *trans*-2) M–H bonds. The experimental H_2 bond dissociation enthalpy is 104.2 kcal/mol (55). The computed values from the BLYP and B3LYP methods nicely bracket this value at 103.9 kcal/mol and 104.5 kcal/mol, respectively. The MP2 value for the H–H bond dissociation enthalpy is only 95.2 kcal/mol; the calculated value improves somewhat at MP4(SDTQ) (99.9 kcal/mol) and CCSD(T) (100.3 kcal/mol). From the computed exothermicities of reaction (5), we can estimate the apparent M–H bond energies in the TBP product. For *trans*-2a, we find apparent Rh–H bond energies (kcal/mol) of 59.6 (BLYP), 60.4 (B3LYP), 63.8 (MP2), 63.4 (MP4(SDTQ)), and 63.6 (CCSD(T)); the corresponding values for the Ir–H bonds in *trans*-2b are 67.3 (BLYP), 69.2 (B3LYP), 70.1 (MP2), 70.7 (MP4(SDTQ)), and 72.9 (CCSD(T)). The computed M–H bond energies are 7–9 kcal/mol higher for Ir than for Rh, larger with the MO-based methods than with DFT, and they show an increasing trend with more extensive correlation treatments (MP2→CCSD(T)). The larger metal–ligand bond strength encountered for $5d$ elements over $4d$ elements may be attributed to the relative

$$Cl-M\overset{PH_3}{\underset{PH_3}{\diagup}} \quad + \quad H\text{-}H \quad \longrightarrow \quad Cl-M\overset{PH_3}{\underset{PH_3}{\mid}}\overset{H}{\underset{H}{\diagup}} \quad (5)$$

cis-1   trans-2
a, M=Rh; b, M=Ir   a, M=Rh; b, M=Ir

radial extension of the $5d$ vs. $4d$ orbitals. The former orbitals are diffuse and destabilized as the core orbitals contract and shield the nuclear charge, a relativistic effect (56). Hence the $5d$ orbitals overlap better with ligands, forming more covalent and stronger bonds. Inspection of Figures 1–3 shows that, when comparisons are appropriate, Ir–X bonds are generally shorter than Rh–X bonds, and this is particularly so for the M–H bonds.

3.2. Oxidative Addition of H_2 to $H_2M(PH_3)_2Cl$, M = Rh and Ir: Reaction Products and Transition States

In this section, we consider species formally obtained by adding H_2 to the most stable five-coordinate Rh and Ir species, *trans*-2. Seven-coordinate Ir(V) polyhydride complexes are known, whereas analogous Rh(V) complexes are not (57). Complexes containing molecular dihydrogen (58) most often show the metal attaining a d^6 electronic configuration (59,60). Hence, a priori, the following isomers appear possible for our seven-coordinate species: (a) a classical isomer with four M–H bonds, $(H)_4M(PH_3)_2Cl$ (3); (b) nonclassical isomers that have one dihydrogen molecule coordinating cis, $(cis-(H_2)-\eta^2-H_2)M(PH_3)_2Cl$ (4), or trans, $(trans-(H_2)-\eta^2-H_2)M(PH_3)_2Cl$ (5), to the Cl atom; (c) nonclassical isomers with two dihydrogen molecules coordinated to the metal, $(\eta^2-H_2)_2M(PH_3)_2Cl$ (6).

3
a, M=Rh; b, M=Ir

4
a, M=Rh; b, M=Ir

5
a, M=Rh; b, M=Ir

6
a, M=Rh; b, M=Ir

When M = Rh, we are unable to locate isomer **3a** as a minimum. With the BLYP and MP2 methods, **3a** possesses one imaginary frequency and is hence a transition state; at the B3LYP level, the structure is a second-order saddle point. All computational methods predict a minimum corresponding to the nonclassical cis isomer **4a** (Fig. 5). In addition, the DFT methods predict a minimum corresponding to trans isomer **5a**, whereas MP2 fails to locate a minimum for the tetrahydride. However, the computed **4a–5a** difference is more than 20 kcal/mol (Table 4) in favor of **4a**. This result may be yet another manifestation of the strong trans influence exerted by H, which renders **5a** with two hydrides as a trans pair disfavored (48,61). Structure **6** appears as a transition state with this B3LYP method; at the BLYP and MP2 levels, any attempt at locating a di-dihydrogen stationary point failed. Although the calculations do not present a fully

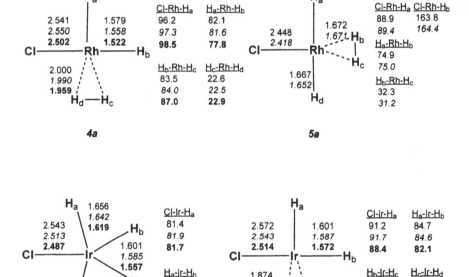

FIGURE 5 Optimized geometries of $H_4M(PH_3)_2Cl$ isomers, M = Rh and Ir. Bond lengths in Å, angles in degrees. Phosphine groups omitted for clarity. BLYP: regular font; B3LYP: *italics* font; MP2: **bold** font.

TABLE 4   Relative Enthalpies (ΔH, kcal/mol) of $H_4M(PH_3)_2Cl$ Species

Species	BLYP	B3LYP	MP2	MP4(SDTQ)	CCSD(T)
M = Rh					
4a	0.0	0.0	0.0		
5a	21.0	22.3	ᵃ		
M = Ir					
4b	0.0	0.0	0.0	0.0	0.0
3b	5.6	7.3	−1.3	1.3	2.0

ᵃ Not a stationary point on the MP2 surface.

uniform picture, they clearly favor nonclassical over classical structures for $H_4Rh(PH_3)_2Cl$. Lin and Hall have pointed out that the presence of contracted metal d orbitals will tend to favor the nonclassical isomers where metal–hydrogen electron transfer is minimized (61,62). Cis isomer **4a** is hardly bound relative to *trans*-**2a** and H_2, with the computed enthalpy for the formation reaction ranging from slightly negative ($\Delta H = -1.3$, -0.5, and -1.2 kcal/mol with MP2, MP4(SDTQ), and CCSD(T), respectively) to positive ($\Delta H = 2.0$ kcal/mol and 2.9 kcal/mol with B3LYP and BLYP, respectively). Since stronger electron-donating phosphines favor H_2 addition, it is likely that the formation enthalpies for **4a** will become more negative by a few kilocalories per mole, when alkylated phosphines are employed. However, ΔG for this bimolecular reaction will remain substantially positive, and the equilibrium for the formation of **4a** will thus lie far toward the reactants (*trans*-**2a**, H_2) under normal experimental conditions.

When M = Ir, we locate the classical, four-hydride isomer, **3b**, and the nonclassical cis isomer, **4b**, as minima with all computational methods. With the singular exception of MP2, the methods agree that **4b** is slightly more stable than **3b**. The **3b**–**4b** enthalpy difference (Table 4) is more than 5 kcal/mol with the DFT methods, but decreases to 2 kcal/mol or less at the highly correlated levels (MP4(SDTQ): 1.3 kcal/mol; CCSD(T): 2.0 kcal/mol). Lin and Hall found that the use of PH_3 rather than PMe_3 in calculations tended to favor the nonclassical isomers (62), but there are no indications of the nonclassical trans isomer **5b** (or of **6**) when the computational method used for geometry optimization includes electron correlation (63). Relativistic effects (destabilization of the $5d$ orbitals) should preferentially favor classical isomers (64), and, indeed, we could not locate the classical tetrahydride when M = Rh (see earlier). There is NMR evidence pointing to a nonclassical structure for $H_4Ir(P^iPr_3)_2Cl$ (65), in accord with the computational results (Table 4). According to the MO-based correlation methods, the seven-coordinate species **4b** is moderately bound with respect to *trans*-**2b**

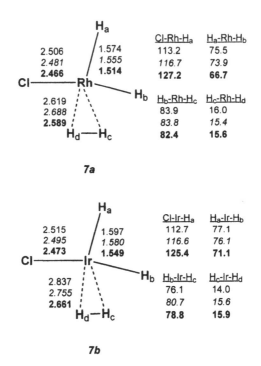

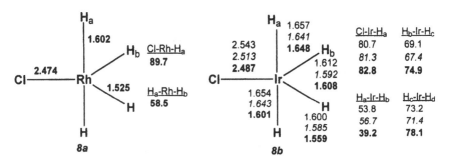

FIGURE 6 Optimized geometries for transition states 7 and 8. Bond lengths in Å, angles in degrees. Phosphine groups omitted for clarity. BLYP: regular font; B3LYP: *italics* font; MP2: **bold** font.

and H_2 (ΔH = -8.6 kcal/mol (MP2), -8.1 kcal/mol (MP4), -6.8 kcal/mol CCSD(T). However, the formation reaction is predicted to be essentially thermoneutral at the DFT levels [ΔH = 0.1 kcal/mol (BLYP), -1.9 kcal/mol (B3LYP)].

The transition state leading to the nonclassical cis isomer (7, Fig. 6) finds H_2 at a large distance ($\sim$2.6 Å) from the metal center and only slightly activated (H–H $\sim$ 0.75 Å). The transition state leading to 4 is only 1–3 kcal/mol above the reactants for both M = Rh and M = Ir. We have been unable to find a transition state, which leads directly to the classical isomer 3b or to the trans nonclassical isomer 5a from the separated reactants. However, 3b should be readily formed by intramolecular rearrangement. Transition state 8b (Fig. 6), which connects 4b and 3b, is located only 2.8 kcal/mol [CCSD(T)] above 4b. The classical tetrahydride 3b forms only a shallow minimum, since 3b and 8b are computed to be very close energetically [0.3 kcal/mol at CCSD(T)] and structurally (cf. Fig. 5 and 6) by all methods. On the MP2 surface for M = Rh, the four-hydride species 8a represents the transition state for the degenerate interconversion of the two equivalent nonclassical cis isomers 4a; 8a is 13.0 kcal/mol higher in enthalpy than 4a.

4. CONCLUSIONS

All three computational methods used here for geometry optimizations (BLYP, B3LYP, and MP2) produce comparable structures for all the isomers. Bond lengths from MP2 are shorter than those obtained from DFT (Figs. 1–3, 5, 6); bond lengths from B3LYP tend to be slightly shorter than those from BLYP, probably reflecting the small admixture of Hartree–Fock exchange present in the B3 functional. There is also general agreement among the methods regarding the relative energies of isomers (Tables 1 and 3). In particular, for M(PH$_3$)$_2$Cl (M = Rh and M = Ir) the singlet T_{PH_3} structure is clearly the preferred isomer. It is noteworthy that the enthalpy differences among the M(I) and M(III) isomers predicted by the B3LYP method are very similar to those predicted by the far more elaborate CCSD(T) method (66). Large differences appear in computed reaction enthalpies for dihydrogen addition, with the MO-based methods [MPn, CCSD(T)] predicting considerably higher exothermicities, which translate into larger M–H bond energies. The MO-based results appear to be closer to the available (limited) experimental data, and the DFT methods thus underestimate the M–H bonding energies, although they do produce the better results for the intrinsic H–H bond enthalpy. The apparent ability of the MPn/CCSD methods to form stronger M–H bonds is on display in the Ir(V) complexes, where a very small enthalpy difference is predicted between classical and nonclassical isomers.

The structural and energetic influences exerted by bulky phosphines continue to be of interest. Unfortunately, the dramatic scaling of MPn/CCSD(T) calculations with molecular size makes it impossible to perform these highly accurate calculations on large systems (67). DFT calculations scale less unfavorably with molecular size and would seem to be the method of choice for further investigations of such "substituent" effects.

ACKNOWLEDGMENTS

We gratefully acknowledge the National Science Foundation for financial support (CHE-9704304) and for a computer equipment grant (DBI-9601851-ARI). We thank Professor A. S. Goldman for stimulating discussions.

NOTES AND REFERENCES

1. PWNM van Leeuwen, JH van Lenthe, K Morokuma, eds. Theoretical Aspects of Homogeneous Catalysis, Applications of Ab Initio Molecular Orbital Theory. Dordrecht, The Netherlands: Kluwer, 1994.
2. DG Truhlar, K. Morokuma, eds. Transition State Modeling for Catalysis. ACS Symposium Series No. 721. Washington, DC: ACS, 1998.
3. L Deng, T Ziegler, TK Woo, P Margl, L Fan. Organometallics 17:3240–3253, 1998.
4. MT Benson, TR Cundari, ML Lutz, SO Sommerer. In: D Boyd, K Lipkowitz, eds. Reviews in Computational Chemistry. Vol. 8. New York: VCH, 1996, pp 145–202.
5. M Springborg, ed. Density-Functional Methods in Chemistry and Materials Science. London: Wiley, 1997.
6. J Burdeniuc, B Jedlicka, RH Crabtree. Chem Ber/Recueil 130:145–154, 1997.
7. K Nomura, Y Saito. Chem Commun 161, 1988.
8. T Sakakura, T Sodeyama, M Tokunaga, M Tanaka. Chem Lett 263–264, 1988.
9. W Xu, GP Rosini, M Gupta, CM Jensen, WC Kaska, K Krogh-Jespersen, AS Goldman. Chem Commun 2273–2274, 1997.
10. S Niu, MB Hall. J Am Chem Soc 121:3992–3999, 1999.
11. K Krogh-Jespersen, M Czerw, M Kanzelberger, AS Goldman. J Chem Info Comput Sci. In press.
12. A Veillard. Chem Rev 91:743–766, 1991.
13. MS Gordon. In: DR Yarkony, ed. Modern Electronic Structure Theory. Singapore: World Scientific, 1994, pp 311–344.
14. CW Bauschlicher Jr, SR Langhoff, H Partridge. In: DR Yarkony, ed. Modern Electronic Structure Theory. Singapore: World Scientific, 1994, pp 1280–1374.
15. RG Parr, W Yang. Density-Functional Theory of Atoms and Molecules. Oxford, UK: Oxford University Press, 1989.
16. WJ Hehre, L Radom, JA Pople, PvR Schleyer. Ab Initio Molecular Orbital Theory. New York: Wiley-Interscience, 1986.
17. JB Foresman, A Frisch. Exploring Chemistry with Electronic Structure Methods. Pittsburgh, PA: Gaussian, 1996.
18. M Head-Gordon, JA Pople, M Frisch. J Chem Phys Lett 153:503–506, 1988.
19. R Krishnan, MJ Frisch, JA Pople. J Chem Phys 72:4244–4245, 1980.
20. JA Pople, M Head-Gordon, K Raghavachari. J Chem Phys 87:5968–5975, 1987.

21. GD Purvis, RJ Bartlett. J Chem Phys 76:1910–1918, 1982.
22. AD Becke. Phys Rev A 38:3098–3100, 1988.
23. C Lee, W Yang, RG Parr. Phys Rev B 37:785–789, 1988.
24. AD Becke. J Chem Phys 98:5648–5652, 1993.
25. PJ Hay, WR Wadt. J Chem Phys 82:270–283, 1985.
26. TH Dunning, PJ Hay. In: HF Schaefer III, ed. Modern Theoretical Chemistry. New York: Plenum, 1976, pp 1–28.
27. R Krishnan, JS Binkley, R Seeger, JA Pople. J Chem Phys 72:650–654, 1980.
28. JS Binkley, JA Pople, WJ Hehre. J Am Chem Soc 102:939–947, 1980.
29. WJ Hehre, RF Stewart, JA Pople. J Chem Phys 51:2657–2664, 1969.
30. HB Schlegel. In: DR Yarkony, ed. Modern Electronic Structure Theory. Singapore: World Scientific, 1994, pp 459–500.
31. DA McQuarrie. Statistical Thermodynamics. New York: Harper and Row, 1973.
32. Gaussian 98 (Revision A.5). MJ Frisch, GW Trucks, HB Schlegel, GE Scuseria, MA Robb, JR Cheeseman, VG Zakrzewski, JA Montgomery, RE Stratmann, JC Burant, S Dapprich, JM Millam, AD Daniels, KN Kudin, MC Strain, O Farkas, J Tomasi, V Barone, M Cossi, R Cammi, B Mennucci, C Pomelli, C Adamo, S Clifford, J Ochterski, GA Petersson, PY Ayala, Q Cui, K Morokuma, DK Mailick, AD Rabuck, K Raghavachari, JB Foresman, J Cioslowski, JV Ortiz, BB Stefanov, G Liu, A Liashenko, P Piskorz, I Komaromi, R Gomperts, RL Martin, DJ Fox, T Keith, MA Al-Laham, CY Peng, A Nanayakkara, C Gonzalez, M Challacombe, PMW Gill, BG Johnson, W Chen, MW Wong, JL Andres, M Head-Gordon, ES Replogle, JA Pople. Pittsburgh, PA: Gaussian, 1998.
33. TA Albright, JK Burdett, M-H Whangbo. Orbital Interactions in Chemistry. New York: Wiley-Interscience, 1985.
34. N Koga, K Morokuma. J Phys Chem 94:5454–5462, 1990.
35. C Daniel, N Koga, J Han, XY Fu, K Morokuma. J Am Chem Soc 110:3773, 1988.
36. N Koga, K Morokuma. J Am Chem Soc 115:6883–6892, 1993.
37. DG Musaev, K Morokuma. J Organomet Chem 504:93–105, 1995.
38. MRA Blomberg, PEM Siegbahn, M Svensson. J Am Chem Soc 114:6095–6102, 1992.
39. J-F Riehl, Y Jean, O Eisenstein, M Pelissier. Organometallics 11:729–737, 1992.
40. P Margl, T Ziegler, PE Bloechl. J Am Chem Soc 117:12625–12634, 1995.
41. M-D Su, S-Y Chu. J Am Chem Soc 119:10178–10185, 1997.
42. YW Yared, SL Miles, R Bau, CA Reed. J Am Chem Soc 99:7076–7078, 1977.
43. A Dedieu, I Hyla-Kryspin. J Organomet Chem 220:115–123, 1981.
44. TR Cundari. J Am Chem Soc 116:340–347, 1994.
45. GP Rosini, F Liu, K Krogh-Jespersen, AS Goldman, C Li, SP Nolan. J Am Chem Soc 120:9256–9266, 1998.
46. K Krogh-Jespersen, AS Goldman. In: DG Truhlar, K. Morokuma, eds. Transition State Modeling for Catalysis. ACS Symposium Series No. 721. Washington, DC: ACS, 1998, pp 151–162.
47. PEM Siegbahn. J Am Chem Soc 118:1487–1496, 1996.
48. JD Atwood. Inorganic and Organometallic Reaction Mechanisms. Monterey, CA: Brooks/Cole, 1985.
49. CJ van Wüllen. J Comput Chem 18:1985–1992, 1997.

50. K Wang, GP Rosini, SP Nolan, AS Goldman. J Am Chem Soc 117:5082–5088, 1995.
51. We have not performed a thorough search of possible dimer configurations. The structures outlined in Figure 2 represent minima at BLYP, B3LYP, and MP2 levels of theory.
52. M Gupta, C Hagen, RJ Flesher, WC Kaska, CM Jensen. Chem Commun 2083–2084, 1996.
53. RG Pearson. Acc Chem Res 4:152–160, 1971.
54. GS Hammond. J Am Chem Soc 77:334–338, 1955.
55. DR Lide, ed. CRC Handbook of Chemistry and Physics. 71st ed. Boca Raton, FL: CRC Press, 1990.
56. WHE Schwarz, EM van Wezenbeek, EJ Baerends, JG Snijders. J Phys B At Mol Opt Phys 22:1515–1523, 1989.
57. FA Cotton, G Wilkinson. Advanced Inorganic Chemistry. 6th ed. New York: Wiley, 1999.
58. G Kubas. Acc Chem Res 21:120–128, 1988.
59. PG Jessop, RJ Morris. Coord Chem Rev 121:155–284, 1992.
60. Z Lin, MB Hall. Inorg Chem 31:4262–4265, 1992.
61. Z Lin, MB Hall. J Am Chem Soc 114:6102–6108, 1992.
62. Z Lin, MB Hall. J Am Chem Soc 114:2928–2932, 1992.
63. Structure 5b is a minimum at the Hartree–Fock level; see Ref. 61.
64. J Li, RM Dickson, T Ziegler. J Am Chem Soc. 117:11482–11487, 1995.
65. M Mediati, GN Tachibana, CM Jensen. Inorg Chem 29:3–5, 1990.
66. RK Szilagyi, G Frenking. Organometallics 16:4807–4815, 1997.
67. BG Johnson, PMW Gill, JA Pople. J Chem Phys 97:7846–7848, 1992.

14

The Electronic Structure of Organoactinide Complexes via Relativistic Density Functional Theory: Applications to the Actinocene Complexes $An(\eta^8\text{-}C_8H_8)_2$ (An = Th–Am)

Jun Li and Bruce E. Bursten
The Ohio State University, Columbus, Ohio

1. INTRODUCTION

The synthesis and characterization of the sandwich complex ferrocene, $Fe(\eta^5\text{-}C_5H_5)_2$, in 1951 stands as the starting point for modern organometallic chemistry (1). Since then, organometallic chemistry has been one of the most rapidly developing new fields in modern chemistry (2). In the development of modern organometallic chemistry, cyclopolyene and cyclopolyenyl systems of $\eta^n\text{-}C_nH_n$ carbocyclics, such as $C_5H_5^-$ (Cp), C_6H_6 (Bz), $C_7H_7^{3-}$ (Cht), and $C_8H_8^{2-}$ (Cot) are among the most frequently used ligands for both transition metal and f-block element organometallic complexes (3,4). Transition metal and f-element sandwich complexes formed by these ligands have played a central role in understanding the fundamental bonding, electronic structures, and chemical properties of organometallic complexes.

Since the synthesis of ferrocene, a multitude of interesting $\eta^n\text{-}C_nH_n$ (n = 4, 5, 6, 7, 8) sandwich complexes of various transition metals and f-elements have been reported. Among these developments, perhaps the most interesting are the syntheses of bis(tetraphenylcyclobutadiene)-nickel and -palladium, $M(\eta^4\text{-}$

$C_4Ph_4)_2$ (M = Ni, Pd) (5,6), bis(benzene)chromium(0), $Cr(\eta^6-C_6H_6)_2$ (7), and bis(cyclooctatetraene)uranium(IV), uranocene or $U(\eta^8-C_8H_8)_2$ (8), the structures of which are shown in Figure 1. The synthesis of uranocene, and its structure and chemistry via-à-vis those of ferrocene, provided a remarkable example of the similarities and differences between organotransition metal chemistry and organo-f-element chemistry.

While sandwich complexes of the Cp, Cp* ($C_5Me_5^-$), and Bz rings are prevalent for transition metals, the development of the chemistry of actinide sandwich complexes has focused mainly on complexes of the Cot ligand; because of the larger size of the actinide metal atoms, complexes with two Cp or Cp* ligands always involve the coordination of additional ligands. Although uranocene and other actinocenes were synthesized in the late 1960s, actinide sandwich complexes of the cycloheptatrienyl (Cht) and benzene (Bz) ligands were unknown until very recently. In 1995, the first bis(cycloheptatrienyl) actinide complex, namely, the $U(Cht)_2^-$ anion, was synthesized and characterized crystallographically by Ephritikhine and coworkers (9). Meanwhile, the bis(arene) actinide cationic complexes $[Bz_2An]^+$ and $[(TBB)_2An]^+$ (An = Th, U) were also observed in the past several years by Pires de Matos, Marshall, and coworkers via mass spectrometry (10), which suggests that it might be possible to synthesize and isolate neutral bis(arene) actinide sandwich complexes. These discoveries have opened a new chapter in the chemistry of organoactinide sandwich complexes, and a comprehensive theoretical investigation is thus needed to provide systematic comparisons of the electronic structures of these unique new complexes.

To date, a great number of experimental and theoretical investigations have been carried out to elucidate the structures and chemical reactivities of transition metal sandwich complexes (11). The parallel interest in investigations of actinide sandwich complexes has revived again recently (4,12). We have found relativistic density functional theory (DFT) to be a superb tool for the elucidation of the

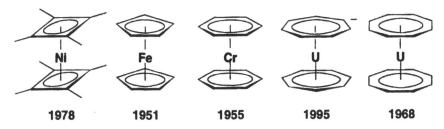

FIGURE 1   Structures of the $M(C_nR_n)_2$ ($n = 4$–8) sandwich complexes. The year in which the first member of each class of sandwich complex was synthesized is given below the structure.

electronic structural aspects of organoactinide complexes. We have recently used this methodology to explore aspects of the structure and bonding of a number of organoactinide sandwich complexes $An(\eta^n\text{-}C_nH_n)_2$ (n = 6, 7, 8) (13,14,15). In this chapter, we will discuss some practical aspects of the application of relativistic DFT to actinide complexes. We will then discuss some specifics of the bonding in organoactinide sandwich complexes, with particular emphasis on some recent detailed studies of the geometries, electronic structures, and vibrational properties of the actinocenes. Our goal is to demonstrate the utility of relativistic DFT in charting the future theoretical and experimental studies of organoactinide complexes.

2. THE CHALLENGES OF THEORETICAL ACTINIDE CHEMISTRY

The application of theoretical electronic structure methods to organoactinide complexes has long been hampered by some well-known challenges. Before we detail the computational methodology used in the present studies, we will briefly summarize some of the particular challenges inherent in the theoretical study of actinide complexes:

- Because of the importance of the An $5f$ orbitals in chemical bonding, the theoretical method chosen must be able to accommodate f orbitals, and, in order to investigate large actinide-containing systems, must do so with good computational efficiency.

- The radial distributions of the An $6s$ and $6p$ "semicore" orbitals for the early actinides lie in the valence region. Thus, the calculations necessarily involve a great number of "valence" orbitals, including at least the An $6s$, $6p$, $6d$, $5f$, $7s$, and $7p$ atomic orbitals in the valence space and a large number of valence electrons in the variational calculations (16,17).

- Dynamical electron correlation effects, i.e., the instantaneous correlation in the motions of electrons at short interelectronic distances, are so important for the heavy-element systems that exclusion of these effects in theoretical calculations of actinide complexes might lead to incorrect conclusions (18).

- The An $5f$ and $6d$ orbitals are very close in energy, so many low-lying, near-degenerate states exist for each given electron configuration. This complication poses a great challenge for conventional correlated ab initio calculations because a multiconfiguration and/or multireference scheme with large active configuration space is generally necessary to account for the nondynamical electron correlation effects arising from the degeneracy or near-degeneracy of different electron configurations.

- Because of their large atomic numbers, the actinide elements are subject to severe relativistic effects (19). Scalar relativistic effects (i.e., mass-velocity and Darwin effects) must be incorporated into the model Hamiltonian in order to predict accurate electronic structures and chemical properties. For example, ThO_2 would be erroneously predicted to be linear had scalar relativistic effects been neglected in the geometry optimization (20). The effects of spin-orbit coupling effects are also important for many of the properties of actinide compounds, especially when involving multiplets in the ground and excited states of complexes.

- Although the An 6d orbitals are strongly split by the presence of a ligand field (as is the case for transition metal complexes), the more contracted An 5f orbitals are generally only slightly split by the ligands (21). The ground configurations of actinide complexes are generally governed by the occupation of these closely spaced An 5f orbitals, which leads to many open-shell states. The determination of the ground state among these multiple open-shell states is therefore a more difficult problem for actinide complexes than for transition metal complexes (22). Additionally, the determination of the ground state is often further complicated by the energetic closeness of different structures.

- All of the foregoing effects, coupled with the generally large size of organoactinide complexes, place extremely high demands on the choice of suitable atomic basis sets, efficient numerical algorithms, and appropriate computational resources in order to enable theoretical calculations on organoactinide molecules of chemical interest.

It is our experience that many of the preceding difficulties can be overcome through the application of advanced density functional theoretical methods. Although DFT still suffers from some disadvantages and is still an evolving methodology, the successful use of DFT for calculating the electronic structure of molecules has increased tremendously in recent years, especially for metal-containing systems. In the next section we will highlight some of the particular advantages of DFT when applied to organoactinide systems.

3. ADVANTAGES OF DENSITY FUNCTIONAL THEORY METHODS

Because of the aforementioned challenges that theoretical actinide chemistry has posed, applications of traditional correlated ab initio methods are very expensive for organoactinide complexes and are thus limited to either small molecules or small basis sets and/or small active configuration spaces. During the past two decades, our group has engaged in the elucidation of the electronic structures of organoactinide complexes via a variety of DFT formalisms, including the Xα-SW and DV-Xα approaches (23,24). Recent advances in density functional theory have greatly expanded the capability of DFT methods (25–27). Of particular importance are the DFT methods that employ the generalized gradient approach (GGA), which involves gradient corrections to the exchange-correlation functionals. These GGA DFT methods, while preserving the molecular orbital description of the electronic structure, have also developed into quite reliable theoretical tools for the prediction of the geometries, vibrational frequencies, molecular energetics, and other properties for organometallic complexes (28).

The results we will report here were obtained using gradient-corrected DFT methods with appropriate relativistic corrections. It is our experience, and that of others, that this approach has several distinct advantages for studies of the electronic structure of actinide complexes:

- As has been demonstrated by numerous studies, the accuracy of properties calculated using the GGA DFT methods is, in most cases, comparable to or better than those from ab initio MP2 (Møller–Plesset second-order perturbation) or CISD (configuration interaction with single and double excitations) methods. In fact, the accuracy of the DFT results in some instances matches those obtained from the much more costly (but, in principle, more exact) CCSD(T) (coupled cluster singles and doubles with a perturbative inclusion of connected triple excitations) method (29) and the ab initio G1 procedure (30).
- The DFT methods achieve their good quantitative accuracy with a computational cost that is much lower than that of traditional ab initio correlated methods because of their better scaling with respect to the size of the atomic basis set (31). For example, one recent study compared the CPU time used in the calculation of the energy hypersurface of organotransition metal systems, by using gradient-corrected DFT, and various correlated ab initio methods. The DFT calculations were found to be 66, 78, and 83 times faster than the ab initio QCISD, CCSD, and QCISD(T) methods, respectively (32).
- Dynamical electron correlation effects are intrinsically included in DFT methods via the local and gradient-corrected correlation functionals. Most importantly, the nondynamical electron correlation effects have been partially accounted for via the exchange functionals (33,34). As a consequence, the multiconfiguration characters that often show up in traditional ab initio calculations using Hartree–Fock references could be slightly reduced in DFT methods. In other words, for approximating the "true" electronic structures of many-electron atoms and molecules

within the one-electron framework, DFT is a better approach than Hartree–Fock theory.

- Because of the inclusion of electron correlation within the one-electron formalism, DFT methods provide chemically meaningful one-electron Kohn–Sham orbitals, which facilitate the analysis of bonding and charge distributions within the usual MO description (35). Because of the inclusion of the effects of correlation in DFT, the energetic ordering of both the occupied and virtual Kohn–Sham orbitals is generally more reasonable than that of Hartree–Fock MOs, especially for metal-containing systems. In addition, the DFT eigenvalues seem to correlate well with experimental ionization energies and excitation energies. The Kohn–Sham orbitals can therefore be used to construct chemically intuitive MO energy-level diagrams for heavy-element systems (35–38).

- Because the convenience of the one-electron formalism is retained, DFT methods can easily take into account the scalar relativistic effects and spin-orbit effects, via either perturbation or variational methods. The retention of the one-electron picture provides a convenient means of analyzing the effects of relativity on specific orbitals of a molecule.

- Spin-unrestricted Hartree–Fock (UHF) calculations usually suffer from spin contamination, particularly in systems that have low-lying excited states (such as metal-containing systems). By contrast, in spin-unrestricted Kohn–Sham (UKS) DFT calculations the spin-contamination problem is generally less significant for many open-shell systems (39). For example, for transition metal methyl complexes, the deviation of the calculated UKS expectation values $\langle \hat{S}^2 \rangle$ ($\hat{S}$ = spin angular momentum operator) from the contamination-free theoretical values are all less than 5% (32).

- For most semiempirical MO methods or uncorrelated ab initio methods, the demand for basis sets is drastically increased, while the accuracy is usually reduced when going from light- to heavy-element systems. However, as we will show, the gradient-corrected DFT methods seem to have smaller basis set truncation error, so they are quite reliable for systems as heavy as those of actinides, and the results seem, at least qualitatively, to be less dependent on the choice of basis sets with high-angular-momentum components (40).

Because modern DFT methods possess the foregoing advantages relative to traditional correlated electronic structure formalisms, DFT has rapidly gained popularity as a tool in computational chemistry applications in recent years. The successive improvements in the exchange-correlation functionals, in concert with increased computational capabilities, has enabled us to calculate the geometries,

vibrational properties, excitation energies, and other chemical properties of organoactinide (as well as organolanthanide) complexes with an accuracy comparable to that achieved for organic and transition metal complexes.

4. COMPUTATIONAL METHODS

All the calculations were carried out using the Amsterdam Density Functional
(ADF) code, Version 2.3 (Theoretical Chemistry, Vrije Universiteit, Amsterdam,
The Netherlands), developed by Baerends et al. (41), which incorporates the relativistic extensions first proposed by Snijders et al. (42). The code was vectorized
by Ravenek (43) and parallelized by Fonseca Guerra et al. (44), and the numerical
integration scheme applied for the calculations was developed by te Velde et al.
(45). The ADF method utilizes Slater-type orbitals (STOs) for basis functions.

For closed-shell and open-shell molecules, spin-restricted Kohn–Sham
(RKS) and spin-unrestricted Kohn–Sham (UKS) density functional calculations
were employed, respectively. Except for the calculations of excited states and
the cases where pure states are sought, we have employed an approximation in
which electron density is "smeared" among the closely spaced orbitals near the
Fermi levels. In this procedure, fractional occupations are allowed for those frontier orbitals with energy difference within 0.01 hartree to avoid the violation of
the Aufbau principle (46).

Unless specified otherwise, all the calculations were carried out using two
approaches, i.e., the local density approach (LDA), where the Slater exchange
functional and Vosko–Wilk–Nusair local correlation functional (parameterization scheme V) were used (47), and the generalized-gradient approach that uses
gradient-corrected Perdew–Wang 91 (PW91) exchange-correlation functionals
(48). It is our experience that the PW91 functional provides consistently accurate
and reliable results on a variety of actinide complexes. For purpose of comparison, we have also employed the Hartree–Fock–Slater (HFS) (49), Becke–Perdew
1986 (BP86) (50), and Becke–Lee–Yang–Parr (BLYP) (51) functionals for some
of the systems discussed later.

In order to account for the relativistic effects in the atomic core orbitals,
the chemically much less important atomic core densities and the core potentials
for all the relevant atoms were generated using the relativistic atomic program
DIRAC (52). These relativistic atomic core densities were kept invariant during
the molecular calculations. This frozen-core approximation, proposed by
Baerends et al. (41a), greatly reduces the computational effort while incorporating
the important relativistic effects for the core electrons. In the present case, the
atomic inner orbitals, [$1s$] for C and [$1s$–$5d$] for An (An = Th–Am), were frozen
during molecular calculations. For fitting the molecular density and accurately
representing the Coulomb and exchange potentials in each SCF cycle, a set of
auxiliary s, p, d, f, and g type of STO functions centered on all nuclei were used

(53). These fitting functions were taken from the standard basis sets library of ADF. The use of these fitting functions has greatly increased the speed of the calculations.

Throughout this chapter, unless specified otherwise, STO valence basis sets of triple-ζ (TZ) quality were used for the actinides Th–Am (54), while valence basis sets used for C and H were of the quality of double-ζ, with $3d$-type or $2p$-type polarization functions (DZP) for C and H, respectively. In order to test the effects of different basis sets, we also used at several places the TZ basis sets with d- and f-type or p- and d-type polarization functions (TZ2P) for C and H. The exponents of the polarization functions were taken from the standard basis sets library of ADF, i.e., $\zeta(3d)$ = 2.20, $\zeta(4f)$ = 3.30 for C, and ζ $(2p)$ = 1.25, $\zeta(3d)$ = 2.50 for H.

Scalar relativistic (mass-velocity and Darwin) effects for the valence electrons were incorporated by using the quasi-relativistic method (55), where the first-order scalar relativistic Pauli Hamiltonian was diagonalized in the space of the nonrelativistic basis sets. The Pauli Hamiltonian used was of the form

$$\hat{H} = \hat{H}_0 + \hat{h}_{MS} + \hat{h}_{DW} + \hat{h}_{SO},$$

where $\hat{H}_0$ is the nonrelativistic one-electron Hamiltonian, $\hat{h}_{MS}$, $\hat{h}_{DW}$, and $\hat{h}_{SO}$ are the first-order relativistic terms, corresponding to mass-velocity, Darwin, and spin-orbit corrections, respectively (56).

Spin-orbit coupling effects were taken into account by including the spin-orbit terms in addition to the scalar relativistic terms in the preceding Pauli Hamiltonian. Because of the coupling of the orbital and spin angular momentum, L and S are no longer good quantum numbers. As a result, double-group symmetry has to be used. These double-group calculations were performed at geometries optimized with scalar relativistic effects included. Although geometry optimization and frequency calculations with inclusion of spin-orbit coupling effects are not readily available yet, from our experience spin-orbit coupling does not significantly affect the ground state geometries and vibrational frequencies of many actinide complexes (57), as will be shown later.

All the geometric structures were fully optimized under the proper symmetry point group via the analytical energy gradient technique implemented in the ADF code. The frequency calculations were performed at the geometries optimized with only scalar relativistic effects included. The frequencies and infrared (IR) absorption intensities were calculated based on numerical differentiation of the energy gradients in slightly displaced (0.01 Å) geometries. In order to account for first-order anharmonicity of the potential energy surface, two-sided displacements were employed in the numerical determination of the force constant matrices. The accuracy of a DFT calculation depends not only on error sources inherent in self-consistent-field calculations, but also on the number of grid points used in numerical integration of the functionals. Thus, we have used stringent criteria for the numerical integration accuracy (INTEGRATION = 8.0) in order to reduce the numerical noise. Further tightening of the numerical integration accuracy generally seems to have minimal effects on the calculated geometries and vibrational frequencies. The convergence thresholds were set as 10^{-8} for energy calculations during self-consistent-field iterations and as 10^{-4} Hartree/Å for the Cartesian energy gradients in geometry optimizations.

5. SELECTION OF THE "BEST" ENERGY FUNCTIONALS FOR ACTINIDE-CONTAINING SYSTEMS

The choice of which is the "best" exchange-correlation functional with gradient (or even higher-order derivative) corrections for any given system is an area of great current interest (58). Many pairings of exchange and correlation functionals are available, and the appropriateness of these to actinide complexes has received little systematic investigation. We have therefore chosen to compare the accuracy and performances of these different GGA methods as applied to organoactinide complexes prior to discussing their electronic structures. We have focused on some of the more popular exchange-correlation functionals that are programmed into the ADF code, namely, the widely used HFS, LDA, BLYP, BP86, and PW91 methods.

A good theoretical method should be able to reproduce appropriately chosen experimental data with sufficient accuracy that it can be used as a predictive tool. Because the molecular structures of many of the systems discussed here have been determined crystallographically, we have used the agreement between the theoretical optimized and experimentally determined geometric parameters as one of the fundamental criteria for the theoretical methods used. We have optimized the geometries of $Pa(Cot)_2$, $U(Cot)_2$, and $U(Cht)_2^-$ using the HFS, LDA, BLYP, BP86, and PW91 methods with the inclusion of scalar-relativistic effects. All of the calculations use the DZP basis sets for C and H, while the TZ2P basis sets were also used for one set of PW91 calculations to examine basis set effects. The optimized U–X (X = centroid of the C_n skeleton of the C_nH_n ring), U–C, C–C, and C–H distances (Å) and ∠HCX angles (°) are compared to the experimental data in Table 1. The experimental geometries of $U(Cot)_2$ and $U(Cht)_2^-$ are taken from the literature (59,9), while the "experimental" geometry of $Pa(Cot)_2$ is interpolated via a quadratic-function fitting of the experimental geometries of $Th(Cot)_2$, $U(Cot)_2$, and $Np(Cot)_2$, (59,60).

The data in Table 1 indicate that the An–X and An–C distances from the HFS and LDA are slightly too short, as is typical for heavy-element systems. The GGA methods lead to bond distances that are longer than those from the HFS and LDA functionals, as is usually the case. The bond lengths predicted by using the BLYP functional are generally longer than the experimental values. In contrast, the PW91 bond lengths are only slightly longer than the experimental

TABLE 1 Optimized DFT Bond Distances (Å) and Bond Angles (°) of Pa(Cot)₂, and U(Cot)₂, and U(Cht)₂⁻

	HFS	LDA	BLYP	BP86	PW91	PW91[b]	Exptl.
				DFT exchange-correlation functional			
Pa(Cot)₂							
Pa–X[a]	1.937	1.915	2.040	1.984	1.975	1.949	1.958 ± 0.010
Pa–C	2.667	2.650	2.755	2.711	2.702	2.683	2.671 ± 0.004
C–C	1.404	1.402	1.417	1.413	1.412	1.411	1.389 ± 0.013
C–H	1.100	1.097	1.092	1.093	1.092	1.088	(1.090)
∠XCH	174.2	174.0	175.0	174.6	174.3	174.2	
U(Cot)₂							
U–X[a]	1.903	1.880	2.013	1.954	1.943	1.928	1.924 ± 0.005
U–C	2.642	2.624	2.734	2.688	2.678	2.669	2.647 ± 0.004
C–C	1.403	1.401	1.416	1.413	1.411	1.413	1.392 ± 0.013
C–H	1.100	1.097	1.092	1.093	1.092	1.089	(1.090)
∠XCH	174.0	173.9	175.0	174.4	174.1	174.2	
U(Cht)₂⁻							
U–X[a]	1.983	1.970	2.063	2.021	2.009	2.002	1.98 ± 0.02
U–C	2.567	2.554	2.641	2.604	2.593	2.589	2.53 ± 0.02
C–C	1.414	1.411	1.429	1.424	1.422	1.425	1.37 ± 0.07
C–H	1.100	1.097	1.092	1.093	1.092	1.090	(1.084)
∠XCH	174.0	173.9	175.3	174.6	174.4	174.6	

[a] X is the centroid of the Cot or Cht rings.
[b] TZ2P basis sets were used for C and H atoms in this column.

values. When the size of the C and H basis sets is increased from DZP to TZ2P, the PW91 An–X and An–C distances are further reduced. At first glance, the optimized HFS and LDA distances for $U(Cht)_2^-$ seem to be in better agreement with the crystal structures than are the PW91 distances. However, the experimental U–C and C–C bond lengths, 1.98 ± 0.02 Å and 1.37 ± 0.07 Å, have large error bars due to significant thermal libration. By comparing the average experimental C–C distances of the C_7H_7 (1.408 Å) and C_8H_8 (1.397 Å) compounds (61), we believe that the actual C–C distance in the $U(Cht)_2^-$ anion, and thus the U–X and U–C ones, should lie near the upper end of the error bar. These ''adjusted'' experimental bond lengths are closer to the PW91 than to the HFS or LDA values. Thus, we believe that the PW91 exchange-correlation functional is the best choice for these organoactinide sandwich complexes, which is expected based on the theoretical analysis of this functional (62). We will show later that the PW91 An–X and An–C distances can be further slightly reduced by including spin-orbit effects, which brings the PW91 distances into even better agreement with the experimental values.

Our conclusion that the PW91 functional is one of the best choices for actinide complexes is also supported by the excellent agreement between the calculated PW91 and experimental vibrational frequencies and infrared intensities of these organoactinide sandwich complexes, as will be presented in the following several sections. In our studies of a variety series of different actinide complexes, ranging from inorganic UF_n ($n = 1$–6) (63) to organoactinide CAnO, OAnCCO, $(\eta^2\text{-}C_2)AnO_2$, $OAn(\eta^3\text{-}CCO)$, and $An(CO)_x$ (An = Th, U) complexes (64), the PW91 geometry optimizations and frequency calculations always lead to excellent agreement with the available experimental data, indicating the broad reliability and quality of this functional for actinide systems. Therefore, all the calculations reported in this chapter were performed using the PW91 functional.

For the remainder of this contribution, we will focus on specific results for the best-known organoactinide sandwich complexes, namely, the actinocene complexes $An(\eta^8\text{-}C_8H_8)_2$. Of organoactinide sandwich complexes, only the actinocenes are common enough to provide some comparative experimental data among different actinide elements. Other classes of organoactinide sandwich complexes are rare. In fact, only one example of an $An(\eta^7\text{-}C_7H_7)_2^-$ system has been well characterized, and no neutral $An(\eta^6\text{-}C_6H_6)_2$ have been characterized at the time of this writing (13,14).

6. ACTINOCENES $An(\eta^8\text{-}C_8H_8)_2$ (An = Th–Am)

Since the theoretical prediction (65) and the subsequent experimental discovery (8) of uranocene, $U(Cot)_2$, a series of actinocenes $An(Cot)_2$ (An = Th, Pa, U, Np, Pu, Am) or their anions have been synthesized and characterized (66). To date, theoretical calculations of the electronic structure of the actinocenes have been performed at various levels of sophistication, from semiempirical MO approaches to ab initio and DFT methods (4c,12,67). Early treatments of actinocene electronic structure included applications of ligand field theory (68) and extended Hückel theory (69). Some recent applications of approximate theory have included the topological approach of King (70) and the INDO approach of Cory et al. (71), although most recent theoretical investigations of actinocenes have used higher-level methodologies. Dolg and coworkers have performed large-scale ab initio calculations of $Th(Cot)_2$ and $U(Cot)_2$ using methods from SCF, MP2, CISD to MCSCF, MRCISD, ACPF (72). Pitzer and coworkers carried out the first spin-orbit CI (SOCI) calculations on the excited states of $U(Cot)_2$ and $Pa(Cot)_2$ (73). With respect to applications of DFT to actinocenes, the approaches have become increasingly sophisticated as computational resources have grown. Starting in the 1970s, Rösch et al. applied the nonrelativistic and quasi-relativistic Xα-SW method to $Th(Cot)_2$ and $U(Cot)_2$ (74). Baerends and coworkers later systematically optimized the geometries and discussed the electronic structures of $An(Cot)_2$ (An = Th–Pu) using the HFS-Xα approach with scalar relativistic and spin-orbit coupling effects included (75). Our group applied the fully relativistic DV-Xα method to the determination of the excitation energies of $Pa(Cot)_2$ (76).

More recently, we have fully optimized the geometry of $Pa(Cot)_2$ using a series of modern GGA DFT methods with scalar relativistic effects included (15). This paper discussed in detail the electronic structures, ionization energies, and electronic transitions of protactinocene, and also calculated and assigned the vibrational frequencies of $Pa(Cot)_2$ for the first time. Since our DFT studies of $Pa(Cot)_2$ agree rather well with the results of other methodologies and with the available experimental data, we will briefly summarize here the major results for $Pa(Cot)_2$ and extend our discussion to the electronic structures, geometries, and vibrational properties of the other actinocenes $An(Cot)_2$ (An = Th–Am).

6.1. Geometries of the Actinocenes

The theoretical determination of metal-ring distances in sandwich complexes such as ferrocene has been a notorious challenge to traditional ab initio methods (77). By contrast, DFT methods seem to provide calculated geometries of metal sandwich complexes that are generally in good accord with experimental structure determinations. We find this to be the case for actinide sandwich complexes as well, which we will illustrate with our PW91 DFT calculations on the early actinide Cot sandwich complexes $An(Cot)_2$ (An = Th–Am). These calculations indicate first that the energy of the eclipsed (D_{8h}) and staggered (D_{8d}) conformers are almost the same for all of the complexes, with the D_{8h} form generally slightly higher in energy. For example, the energy differences of the two conformers of

Pa(Cot)$_2$ and U(Cot)$_2$ are less than 0.037 and 0.044 kcal/mol, respectively, which is much less than the thermal translation energy (kT = 0.59 kcal/mol) at ambient temperature. Thus, the Cot rings of these actinocenes will essentially have a free rotation in gas phase. We therefore focus only on the D$_{8h}$ structures here.

Table 2 lists the An–X (X = centroid of Cot), An–C, C–C, C–H distances and the $\angle HCX$ of the D$_{8h}$ actinocenes obtained by using the scalar relativistic PW91 and LDA methods. These geometries are optimized for the high-spin electronic states with $(3e_{2g})^4 (3e_{2u})^4 f^n$ (n = 0, 1, 2, 3, 4, 5 for An = Th–Am) configurations.

The An–X distances from experimental crystallographic geometries are 2.004 Å for Th(Cot)$_2$, 1.924 Å for U(Cot)$_2$ (59), and 1.9088 Å for Np(Cot)$_2$ (60); i.e., they decrease smoothly as the atomic number of An increases. We have therefore fitted the experimental data of these three actinocenes to a parabolic function, which enables us to predict the "experimental" An–X distances of Pa(Cot)$_2$, Pu(Cot)$_2$, and Am(Cot)$_2$ as 1.956 Å, 1.911 Å, and 1.910 Å, respectively. Although this fitting is based on crystallographic data, solid-state effects are expected to be small for these neutral molecules, and we believe that these estimates should be close to the distances for the free molecules in the gas phase (78).

All the optimized PW91 An–X distances listed in Table 2 are about 0.02 Å (0.04 Å for Th(Cot)$_2$) too long as compared to the "experimental" distances. However, as was shown in Table 1, the PW91 An–X distances will decrease with increasing quality of the basis sets for the C and H atoms. For example, increasing the size of the C and H atomic basis sets from DZP to TZ2P decreases the Pa–X and U–X distances by 0.024 Å and 0.015 Å, respectively (Table 1), which brings the PW91 bond distances into excellent agreement with the experimental distances.

The trends in the calculated distances in the An(Cot)$_2$ molecules reflect the expected changes upon varying the actinide atom. As expected from the *actinide contraction* (79), the An–X and An–C distances all decrease from Th to Am. At the same time, the corresponding C–C distances decrease slightly, reflecting the fact that the An–Cot interactions become slightly weaker as one progresses from left to right in the actinide series (vide infra).

It is also interesting to note that the H atoms do not lie in the same plane as the C atoms in the C$_8$H$_8$ ligands. As shown in Table 2, the calculated $\angle XCH$ angles in the An(Cot)$_2$ deviate significantly from the planar value of 180°. In fact, a neutron-diffraction study of ferrocene has shown that the H atoms of the C$_5$H$_5$ rings are tilted toward the metal by 1.6° ± 0.4° (80). In our geometry optimizations, the H atoms in An(Cot)$_2$ molecules are found to be bent ca. 5°–6° in the same direction. This phenomenon, which is common for bis(cyclopentadienyl) sandwich compounds, has been rationalized by the hybridization caused by reorientation of the $p\pi$ orbitals toward the central metal (81) and the repulsion between the electron clouds of the C–H bonds and the π orbitals of the C$_n$H$_n$

TABLE 2 Optimized PW91 and LDA (in parentheses) Bond Distances (Å) and Bond Angles (°) of An(Cot)$_2$ (An = Th, Pa, U, Np, Pu, Am)

	An–X[a]	An–C	C–C	C–H	∠HCX
Th(Cot)$_2$	2.047 (2.001)	2.756 (2.713)	1.412 (1.402)	1.092 (1.096)	174.4 (175.0)
Pa(Cot)$_2$	1.974 (1.915)	2.702 (2.650)	1.412 (1.402)	1.092 (1.097)	174.4 (174.0)
U(Cot)$_2$	1.941 (1.880)	2.677 (2.624)	1.411 (1.401)	1.092 (1.097)	174.2 (173.9)
Np(Cot)$_2$	1.931 (1.860)	2.669 (2.609)	1.410 (1.400)	1.092 (1.097)	174.4 (173.9)
Pu(Cot)$_2$	1.932 (1.851)	2.670 (2.602)	1.410 (1.400)	1.092 (1.097)	174.7 (174.2)
Am(Cot)$_2$	1.927 (1.842)	2.666 (2.596)	1.409 (1.400)	1.092 (1.097)	175.2 (174.5)

[a] X is the centroid of the C$_8$ plane of the Cot rings.

ring (82). Based on these arguments, large rings will tend to tilt inward (toward the metal), while smaller rings will tilt outward (away from the metal) in order to achieve the best overlap with the central metals. Because of these effects, the assumption of planar C_nH_n rings is not appropriate; we have allowed the C_8H_8 rings to be nonplanar in the geometry optimization of these actinide sandwich complexes.

Although spin-orbit coupling effects are not included in these optimizations, the good agreement between the PW91 bond distances and the experimental values indicates that these effects are not important on the geometry and thus on the vibrational frequencies. In fact, inclusion of spin-orbit effects has only very small effects on the calculated geometries (83). It is our experience that, with respect to geometries, the scalar relativistic effects account for the major relativistic effects, while spin-orbit coupling effects are not essential. This conclusion is qualitatively supported by the fact that spin-orbit interactions of the radially contracted f-electrons have only minimal effects on the bond strengths and the geometric structures.

6.2. Electronic Structure of the Actinocenes

The qualitative energy-level diagram for $An(Cot)_2$ under the D_{8h} single group is shown in Figure 2, where the principal interactions of the An $5f$ and $6d$ orbitals with the π orbitals of the two Cot rings are also indicated. The eight C $2p\pi$ orbitals of a planar D_{8h} C_8H_8 ligand form eight π molecular orbitals, which, in order of increasing energy, are bases for the a_{2u}, e_{1g}, e_{2u}, e_{3g}, and b_{1u} representations. We will denote these π MOs as π_0, π_1, π_2, π_3, and π_4, respectively. In a D_{8h} $An(Cot)_2$ complex, these π MOs form gerade π_{ng} and ungerade π_{nu} ($n = 0$–4) ligand group orbitals of $(C_8H_8)_2$. In the absence of spin-orbit coupling, the $6d$ orbitals are split in a D_{8h} field as a_{1g} ($d\sigma$) $\ll$ e_{2g} ($d\delta$) $<$ e_{1g} ($d\pi$); the $d\delta$ and $d\pi$ orbitals are considerably destabilized by strong interaction with the ligand orbitals. Similarly, the An $5f$ orbitals are split as a_{2u} ($f\sigma$) $\sim$ e_{3u} ($f\phi$) $<$ e_{1u} ($f\pi$) $\ll$ e_{2u} ($f\delta$), where the strong destabilization of the $f\delta$ orbitals from the f-manifold is an indication of their strong interaction with the filled π_{2u} orbitals of the $(Cot)_2$.

The 16 $p\pi$-electrons from the Cot rings and 4 electrons from the An atom will fill all the MOs up to the e_{2g} and e_{2u} pairs (i.e., π_{2g} and π_{2u}) derived from the π_2 MOs of the Cot ring. As expected, the orbital interactions involving the radially diffuse An $6d$ orbitals are stronger than those involving the more contracted An $5f$ orbitals (16). As such, the a_{1g}, e_{1g}, and e_{2g} ligand group orbitals of $(Cot)_2$ interact more strongly than do the a_{2u}, e_{1u}, and e_{2u} ligand group orbitals. For example, in $Pa(Cot)_2$, the MOs derived from the $(Cot)_2$ a_{1g}, e_{1g}, and e_{2g} orbitals all lie lower energetically than their a_{2u}, e_{1u}, and e_{2u} counterparts. Our calculated overlap integrals for the $d\delta$–π_{2g} interaction are all about 0.35–0.33, while those for $f\delta$–π_{2u} decrease from 0.11 to 0.08 from $Th(Cot)_2$ to $Am(Cot)_2$,

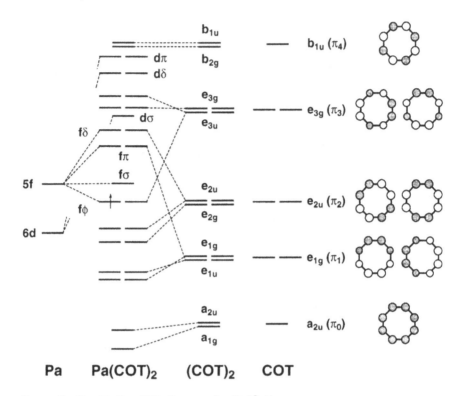

Pa   Pa(COT)₂   (COT)₂   COT

FIGURE 2   Qualitative MO diagram for Pa(Cot)₂.

in agreement with the foregoing conclusion. Because of their greater interaction, the $6d$-based MOs are destabilized relative to the $5f$-based orbitals. Therefore, all the "extra" metal-based electrons (from 0 for An = Th to 5 for An = Am) are localized in the $5f$-based MOs in the An(Cot)₂ (An = Th–Am) complexes.

The net charges and net spin densities obtained from Mulliken population analysis are listed in Table 3. The net charges of An are all ca. +2.9 except for Pa, for which the charge is +3.5. For all of the complexes, the charges on the C atoms are all about −0.2, and the H atoms are essentially neutral, consistent with the picture that the charge transfer occurs primarily between the $Cp\pi$ orbitals to the An orbitals, as expected. The net spin density increases from Th to Am, consistent with the increase in the number of metal-localized electrons in the series of actinocenes.

From Th to Am, both the increased radial contraction of the $5f$ orbitals and the increase of the $6d$ orbital energies from the early to late actinides (84,85)

TABLE 3 Mulliken Net Charges and Spin Densities for the An, C, and H Atoms in An(Cot)$_2$

	Charges			ρ_{spin}		
	An	C	H	An	C	H
Th(Cot)$_2$	2.881	−0.214	0.034	0.000	0.000	0.000
Pa(Cot)$_2$	3.519	−0.243	0.023	1.000	−0.001	0.001
U(Cot)$_2$	2.925	−0.214	0.031	2.197	−0.014	0.002
Np(Cot)$_2$	2.931	−0.215	0.032	3.458	−0.031	0.003
Pu(Cot)$_2$	2.899	−0.213	0.032	4.767	−0.052	0.004
Am(Cot)$_2$	2.909	−0.213	0.031	6.047	−0.071	0.005

help to decrease the $\pi_{2u} \rightarrow f\delta$ and $\pi_{2g} \rightarrow d\delta$ back-donations. As a result, the covalent bonding between An and the Cot rings is decreased from the early to the late actinide elements. The calculated overlap integrals between the actinide and the ligand group orbitals support this conclusion.

The predicted decrease from Th to Am in the covalent interaction between An and the Cot ligands should be reflected in the overall energy of interaction between the rings and the An atom. To test this notion, we have estimated the average An–C_8H_8 bond energies by calculating the change in total energy for the ligand-dissociation reaction

$$An(C_8H_8)_2 \rightarrow An + 2\ C_8H_8 \qquad \Delta E \approx 2\ \Delta E(An–C_8H_8)$$

where the UKS-optimized scalar relativistic energies of An(C_8H_8)$_2$ and C_8H_8 (planar $^3A_{2g}$) have been employed, while the average-of-configuration (AOC) UKS atomic energies calculated with spherical density were utilized for the An atoms. The resultant bond energies are depicted in Figure 3. Because of the complications associated with multiplets (86,87), spin-orbit effects have not been explicitly included in these calculations. As a result, the calculated bond energies are too high relative to the available experimental An–C_8H_8 bond energies: 98.0 and 82.9 kcal/mol for Th and U, respectively (88). Nevertheless, the difference in the calculated bond energies of Th–C_8H_8 and U–C_8H_8 (15.1 kcal/mol) is in excellent agreement with the measured difference (15.06 kcal/mol). Therefore the bond energies determined here should reflect accurately the trend for this series of actinocene complexes. The bond energies clearly show that the covalent interaction between An and Cot ring decreases as one proceeds to the later actinides, which is in agreement with the well-known fact that complexes of the late actinide elements are predominantly ionic in nature, similar to lanthanide complexes.

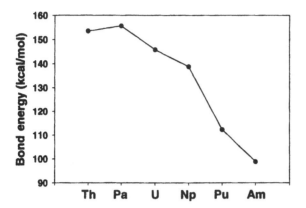

FIGURE 3   Calculated An–Cot bond energies for An(Cot)$_2$ (without multiplet and spin-orbit effects).

6.3. Spin-Orbit Effects and Excited States of Pa(Cot)$_2$

While spin-orbit effects do not significantly affect the geometries and vibrational properties of actinide compounds, these effects are very important for describing the excited-state energies of the complexes. We shall now discuss spin-orbit coupling in the actinocene complexes. Because of the large atomic numbers of the actinides, spin-orbit effects are as important as ligand-field effects in determining the electronic states of actinide compounds. When spin-orbit coupling is taken into account, all the doubly degenerate spatial MOs (e_{iu} and e_{ig}, $i = 1, 2, 3$) of An(Cot)$_2$, which are fourfold spin degenerate under the single group, are reduced to twofold degenerate spin-orbitals (spinors) because of the half-integer angular momentum ($s = 1/2$) (89). The correlation of the irreducible representations of the D$_{8h}$ single group and the D$_{8h}$* double group can be derived by taking the direct product of the single-group representation with the one-electron spin representation ($e_{1/2g}$). Table 4 lists the derived correlation of the symmetry species of the D$_{8h}$ single group and the D$_{8h}$* double group (90).

In order to show the quantitative splitting of these MOs by the spin-orbit effects, Figure 4 depicts the single-group and double-group energy levels calculated for Pa(Cot)$_2$. Although in a π-only picture (Fig. 2) the $f\phi$ orbitals are predicted to be the lowest among the f-manifold because of the interaction with Cot π_3 orbitals, our PW91 calculations reveal that the $5f$ orbitals in the D$_{8h}$ ligand field will be split as $f\sigma < f\phi < f\pi \ll f\delta$, as shown in Figure 4. The fact that the $f\sigma$ orbital becomes even lower in energy than the $f\phi$ orbitals indicates that the σ-type orbitals also play a role in these orbital interactions. The interactions between $f\delta$ and the e_{2u} ligand group orbitals are so strong that these orbitals are

TABLE 4 Correlation of the Symmetry Species of Single D_{8h} Group and Double $D_{8h}*$ Group

Γ_i (D_{8h})	An basis	(Cot)$_2$ basis	$\Gamma_i \times e_{1/2g}$ ($D_{8h}*$)
a_{1g}	s, $d\sigma$	π_{0g}	$e_{1/2g}$
a_{2g}			$e_{1/2g}$
b_{1g}		π_{4g}	$e_{7/2g}$
b_{2g}			$e_{7/2g}$
e_{1g}	$d\pi$	π_{1g}	$e_{1/2g} + e_{3/2g}$
e_{2g}	$d\delta$	π_{2g}	$e_{3/2g} + e_{5/2g}$
e_{3g}		π_{3g}	$e_{5/2g} + e_{7/2g}$
a_{1u}			$e_{1/2u}$
a_{2u}	$p\sigma$, $f\sigma$	π_{0u}	$e_{1/2u}$
b_{1u}			$e_{7/2u}$
b_{2u}		π_{4u}	$e_{7/2u}$
e_{1u}	$p\pi$, $f\pi$	π_{1u}	$e_{1/2u} + e_{3/2u}$
e_{2u}	$f\delta$	π_{2u}	$e_{3/2u} + e_{5/2u}$
e_{3u}	$f\phi$	π_{3u}	$e_{5/2u} + e_{7/2u}$

pushed even energetically close to the $d\sigma$ orbitals. As a result, in the absence of spin-orbit effects, the energies of the low-lying electronic states of Pa(Cot)$_2$ increase as $^2A_{2u}$ ($f\sigma$) < $^2E_{3u}$ ($f\phi$) < $^2E_{1u}$ ($f\pi$) < $^2A_{1g}$ ($d\sigma$) $\ll$ $^2E_{2u}$ ($f\delta$) in the single group. Thus, we predict that the $^2A_{2u}$ state, corresponding to the ($f\sigma$)^1 configuration, is the ground state of the complex. However, the near-degeneracy of the $f\phi$ and $f\sigma$ orbitals suggests that the ground state could be strongly dependent on the spin-orbit coupling effects as well.

When spin-orbit coupling is included, the ordering of the low-lying states is found to increase as $E_{5/2u}$ ($f\phi$) < $E_{1/2u}$ ($f\sigma + \pi$) $\ll$ $E_{3/2u}$ ($f\pi$) ~ $E_{7/2u}$ ($f\phi$) < $2E_{1/2u}$ ($f\pi + \sigma$) < $E_{1/2g}$ ($d\sigma$) $\ll$ $2E_{3/2u}$ ($f\delta$) < $2E_{5/2u}$ ($f\delta$) in the double-group representation. Therefore, the ground state is now identified as an $E_{5/2u}$ ($f\phi$) Kramers doublet. From the symmetry correlation listed in Table 4, the $E_{5/2u}$ ($f\phi$) double-group state can only come from the $^2E_{2u}$ and $^2E_{3u}$ single-group states. In Pa(Cot)$_2$, the lowest available $^2E_{2u}$ ($f\delta$) state lies too high in energy to contribute to the state mixing. Thus, the spin-orbit-coupled ground state corresponds to an almost pure ($f\phi$)^1 configuration (instead of the single-group ground state with an ($f\sigma$)^1 configuration), which is accidentally just the ground state expected from the simple π-only picture (Fig. 2). This example is typical, and it illustrates an important aspect of actinide quantum chemistry: the inclusion of spin-orbit coupling, while insignificant for the calculation of geometries, can change the nature of the ground state.

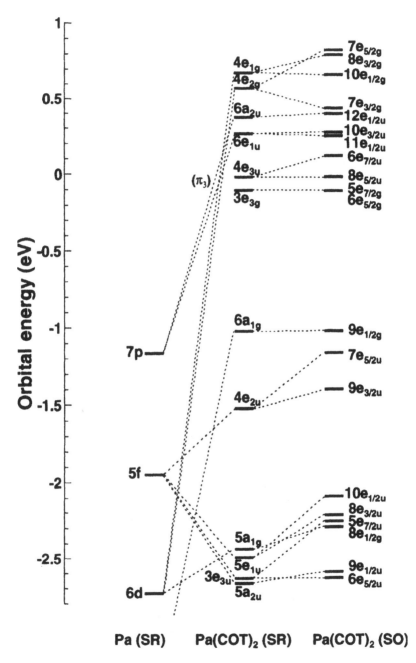

FIGURE 4 Single-group and double-group energy level correlation for Pa(Cot)₂.

Table 5 compares the relative energies of the low-lying states of Pa(Cot)$_2$ calculated using the PW91 method, with spin-orbit effects included, and those calculated from the ab initio spin-orbit CI method (91). The electronic transition energies labeled PW(TS) were determined by using Slater's transition-state method (92), whereas the other PW91 energies were calculated as the difference in total energy between excited states and the ground state. Because the SOCI calculation used a nonoptimized (Nopt) geometry, we also performed a PW91 calculation of these excited states using this assumed geometry in order to allow a direct comparison between the two methods at the same geometry.

Several features in Table 5 are notable. We see excellent agreement between the excitation energies calculated by using Slater's transition-state method and those from the state energy differences (within ± 0.08 eV). Further, when using the same geometry as used in the SOCI calculations, the PW91 method reproduces the SOCI excitation energies very well. The differences between the calculated energies by the two methods are all less than 0.08 eV, except for the d^1-state, where a larger difference (0.27 eV) exists. A similar difference in the energy of this state exists when the geometry is improved by increasing the C and H basis sets in the PW91 calculations from DZP to TZ2P. We therefore believe that the difference between the results of the two methods might be partly due to the smaller basis sets (DZ) used in the SOCI calculations; because of the diffuse nature of the $6d$ orbitals, it seems likely that a more extensive basis set is needed to describe them accurately.

It is remarkable that the PW91 DFT method can reproduce the results of the much more expensive SOCI method so well for the actinide excited states. Given the huge difference in the amount of computer time demanded by these two methods, the application of the DFT method to excited states of other actinide compounds with f^n ($n > 1$) configurations promises to be a challenging venture (because of the problem of state multiplets) but potentially a very fruitful one.

Based on the double-group PW91 calculations of Pa(Cot)$_2$, we have determined all the energies for the transitions from the $6e_{5/2u}$ level to the virtual levels up to $10e_{3/2g}$ (cf. Fig. 4). Because of the centrosymmetry of the molecule, allowed transitions must involve a parity change (93). In the D_{8h}^* double group, the excited state has to be one of the $E_{3/2g}$, $E_{5/2g}$, or $E_{7/2g}$ states for an electric-dipole-allowed transition from the $E_{5/2u}$ ground state. Based on this selection rule, only some f–d, f–π_3, and LMCT (ligand-to-metal charge transfer) transitions of types $E_{5/2u} \rightarrow E_{3/2g}$ (x, y polarization), $E_{5/2u} \rightarrow E_{5/2g}$ (z polarization), and $E_{5/2u} \rightarrow E_{7/2g}$ (x, y polarization) are allowed.

Among the dipole-allowed transitions, the lowest-energy f–π_3 transitions, $E_{5/2u} \rightarrow E_{5/2g}$ and $E_{5/2u} \rightarrow E_{7/2g}$, are both predicted to occur at 397 nm, while the lowest-energy f–d transition, $E_{5/2u} \rightarrow E_{3/2g}$, is predicted to occur at 27,200 cm^{-1}, or 368 nm. The latter is in near-perfect agreement with the experimental estimate of 365 nm by Streitwieser and coworkers (94). We find only two π_2–$d\sigma$ LMCT

TABLE 5 Comparison of Calculated PW91 and SOCI State Energies (eV) for Some Low-Lying States of Pa(Cot)$_2$[a]

State	Configuration	PW(TS)[b]	PW91	PW(TZ2P)	PW(Nopt)[c]	SOCI[d]	ΔE[e]
$E_{5/2u}$	$e_{3u}(f^1)$	0.000	0.000	0.000	0.000	0.000	0.000
$E_{1/2u}$	$a_{2u}+e_{1u}(f^1)$	0.041	0.049	0.131	0.101	0.166	0.065
$E_{3/2u}$	$e_{1u}(f^1)$	0.375	0.369	0.448	0.395	0.477	0.082
$E_{7/2u}$	$e_{3u}(f^1)$	0.378	0.379	0.357	0.362	0.362	0.000
$2E_{1/2u}$	$e_{1u}+a_{2u}(f^1)$	0.523	0.541	0.613	0.567	0.569	0.002
$E_{1/2g}$	$a_{1g}(d^1)$	0.598	0.685	0.680	0.651	0.925	0.274
$2E_{3/2u}$	$e_{2u}(f^1)$	1.141	1.122	1.206	1.227	1.222	−0.005
$2E_{5/2u}$	$e_{2u}(f^1)$	1.383	1.350	1.431	1.463	1.427	−0.036
$3E_{5/2u}$	$(e_{2u})^3 (e_{3u})^2 (\pi^3 f^1)$	2.956	2.896	—	2.873	—	—
$3E_{3/2u}$	$(e_{2u})^3 (e_{3u})^2 (\pi^3 f^2)$	2.982	2.924	—	2.901	—	—

[a] All energies are calculated by using the DZP basis sets for C and H, except for PW(TZ2P), where the TZ2P basis sets are used for C and H.

[b] The PW(TS) values are calculated by using Slater's transition-state method.

[c] The PW(Nopt) energies are calculated at the same geometry assumed in the SOCI calculations.

[d] The spin–orbit CI results are taken from Ref. 91.

[e] ΔE is the energy difference between SOCI and PW(Nopt).

transitions that should occur in the visible region (430–435 nm), corresponding to the transitions from $3e_{2u}$-based ligand orbitals ($5e_{5/2u}$ and $7e_{3/2u}$) to the Pa $d\sigma$ orbital ($8e_{1/2g}$). These two LMCT transitions at 430–435 nm are the likely origin for the low-energy shoulder reported to occur at 490 nm in the spectrum of $Pa(TMCot)_2$ (94). These transitions are in the violet portion of the visible spectrum and may thus be responsible for the characteristic golden-yellow color of protactinocene.

As discussed earlier, $Pa(Cot)_2$ has an $E_{5/2u}$ ($f\phi$), i.e., $|M_J| = 5/2$, ground state, as a consequence of substantial contributions of both spin-orbit and ligand-field effects. The ground magnetic properties of protactinocene will be dominated by the characteristics of the $E_{5/2u}$ ground state. Based on the magnetic dipole transition selection rule ($\Delta M_J = \pm 1$) (95), the $E_{5/2u}$ ground state is expected to be ESR silent: When the M_J components of the $E_{5/2u}$ state are split in a magnetic field, the predicted ESR transition from $M_J = -5/2$ to $M_J = +5/2$ is not allowed.

Besides the ground state, the $E_{1/2u}$ and $E_{3/2u}$ excited states are possible contributors to the observed magnetic moment of $Pa(Cot)_2$. Using our calculated PW91 state energies and Warren's formalism for an f^1 axial system (96), we have determined the anisotropic room-temperature magnetic moments, μ_z and $\mu_{x,y}$, as 3.33 and 0.99 BM, respectively, which will give rise to an average magnetic moment $\langle \mu \rangle = 2.09$ BM. This calculated room-temperature magnetic moment is fairly close to the value of 1.96 BM that was obtained via the spin-orbit CI calculations on $Pa(Cot)_2$ (91). Both values are close to the experimental room-temperature value $\langle \mu \rangle = 1.88$ BM for the $4f^1$ sandwich complex $Ce(Cot)_2^-$ (97). These calculated excitation energies and magnetic properties show that the gradient-corrected DFT methods, especially PW91, can be used as a reliable theoretical method for large actinide molecules, for which a spin-orbit CI calculation is too expensive to be carried out.

6.4. Vibrational Analysis of Actinocenes

Not surprisingly, vibrational spectra have proven to be an invaluable tool for experimental chemists in the characterization of transition metal and actinide sandwich compounds (98). Most known actinocenes have been characterized early on by vibrational spectroscopy (99). The IR and Raman spectra of thorocene and the IR spectra of protactinocene and uranocene were reported in the 1970s (100,101). However, normal coordinate analysis of these vibrational spectra is difficult because of the large number of vibrational modes involved. So far only a tentative assignment of the vibrational spectra of thorocene and uranocene, based on a qualitative group theory analysis, has been advanced (102).

To date, the lack of appropriate theoretical computational methods has hampered a comprehensive first-principle analysis of the detailed vibrational properties of large molecules such as the actinocenes. Because the theoretical calculation of vibrational properties requires full geometry optimizations to a high degree of accuracy, the PW91 DFT method is anticipated to be an appropriate theoretical tool, especially for metal-containing systems. We have recently shown that this method can be used with good success to calculate the vibrational modes of $Pa(Cot)_2$ (15). Here we will extend these studies to the calculations and assignments of the vibrational spectra to the other $An(Cot)_2$ ($An = Th-Am$) systems, with comparisons to the available experimental data. The present work represents the first comprehensive theoretical study of the vibrational spectra and assignments of the actinocenes.

Based on standard group theoretical analysis (103), among the 93 vibrational modes of actinocenes with D_{8h} symmetry, only the 4 A_{2u} and 6 E_{1u} modes are IR active, whereas 4 A_{1g}, 5 E_{1g}, and 6 E_{2g} modes are Raman active. Table 6 lists the calculated LDA and PW91 frequencies, infrared intensities, and assignments for the vibrational modes of $Th(Cot)_2$. The vibrational frequencies and IR intensities for the other actinocenes are listed in Table 7.

The simple linear-three-mass model for XYX systems (104) can be applied to the $An(Cot)_2$ sandwich complexes when treating the Cot rings as rigid mass points. Using this model, three vibrational modes involving An–Cot interactions are apparent: (1) the symmetric ring–metal–ring stretching mode (A_{1g}), (2) the asymmetric ring–metal–ring stretching mode (A_{2u}), and (3) the ring–metal–ring bending mode (E_{1u}). These three modes represent the lowest-frequency Raman and IR vibrations of actinocenes. As expected, the symmetric ring–metal–ring stretching causes a large change in the molecular polarizability, leading to a strong Raman absorption, observed at 225 cm^{-1} for $Th(Cot)_2$. The asymmetric ring–metal–ring stretching mode of $Th(Cot)_2$ is observed as a strong IR absorption at 250 cm^{-1}. For the ring–metal–ring bending mode, which has not been observed experimentally for any actinocene, the potential energy surface is so flat that the calculated frequency of this mode becomes imaginary even when a very stringent numerical integration accuracy (INTEGRATION = 8.0) is employed. By increasing the accuracy of the numerical integration even more (to INTEGRATION = 10.0), we have obtained this frequency as 57 cm^{-1} for $Th(Cot)_2$. Based on the experimental estimation, this band should appear around 125 cm^{-1} (102).

The C_8H_8 rings are, of course, neither structureless point masses nor rigid rings. In addition to the modes involving the metal–ring interactions (for which the rings are largely rigid), there are numerous modes corresponding to C–C and C–H stretching and in-plane ($\parallel$) and out-of-plane ($\perp$) C–H bending. We will not detail the descriptions of these modes but only summarize the results in Table 6. We will examine more closely the ring-centered modes that lead to experimentally observed IR and Raman bands with strong intensities.

In addition to the previously mentioned strong absorption at 250 cm^{-1}, the IR spectrum of $Th(Cot)_2$ exhibits strong absorptions at 695, 742, and 895 cm^{-1}.

TABLE 6 Calculated LDA and PW91 DFT Vibrational Frequencies (cm^{-1}) and Absolute IR Intensities (km/mol, in parenthesis) for the IR and Raman Active Modes of Th(Cot)$_2$

Mode	Exptl.[a]	LDA	PW91	Assignment[b]
A_{2u}	250 (s)	246 (72.7)	220 (76.9)	Asymmetric ring–metal–ring stretching
	695 (vs)	666 (457)	633 (446)	Asymmetric C–H bending ($\perp$)
	742 (s)	762 (17.2)	747 (0.7)	Asymmetric in-plane C–C stretching
	3005 (m)	3059 (0.4)	3079 (0.1)	Asymmetric C–H stretching
E_{1u}	(125)	8 (7.7)	57 (8.9)	Asymmetric ring–metal–ring bending
		225 (0.0)	235 (0.2)	Asymmetric in-plane C–C stretching
		743 (35.1)	745 (23.2)	Asymmetric C–H bending ($\perp$)
	895 (s)	915 (69.4)	895 (84.1)	Asymmetric ring–metal–ring tilting
	1315 (m)	1406 (1.5)	1430 (0.2)	Asymmetric C–H bending ($\parallel$)
	2920 (m)	3052 (0.1)	3072 (8.3)	Asymmetric C–H stretching
A_{1g}	225 (s)	227	138	Symmetric ring–metal–ring stretching
		680	683	Symmetric C–H bending ($\perp$)
		770	809	Symmetric in-plane C–C stretching
	3045 (m)	3059	3058	Symmetric C–H stretching
E_{1g}	242 (s)	265	260	Symmetric ring–metal–ring tilting
		768	762	Symmetric C–H bending ($\perp$)
		911	898	Symmetric C–C stretching
		1405	1426	Symmetric C–H bending ($\parallel$)
	3022 (m)	3052	3068	Symmetric C–H stretching
E_{2g}		237	235	Symmetric CCC bending ($\perp$)
	391 (m)	388	365	Symmetric in-plane C–C stretching
	775 (s)	817	819	Symmetric C–H bending ($\perp$)
		1187	1177	Symmetric C–H bending ($\parallel$)
		1498	1500	Symmetric in-plane C–C stretching
	2905 (m)	3038	3066	Symmetric C–H stretching

[a] The experimental intensities are shown as very strong (vs), strong (s), and medium (m) based on Ref. 100.
[b] The $\perp$ and $\parallel$ are referred to the Cot plane.

TABLE 7 PW91 Vibrational Frequencies (cm⁻¹) and Absorption Intensities
(km/mol) for IR and Raman Active Modes of An(Cot)₂ (An = Pa–Am)

Mode	Pa(Cot)₂	U(Cot)₂	Np(Cot)₂	Pu(Cot)₂	Am(Cot)₂
A_{2u}	239	232	216	203	195
	683	735	741	734	722
	771	751	748	741	746
	3065	3072	3073	3076	3072
E_{1u}	38	36	45	67	72
	220	236	235	235	235
	761	760	759	757	755
	895	908	910	907	909
	1438	1438	1438	1438	1439
	3064	3066	3066	3070	3064
A_{1g}	328	385	252	192	87
	530	586	700	705	699
	839	756	716	708	765
	3090	3133	3125	3125	3072
E_{1g}	253	233	221	221	216
	787	783	780	778	772
	902	902	903	901	903
	1427	1430	1432	1433	1435
	3063	3063	3065	3067	3065
E_{2g}	219	143	235	250	253
	410	404	380	376	371
	823	829	828	831	824
	1161	1171	1174	1173	1176
	1513	1516	1518	1516	1516
	3046	3040	3041	3045	3051

All of these vibrational frequencies are reproduced with reasonable accuracy in the DFT calculations. The 695-cm⁻¹ band is the strongest in the spectrum. The calculated frequencies, 666 cm⁻¹ at the LDA level and 633 cm⁻¹ at the PW91 level, are only in fair agreement with the experimental value. However, the mode is calculated to have the strongest absolute intensity, which is in complete accord with the experimental observation. We are therefore confident in assigning this mode as asymmetric C–H bending, which differs from a previous assignment of it as the asymmetric ring–metal–ring tilting mode (102). The 742-cm⁻¹ and 895-cm⁻¹ modes are reproduced with remarkable accuracy at the PW91 level (Table 6). These bands are due to C–C stretching and asymmetric ring–metal–ring tilting, respectively.

In addition to the strong band at 225 cm^{-1}, the Raman spectrum of $Th(Cot)_2$ shows two other strong absorptions at 242 and 775 cm^{-1}, which are assigned as symmetric ring–metal–ring tilting and a C–C bending. These vibrational frequencies of these bands have been reproduced with reasonable accuracy at both the LDA and PW91 levels of calculation.

Table 6 demonstrates that the PW91 frequencies and IR intensities are overall in fair agreement with the experimental spectra. These theoretical results can be further improved by increasing the quality of the optimized geometry via the use of larger basis sets. By combining the calculated IR intensities and the frequencies, the PW91 theoretical calculations can provide reliable predictions of the positions and absorbency of the IR transitions. Therefore, theoretical calculations of IR spectra can be very useful in helping experimentalists to identify the IR bands and their microscopic origins, even for molecules as large as the actinocenes.

Because the geometric structures and bonding are very similar for the entire series of actinocenes, we might expect their vibrational frequencies to be very similar as well. Table 7 lists the PW91 calculated frequencies for the $An(Cot)_2$ complexes, with An = Pa–Am. We see a smooth trend in the vibrational frequencies across the series, with similar values to those discussed earlier for $Th(Cot)_2$. The ability to calculate the predicted changes in the vibrational frequencies across a series of homologous complexes is certainly a strength of the application of DFT to vibrational problems.

From Tables 6 and 7, we see that the lowest A_{1g} vibration (i.e., the symmetric Cot–An–Cot stretch) and the lowest A_{2u} vibration (the asymmetric Cot–An–Cot stretch) show nearly the same trend across the actinide series: the frequencies increase slightly from Th to Pa and then decrease from U to Am. This trend is entirely consistent with the trends in the group orbital overlap integrals mentioned earlier and the $An–C_8H_8$ bond energies. Because the frequencies of fully symmetric vibrations are rigorously independent of the mass of An, the changes of the vibrational frequencies of the A_{1g} mode from Th to Am directly reflects the An–Cot bonding strength. For the asymmetric Cot–An–Cot stretching mode, the frequency changes from Th to Am agree well with the increase of the atomic mass from Th to Am and the decrease of the symmetric stretching frequencies.

In conclusion, the calculated PW91 vibrational frequencies are in reasonably good agreement with the experimental data. As we discussed earlier, augmentation of the basis sets can improve the molecular geometries of the actinocenes. We can therefore expect that better agreement with experiments can be achieved if the vibrational frequencies are calculated with larger basis sets. The assignments of the IR and Raman vibrational modes will help us to understand the vibrational spectra of other actinocenes and their microscopic origins. Our calculations in this work indicate that gradient-corrected density functional methods, especially the PW91 functional, not only can be used to elucidate the bonding and electronic structure, but can also be of great value in interpreting and predicting the vibrational properties of actinide complexes. Because the symmetric vibrational frequencies are not sensitive to the mass of the central metal, the present vibrational analysis and mode assignments will also be useful in understanding the vibrational spectra of lanthanocenes (105).

7. CONCLUSIONS

Relativistic density functional theory, especially with the inclusion of nonlocal exchange and correlation corrections, has become a powerful predictive tool in actinide chemistry. The methodology is sufficiently efficient to allow experimentally important properties, such as the geometry, vibrational frequencies, and infrared absorption intensities, to be calculated even for large organoactinide systems such as those discussed here. Inasmuch as many aspects of actinide chemistry are experimentally challenging because of the difficulty in handling of the elements, reliable theoretical calculations provide a valuable adjunct to experimental studies.

State-of-the-art DFT methods can provide theoretical interpretations of experimental results and are becoming more and more reliable in predicting physicochemical properties of actinide compounds. In spite of the current shortcomings of the method, new developments and future advances in density functional theory, such as the hybrid exchange-correlation functionals (106), meta-GGA functionals with high-order gradient corrections and kinetic energy density included (107), and time-dependent DFT (108), promise to provide even greater utility with respect to the study of ground-state and excited-state properties of actinide and organoactinide complexes (109).

ACKNOWLEDGMENTS

We gratefully acknowledge support for this research from the Division of Chemical Sciences, U.S. Department of Energy (Grant DE-FG02-86ER13529 to BEB), from Los Alamos National Laboratory, and from the Ohio Supercomputer Center and the Environmental Molecular Sciences Laboratory at Pacific Northwest National Laboratory for grants of computer time.

ABBREVIATIONS

An	actinide element
BLYP	Becke 1988–Lee-Yang-Parr
BP86	Becke 1988–Perdew 1986
Bz	benzene, C_6H_6

Cht	cycloheptatrienyl, C_7H_7
Cot	cyclooctatetraene, C_8H_8
Cp	cyclopentadienyl, C_5H_5
Cp*	pentamethylcyclopentadienyl, C_5Me_5
DFT	density functional theory
DZP	double-ζ basis set with polarization function
GGA	generalized gradient approach
HFS	Hartree–Fock–Slater
LDA	local density approach
NR	nonrelativistic
PW91	Perdew–Wang functional (1991)
RKS	spin-restricted Kohn–Sham
SO	spin-orbit
SOCI	spin-orbit configuration interaction
SR	scalar relativistic
TBB	1,3,5-tri-*tert*-butylbenzene, C_6H_3-$^tBu_3$
TZ2P	triple-ζ basis set with two polarization functions
UKS	spin-unrestricted Kohn–Sham

NOTES AND REFERENCES

1. (a) TJ Kealy, PL Pauson. Nature 168:1039–1040, 1951. (b) SA Miller, JA Tebboth, JF Tremaine. J Chem Soc 632–635, 1952. (c) EO Fischer, W Phab. Z Naturforsch 76:377–379, 1952. (d) G Wilkinson, M Rosenblum, MC Whiting, RB Woodward. J Am Chem Soc 74:2125–2126, 1952.
2. RH Crabtree. The Organometallic Chemistry of the Transition Metals. 2nd ed. New York: Wiley, 1994.
3. For cyclopolyene and cyclopolyenyl complexes of transition metals, see: (a) G Deganello. Transition Metal Complexes of Cyclic Polyolefins. London: Academic Press, 1979, Chap 1. (b) MLH Green, DKP Ng. Chem Rev 95:439–473, 1995.
4. For recent studies of f-element C_6H_6, C_7H_7, and C_8H_8 sandwich complexes, see, for example: (a) WA King, S Di Bella, G Lanza, K Khan, DJ Duncalf, FGN Cloke, IL Fragala, TJ Marks. J Am Chem Soc 118:627–635, 1996. (b) D Gourier, D Caurant, T Arliguie, Ephritikhine. J Am Chem Soc 120:6084–6092, 1998. (c) M Dolg, P Fulde. Eur J Chem 4:200–204, 1998.
5. (a) H Hoberg, R Krause-Göing, R Mynott. Angew Chem Int Ed Engl 17:123–124, 1978. (b) H Hoberg, C Fröhlich. J Organomet Chem 168:C52–C52, 1979.
6. H Hoberg, C Fröhlich. J Organomet Chem 197:105–109, 1980.
7. EO Fischer, W Hafner. Z Naturforsch B10:665–667, 1955.
8. U Müller-Westerhoff, A Streitwieser Jr. J Am Chem Soc 90:7364–7364, 1968.
9. T Arliguie, M Lance, M Nierlich, J Vigner, M Ephritikhine. J Chem Soc Chem Commun 183–184, 1995.
10. (a) Z Liang, AG Marshall, A Pires de Matos, JD Spirlet. In: LR Morss, J Fuger, eds. Transuranium Elements: A Half Century. Washington, DC: American Chemical

Society, 1992, pp 247–250. (b) WW Yin, AG Marshall, J Marcalo, A Pires de Matos. J Am Chem Soc 116:8666–8672, 1994. (c) J Marcalo, JP Leal, A Pires de Matos, AG Marshall. Organometallics 16:4581–4588, 1997.

11. A large number of review papers and books about ferrocenes have been published. See, for example: (a) A Togni, T Hayashi, eds. Ferrocenes. Weinheim, Germany: VCH, 1995. (b) WP McNutt, CU Pittman Jr. Ferrocene and Ferrocene Derivatives. Redstone Arsenal, Redstone Scientific Information Center, 1966.

12. (a) M Pepper, BE Bursten. Chem Rev 91:719–741, 1991. (b) G Schreckenbach, PJ Hay, RL Martin. J Comp Chem 20:70–90, 1999.

13. J Li, BE Bursten. J Am Chem Soc 121:10243–10244, 1999.

14. J Li, BE Bursten. J Am Chem Soc 119:9021–9032, 1997.

15. J Li, Be Bursten. J Am Chem Soc 120:11456–11466, 1998.

16. A numerical relativistic Dirac–Fock calculation for the uranium atom indicates that the spin-orbit averaged radii of the maximum radial density are 0.72 Å, 0.86 Å, 0.56 Å, and 1.30 Å, respectively, for $6s$, $6p$, $5f$, and $6d$ orbitals. See: JP Desclaux. Atomic Data and Nuclear Data Tables 12:311–406, 1973.

17. PJ Hay, WR Wadt, LR Kahn, RC Raffenetti, DH Phillips. J Chem Phys 71:1767–1779, 1979.

18. For a general discussion of dynamic and nondynamic electron correlation, see, for example: (a) DKW Mok, R Neumann, NC Handy. J Phys Chem 100:6225–6230, 1996. (b) MA Buijse, EJ Baerends. In: ED Ellis, ed. Density Functional Theory of Molecules, Clusters, and Solids. Dordrecht, The Netherlands: Kluwer Academic, 1995, pp 1–46.

19. For qualitative and quantitative discussions of scalar and spin-orbit relativistic effects, see, for example: (a) KS Pitzer. Acc Chem Res 12:271–275, 1979. (b) P Pyykkö, JP Desclaux. Acc Chem Res 12:276–281, 1979. (c) P Pyykkö. Chem Rev 88:563–594, 1988. (d) WHE Schwarz, EM van Wezenbeek, EJ Baerends, JG Snijders. J Phys B22:1515–1530, 1989.

20. J Li, BE Bursten. Unpublished results.

21. CJ Burns, BE Bursten. Comments Inorg Chem 9:61–69, 1989.

22. ER Davidson. Chem Rev 100:351–352, 2000, and references therein.

23. BE Bursten, RJ Strittmatter. Angew Chem Int Ed Engl 30:1069–1085, 1991.

24. WF Schneider, RJ Strittmatter, BE Bursten, DE Ellis. In: JK Labanowski, JW Andzelm, eds. Density Functional Methods in Chemistry. New York: Springer-Verlag, 1991, pp 247–260.

25. (a) RG Parr, W Yang. Density Functional Theory of Atoms and Molecules. New York: Oxford University Press, 1989. (b) RM Dreizler, EKU Gross. Density Functional Theory. Berlin: Springer-Verlag, 1990. (c) ES Kryachko, EV Ludena. Energy Density Functional Theory of Many-Electron Systems. Dordrecht, The Netherlands Kluwer, 1990. (d) JK Labanowski, JW Andzelm, eds. Density Functional Methods in Chemistry. New York: Springer-Verlag, 1991.

26. (a) EKU Gross, RM Dreizler. Density Functional Theory. New York: Plenum Press, 1995. (b) JM Seminario, P Politzer, eds. Modern Density Functional Theory: A Tool for Chemistry. Amsterdam: Elsevier, 1995. (c) RF Nalewajski, ed. Density Functional Theory. In: Topics in Current Chemistry. Vols 180–183. Berlin:

Springer-Verlag, 1996. (d) BB Laird, RB Ross, T Ziegler. Chemical Applications of Density-Functional Theory. In: ACS Symposium Series, Vol 629. Washington, DC: American Chemical Society, 1996. (e) JF Dobson, G Vignale, MP Das, eds. Electronic Density Functional Theory: Recent Progress and New Directions. New York: Plenum Press, 1998.

27. (a) T Ziegler. Chem Rev 91:651–667, 1991. (b) SB Trickey. In: ES Kryachko, JL Calais, eds. Conceptual Trends in Quantum Chemistry. Dordrecht, The Netherlands: Kluwer Academic, 1994, pp. 87–100. (c) RG Parr, W Yang. Annu Rev Phys Chem 46:701–728, 1995. (d) W Kohn, AD Becke, RG Parr. J Phys Chem 100: 12974–12980, 1996. (e) A Nagy. Phys Reports 298:1–79, 1998.

28. (a) J Li, G Schreckenbach, T Ziegler. J Am Chem Soc 117:486–494, 1995. (b) L Fan, T Ziegler. In: DE Ellis, ed. Density Functional Theory of Molecules, Clusters, and Solids. Dordrecht, The Netherlands: Kluwer Academic, 1995. (c) TV Russo, RL Martin, PJ Hay. J Chem Phys 102:8023–8028, 1995. (d) AP Scott, L Radom. J Phys Chem 100:16502–16513, 1996.

29. For discussions of the performance of DFT methods, see, for example: (a) BG Johnson, PMW Gill, JA Pople. J Chem Phys 98:5612–5626, 1993. (b) B Delley, M Wrinn, HP Lüthi. J Chem Phys 100:5785–5791, 1994. (c) SG Wang, DK Pan, WHE Schwarz. J Chem Phys 102:9296–9308, 1995. (d) CW Bauschlicher Jr. Chem Phys Lett 246:40–44, 1995. (e) M Kaupp, OL Malkina, VG Malkin. J Chem Phys 106:9201–9212, 1997. (f) M Kaupp. Eur J Chem 4:2059–2071, 1998. (g) ON Ventura, M Kieninger, RE Cachau. J Phys Chem A 103:147–151, 1999.

30. For lighter molecules, see: AD Becke. J Chem Phys 96:2155–2160, 1992. Good accuracy can also be reached for heavier systems, such as transition metals and lanthanides. See, for example: (a) A Ricca, CW Bauschlicher Jr. Theor Chim Acta 92:123–131, 1995. (b) LA Eriksson, LGM Pettersson, PEM Siegbahn, U Wahlgren. J Chem Phys 102:872–878, 1995. (c) SG Wang, WHE Schwarz. J Phys Chem 99: 11687–11695, 1995.

31. BG Johnson, PMW Gill, JA Pople. J Chem Phys 97:7846–7848, 1992.

32. MC Holthausen, C Heinemann, HH Cornehl, W Koch, H Schwarz. J Chem Phys 102:4931–4940, 1995.

33. PRT Schipper, OV Gritsenko, EJ Baerends. J Phys Chem 111:4056–4067, 1999.

34. AD Becke. In: DR Yarkony, ed. Modern Electronic Structure Theory. Singapore: World Scientific, 1995, pp 1022–1046.

35. For the meaning of KS orbitals and eigenvalues, see: (a) EJ Baerends, OV Gritsenko. J Phys Chem A 101:5383–5403, 1997. (b) R Stowasser, R Hoffmann. J Am Chem Soc 121:3414–3420, 1999.

36. We have found a good linear correlation between the calculated ionization energies and the orbital energies of $Pa(Cot)_2$, $IE_i = 2.106 - 0.990\ \epsilon_i$; see Ref. (15).

37. It has been discovered that excitations from the highest occupied orbital are in agreement with differences of Kohn–Sham eigenvalues. See: CJ Umrigar, A Savin, X Gonze. In: JF Dobson, G Vignale, MP Das, eds. Electronic Density Functional Theory: Recent Progress and New Directions. New York: Plenum Press, 1998, pp 167–176.

38. P Politzer, F Abu-Awwad. Theor Chem Acc 99:83–87, 1998.

39. See, for example: CJ Cramer, FJ Dulles, DJ Giesen, J Almlöf. Chem Phys Lett 245:165–170, 1995, and references cited therein.
40. For examinations of basis set effects in DFT calculations, see: (a) KS Raymond, RA Wheeler. J Comp Chem 20:207–216, 1999. (b) AC Scheiner, J Baker, JW Andzelm. J Comp Chem 18:776–795, 1997.
41. (a) EJ Baerends, DE Ellis, P Ros. Chem Phys 2:41–51, 1973. (b) EJ Baerends, P Ros. Chem Phys 2:52–59, 1973. (c) EJ Baerends, P Ros. Int J Quantum Chem Quantum Chem Symp 12:169–190, 1978.
42. (a) JG Snijders, EJ Baerends. J Mol Phys 36:1789–1804, 1978. (b) JG Snijders, EJ Baerends, P Ros. Mol Phys 38:1909–1929, 1979.
43. W Ravenek. In: HJJ te Riele, ThJ DeDekker, HA van de Vorst, eds. Algorithms and Applications on Vector and Parallel Computers. Amsterdam: Elsevier, 1987.
44. C Fonseca Guerra, O Visser, JG Snijders, G te Velde, E J Baerends. In: E Clementi, G Corongiu, eds. Methods and Techniques for Computational Chemistry. Cagliari, Italy: STEF, 1995, p 305.
45. (a) PM Boerrigter, G te Velde, E J Baerends. Int J Quantum Chem 33:87–113, 1988. (b) G te Velde, EJ Baerends. J Comput Phys 99:84–98, 1992.
46. For discussions of the fractional occupation number approach, see: (a) SG Wang, WHE Schwarz. J Chem Phys 105:4641–4648, 1996. (b) FW Averill, GS Painter. Phys Rev B46:2498–2502, 1992. (c) BI Dunlap. In: KP Lawlay, ed. Ab initio Methods in Quantum Chemistry II. New York: Wiley, 1987. (d) JF Janak Phys Rev B18:7165–7168, 1978. (e) JC Slater, JB Mann, TM Wilson, JH Wood. Phys Rev 184:672–694, 1969.
47. (a) JC Slater. Quantum Theory of Molecular and Solids. Vol 4. New York: McGraw-Hill, 1974. (b) SH Vosko, L Wilk, M Nusair. Can J Phys 58:1200–1211, 1980.
48. (a) JP Perdew, Y Wang. Phys Rev B45:13244–13249, 1992. (b) JP Perdew, JA Chevary, SH Vosko, KA Jackson, MR Pederson, DJ Singh, C Foilhais. Phys Rev B46:6671–6687, 1992.
49. The HFS functional uses Slater's $X\alpha(\alpha = 0.7)$ pure-exchange electron gas formulation with no correlation functional.
50. BP86 exchange functional: AD Becke. Phys Rev A38:3098–3100, 1988. BP86 correlation functional: (a) JP Perdew. Phys Rev B33:8822–8824, 1986. (b) JP Perdew. Phys Rev B34:7406–7406 (erratum), 1986.
51. BLYP correlation functional: C Lee, W Yang, RG Parr. Phys Rev B37:785–789, 1988.
52. The fully relativistic Dirac–Slater atomic code, DIRAC, is one of the auxiliary programs distributed with the ADF code.
53. J Krijn, EJ Baerends. Fit Functions in the HFS Method (Internal Report). Free University of Amsterdam, 1984.
54. (a) JG Snijders, EJ Baerends, P Vernooijs. At Nucl Data Tables 26:483–509, 1981. (b) P Vernooijs, GJ Snijders, EJ Baerends. Slater-Type Basis Functions for the Whole Periodic System (Internal Report). Free University of Amsterdam, 1984.
55. T Ziegler, EJ Baerends, JG Snijders, W Ravenek. J Phys Chem 93:3050–3056, 1989.
56. For definitions of the mass-velocity, Darwin, and spin-orbit coupling terms, see:

JG Snijders, EJ Baerends. In: P Coppens, MB Hall, eds. Electron Distribution and the Chemical Bond. New York: Plenum Press, 1981.

57. J Li, BE Bursten. Unpublished results.

58. For exchange-correlation functionals beyond first-order gradient-correction, see, for example: HL Schmider, AD Becke. J Chem Phys 109:8188–8199, 1998.

59. A Avdeef, KN Raymond, KO Hodgson, A Zalkin. Inorg Chem 11:1083–1088, 1972.

60. DJA De Ridder, J Rebizant, C Apostolidis, B Kanellakopulos, E Dornberger. Acta Cryst C52:597–600, 1996.

61. AG Orpen, L Brammer, FH Allen, O Kennard, DG Watson, R Taylor. J Chem Soc Dalton Trans S1–S83, 1989 (Suppl).

62. For a theoretical analysis of the advantages of the PW91 functional, see: (a) JP Perdew, K Burke. Int J Quantum Chem 57:309–319, 1996. (b) K Burke, JP Perdew, M Ernzerhof. Int J Quantum Chem 61:287–293, 1997.

63. J Li, BE Bursten. Relativistic Density Functional and Ab initio Theoretical Studies of Actinide Fluorides. Book of Abstracts, 217th National ACS Meeting, Anaheim, CA, 1999, NUCL 174.

64. (a) MF Zhou, L Andrews, J Li, BE Bursten. J Am Chem Soc 121:9712–9721, 1999. (b) MF Zhou, L Andrews, J Li, BE Bursten. J Am Chem Soc 121:12188–12189, 1999. (c) L Andrews, B Liang, J Li, BE Bursten. Angew Chem Int Ed. In press.

65. RD Fischer. Theor Chim Acta (Berlin) 1:418–431, 1963.

66. (a) A Streitwieser Jr, SA Kinsley. NATO ASI Ser (Ser C). 155:77–114, 1985, and references cited therein. (b) A Streitwieser Jr. Inorg Chim Acta 94:171–177, 1984.

67. For recent reviews, see: (a) P Pyykkö. Inorg Chim Acta 139:243–245, 1987. (b) K Balasubramanian. In: KA Gschneider Jr, L Eyring, eds. Handbook on the Physics and Chemistry of Rare Earths. Vol 18. Amsterdam: North-Holland, 1994, pp 29–158. (c) M Dolg, H Stoll. In: KA Gschneider Jr, L Eyring, eds. Handbook on the Physics and Chemistry of Rare Earths. Vol 22. Amsterdam: Elsevier Science, 1996, pp 607–729.

68. (a) KD Warren. Struct Bond 33:97–138, 1975. (b) DW Clack, KD Warren. J Organomet Chem 122:C28–C30, 1976.

69. (a) P Pyykkö, LL Lohr. Inorg Chem 20:1950–1959, 1981. (b) P Pyykkö, LJ Laaksonen, K Tatsumi. Inorg Chem 28:1801–1805, 1989.

70. RB King. Inorg Chem 31:1978–1980, 1992.

71. MG Cory, S Köstlmeier, M Kotzian, N Rösch, MC Zerner. J Chem Phys 100:1353–1365, 1994.

72. (a) M Dolg, P Fulde, H Stoll, H Preuss, AHH Chang, RM Pitzer. Chem Phys 195:71–82, 1995. (b) W Liu, M Dolg, P Fulde. J Chem Phys 107:3584–3591, 1997. (c) W Liu, M Dolg, P Fulde. Inorg Chem 37:1067–1072, 1998.

73. (a) AHH Chang, RM Pitzer. J Am Chem Soc 111:2500–2507, 1989. (b) AHH Chang, K Zhao, WC Ermler, RM Pitzer. J Alloys Comp 213/214: 191–195, 1994.

74. (a) N Rösch, A Streitwieser Jr. J Organomet Chem 145:195–200, 1978. (b) N Rösch, A Streitwieser Jr. J Am Chem Soc 105:7237–7240, 1983. (c) N Rösch. Inorg Chim Acta 94:297–299, 1984.

75. PM Boerrigter, EJ Baerends, JG Snijders. Chem Phys 122:357–374, 1988.

76. N Kaltsoyannis, BE Bursten. J Organomet Chem 528:19–33, 1997.
77. (a) H Koch, P Jørgensen, T Helgaker. J Chem Phys 104:9528–9530, 1996. (b) K Pierloot, BJ Persson, BO Roos. J Phys Chem 99:3465–3472, 1995. (c) C Park, JJ Almlöf. Chem Phys 95:1829–1833, 1991. (d) HP Lüthi, PEM Siegbahn, J Almlöf, K Faegri Jr, A Heiberg. Chem Phys Lett 111, 1–6, 1984. (e) TE Taylor, MB Hall. Chem Phys Lett 114:338–342, 1985.
78. For discussion of the effects of Madelung potentials of the crystal field on the bond lengths of molecules in crystals, see, for example: J Li, S Irle, WHE Schwarz. Inorg Chem 36:100–109, 1996.
79. (a) See, for example: JE Huheey, EA Keiter, RL Keiter. Inorganic Chemistry: Principles of Structure and Reactivity. 4th ed. New York: Harper Collins, 1993, Chap 14. (b) For a recent theoretical analysis of actinide contraction and its comparison with the lanthanide contraction, see: M Seth, M Dolg, P Fulde, P Schwerdtfeger. J Am Chem Soc 117:6597–6598, 1995, and references therein.
80. F Takusagawa, TF Koetzle. Acta Crystallogr B35:1074–1081, 1979. An electron-diffraction study determines even a greater bending (3.7°): A Haaland, J Lusztyk, DP Novak, J Brunvoll, KB Starowieyski. J Chem Soc Chem Commun 54–55, 1974.
81. M Elian, MML Chen, MP Mingos, R Hoffmann. Inorg Chem 15:1148–1155, 1976.
82. IA Ronova, DA Bochvar, AL Chistyakov, Yu T Struchkov, NV Alekseev. J Organomet Chem 18:337–344, 1969.
83. We have examined the effects of spin-orbit coupling on the geometry structures and vibrational frequencies of $U(C_7H_7)_2^-$ and found them to be negligible ($<1\%$) for the f^1 actinide system. J Li, BE Bursten. To be submitted.
84. Experimental results: (a) L Brewer. J Opt Soc Am 61:1666–1682, 1971. (b) MSS Brooks, B Johansson, HL Skriver. In: AJ Freeman, GH Lander, eds. Handbook on the Physics and Chemistry of the Actinides. Vol 1. Amsterdam: North-Holland, 1984, Chap 3. (c) MS Fred. In: JJ Katz, GT Seaborg, LR Morss, eds. The Chemistry of the Actinides Elements. Vol 2. 2nd ed. New York: Chapman and Hall, 1986, Chap 15. (d) WT Carnall, HM Crosswhite. In: JJ Katz, GT Seaborg, LR Morss, eds. The Chemistry of the Actinides Elements. Vol 2. 2nd ed. New York: Chapman and Hall, 1986, Chap 16.
85. For theoretical results, see, for example: (a) P Pyykkö, LJ Laakkonen, K Tatsumi. Inorg Chem 28:1801–1805, 1989. (b) KW Bagnall. The Actinide Elements. Amsterdam: Elsevier, 1972, Chap 1.
86. The multiplet effects are estimated to lower the Th and U atomic energies by 4.3 and 19.2 kcal/mol following the procedure described in Ref. 87.
87. Without spin-orbit coupling, atomic ground state multiplet energies can be calculated in DFT from $D_{\infty h}$ determinants; see: EJ Baerends, V Branchadell, M Sodupe. Chem Phys Lett 265:481–489, 1997.
88. NT Kuznetsov, KV Kir'yanov, VA Mitin, VG Sevast'yanov, VA Bogdanov. Radio-khimiya 28:709–7121, 1986.
89. For a discussion of spin-orbit coupling of f electrons, see, for example: A Abragam, B Bleaney. Electron Paramagnetic Resonance of Transition Ions. Oxford: Clarendon Press, 1970.
90. Extensive compilations of all the single- and double-point groups can be found in: (a) JA Salthouse, MJ Ware. Point Group Character Tables and Related Data. Lon-

don: Cambridge University Press, 1972. (b) SL Altmann, P Herzig. Point-Group Theory Tables. Oxford: Clarendon Press, 1994.

91. K Zhao. Electronic Structure of Heavy-Element Molecules: $Pa@C_{28}$, $Pa(C_8H_8)_2$, and the Jahn–Teller Effect in VCl_4. PhD dissertation, The Ohio State University, Columbus, Ohio, 1996.

92. JC Slater. Adv Quantum Chem 6:1–92, 1972.

93. See, for example: BE Douglas, CA Hollingsworth. Symmetry in Bonding and Spectra: An Introduction. Orlando, Fl: Academic Press, 1985.

94. JP Solar, HPG Burghard, RH Banks, A Streitwieser Jr, D Brown. Inorg Chem 19: 2186–2188, 1980.

95. NM Edelstein, J Goffart. In: JJ Katz, GT Seaborg, LR Morss, eds. The Chemistry of the Actinides Elements. Vol 2. 2nd ed. New York: Chapman and Hall, 1986, Chap 18.

96. KD Warren. Inorg Chem 14:3095–3103, 1975.

97. KO Hodgson, F Mares, DF Starks, A Streitwieser Jr. J Am Chem Soc 95:8650–8658, 1973.

98. For a comprehensive review of vibrational spectra of transition metal π-complexes of C_nH_n ($n = 3–8$) rings, see: HP Fritz. In: FGA Stone, R West, eds. Advances in Organometallic Chemistry. Vol 1. New York: Academic Press, 1964, pp 239–316.

99. For a summary, see: DG Karraker, JA Stone, ER Jones Jr, N Edelstein. J Am Chem Soc 92:4841–4845, 1970.

100. J Goffart, J Fuger, B Gilbert, B Kanellakopulos, G Duyckaerts. Inorg Nucl Chem Lett 8:403–412, 1972.

101. J Goffart, J Fuger, D Brown, G Duyckaerts. Inorg Nucl Chem Lett 10:413–419, 1974.

102. L Hocks, J Goffart, G Duyckaerts, P Teyssie. Spectrochim Acta 30A:907–914, 1974.

103. FA Cotton. Chemical Applications of Group Theory. 3rd ed. New York: Wiley, 1990.

104. G Herzberg. IR and Raman Spectra. New York: Van Nostrand, 1945, p 154.

105. F Mares, K Hodgson, A Streitwieser Jr. J Organomet Chem 24:C68–C70, 1970.

106. (a) C Adamo, V Barone. J Chem Phys 110:6158–6170, 1999. (b) AD Becke. J Chem Phys 107:8554–8555 1997. (c) PJ Stephens, F Devlin, CF Chabalowski, MJ Frisch. J Phys Chem 98:11623–11627, 1994. (d) AD Becke. J Chem Phys 98: 5648–5652, 1993.

107. (a) AD Becke. J Comp Chem 20:63–69, 1999. (b) JP Perdew, S Kurth, A Zupan, P Blaha. Phys Rev Lett 82:2554–2547, 1999. (c) AD Becke. J Chem Phys 109: 2092–2098, 1998. (d) T Van Voorhis, GE Scuseria. J Chem Phys 109:400–410, 1998.

108. (a) EKU Gross, JF Dobson, M. Petersilka. Top Curr Chem 181:81–172, 1996. (b) R Singh, BM Deb. Phys Rep 311:47–94, 1999.

109. Assessments of the performances of various new exchange-correlation functionals have been carried out recently. See, for example: (a) AJ Cohen, NC Handy. Chem Phys Lett 316:160–166, 2000. (b) S Kurth, JP Perdew, P Blaha. Int J Quantum Chem 75:889–909, 1999. (c) AD Rabuck, GE Scuseria. Chem Phys Lett 309:450–456, 1999.

15

Pi Bonding in Group 13–Group 15 Analogs of Ethene

**Ashalla McGee, Freida S. Dale,
Soon S. Yoon, and Tracy P. Hamilton**
The University of Alabama at Birmingham, Birmingham, Alabama

1. INTRODUCTION

The study of pi bonding in main group chemistry has been primarily concerned with organic molecules or analogs within Group 14, such as disilene. The latter was quite challenging, and much experimental and theoretical work went into its study and preparation (1). The difficulty of making pi bonds in the heavier elements is due to their weakness and reactivity (which necessitates steric protection). However, Power's group has made a great deal of progress in synthesizing molecules that are isoelectronic to derivatives of famous organics, such as ethene and benzene, using heavy main group elements from families other than Group 14 (2). We will concentrate only on those that have two groups bonded to each atom, just as in ethene.

The pi bond strength in the case of ethene (and most other olefins) is 64 kcal/mol and is determined by several methods. We will discuss some of the most common and evaluate their prospects for being meaningful in Group 13–Group 15 pi bonds. The conceptually simplest is to twist the molecule so that the $2p$ orbitals no longer overlap. This has the advantage that the number obtained is able to be determined by kinetics; i.e., the pi bond strength is the activation barrier for rotation. For ethene, rotation produces an unpaired electron at each carbon (a diradical). This is poorly described by a single electronic configuration, and an electronic correlation method based on a single Slater determinant will have problems (3). Fortunately, in the study here, the pair of electrons that make the pi bond remains on the Group 15 atom. This enables the use of low-level theories for accurate rotation barriers.

Another procedure involves the hydrogenation of ethene. Assuming that the C–C sigma bond does not significantly change, then the energy of the reaction is the energy of the two newly formed C–H bonds minus the H–H bond energy and the C–C pi bond energy. Hydrogenation energies are typically 28 kcal/mol (ethene's is 32.8 kcal/mol) (4), the C–H bond is 98 kcal/mol and the H–H bond energy is 103 kcal/mol (5). This gives 60 kcal/mol for the pi bond energy. The agreement between this and the foregoing is good. For the compounds we will consider, there is a serious problem, because hydrogenation produces a compound that has a significantly different sigma bond—an adduct!

A third option, closely related to the second, involves computing pi bond energies from a reaction and standard or assumed bond energies. Two ethenes could in principle dimerize to form cyclobutane so that the energy of the forma-tion reaction is the two newly formed C–C bonds plus the strain energy minus the two C–C pi bond energies. Using the strain energy of 26.3 kcal/mol (5), the C–C sigma bond energy of 81 kcal/mol and the dimerization energy of − 18.3 kcal/mol (from heats of formation of ethene and cyclobutane) (6), the computed pi bond energy is 59 kcal/mol. In this case, the prospect for obtaining a check on the pi bond energy is good, for dimerization will create what are definitely sigma bonds, but the strain energy of the resulting rings still needs to be known. This is apparent when one examines the variety of rings and cages, some with quite acute bond angles, as in the chemistry of Al-N compounds (7).

This chapter will present a comprehensive study of the ethene analogs pro-duced by taking an element from the set M = B, Al, Ga and one from Group 15 E = N, P, As. Most of this work was performed by undergraduates in this lab over the years. It was ideally suited for an undergraduate research project because of the simplicity of the bonding concepts and of the absence of many complicated theoretical issues. Many of the systems here have been studied by other groups in the meantime, with results scattered throughout the literature. However, this chapter will include new material, such as an examination of the performance of relativistic effective core potentials.

Some of the early explanations of the weakness of pi bonding in the heavier elements rationalized it as being due to the weaker overlap between p orbitals. Now it is better understood to result from energetic effects. For the heavier ele-ments there is an energetic preference to form lone pairs. Lappert first explained

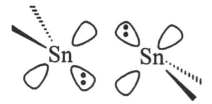

FIGURE 1   Dative bonding model proposed by Lappert. (From Ref. 8.)

the structure of a distannene as being due to the double dative interaction, as shown in Figure 1 (8).

Another way of rationalizing the bending associated with lack of pi bonding is by a second-order Jahn–Teller effect that mixes the pi and a σ* orbital (9). A third explanation, by Trinquier and Malrieu, is an extension of Lappert's work. It predicts whether the Lappert model or the classic σ, π formation will take place based on the singlet–triplet splitting in the fragments created by dissociating the bond (10). As a consequence, electropositive substituents, which stabilize the triplet relative to the singlet, increase the pi bond strength. A very good review is given by Grev (11). No theoretical studies to date have attempted to see how these concepts work for the Group 13–Group 15 molecules here. It is reasonable to assume that the orbital mixing will work, since it explains pyramidalization, which is to be expected for P and As (and is seen in these compounds). The fragment electronic states are going to be doublets and quartets upon homolytic

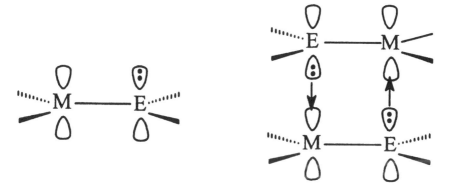

FIGURE 2   Simple model for the pi bonding and dimerization in $R_2M{=}ER_2$ molecules.

bond dissociation into fragments, so how these apply is unclear. One thing we will briefly look at later is the effect of an electropositive substituent on the pi bond energy.

A good model of the bonding in the $R_2M{=}ER_2$ compounds is given in Figure 2. This model, in which the weak pi bond is most like a dative donation of the lone pair on the Group 15 atom to the empty orbital on the Group 13 atom, is one reason why the pi bond is expected to be even weaker than between two Group 14 elements from the same rows. Elimination reactions that produce pi bonds rarely give end products with pi bonds, because the bonding allows easy dimerization (also shown in Fig. 2).

2. THEORETICAL METHODS

Geometries were initially optimized at the Hartree–Fock (HF) level of theory with the 3-21G basis set. Even though high levels of theory are not required to study these systems, the 3-21G basis set lacks d polarization functions. Because of that, the lone pairs of the Group 15 element are inadequately described. It is well known that a basis set that is saturated with s and p functions will predict that ammonia is planar. Subsequent geometry optimizations were carried out at the HF/6-31G** and MP2/6-31G** levels of theory.

For the heavier elements, relativistic effects due to the core may become important. To account for this in the simplest way, the electrons in the core can be replaced by a potential that produces the same valence electron distribution as an all-electron relativistic computation. This also reduces the computer time needed as well, since the number of functions is reduced. Another hazard of doing all-electron calculations with small basis sets on lower-row elements is that the bond lengths have large error. The relativistic effective core potential (RECP) that we employed was CEP-121G** (12). For this RECP, the geometry was optimized at the MP2 level of theory, and a single-point energy was computed at the CCSD(T) level of theory (13).

HF and MP2 optimized structures with the 6-31G** basis set were characterized by computing the harmonic vibrational frequencies. Given the popularity and success of the hybrid density functional method B3LYP, attempts were made to compare it with the MP2 results using the CEP-121G** basis. Since the capability for doing frequencies with RECP basis sets is not yet programmed into Gaussian, the frequencies were unable to be computed. Given the consistency of the structures for every level of theory and the clear understanding of whether the structures should be minima or transitions state or have two imaginary frequencies, the results will be reported as having the same hessian signature as the other levels of theory. Calculations were performed using various versions of the Gaussian program, the latest being Gaussian 98 (14).

3. RESULTS AND DISCUSSION

3.1. Monoamides

The results break into two classes: the molecules with nitrogen as the Group 15 element, and those with heavier Group 15 elements. We will discuss the more interesting nitrogen-containing compounds first. Energies for the monoamides are given in Table 1, and the optimized geometrical parameters are given in Table 2.

The minimum structures are all planar (see Fig. 3). This alone indicates significant pi bonding, because there must be significant conjugation of the "lone pair" of electrons. This was previously computed by other groups for H_2BNH_2 and H_2AlNH_2 (15–18). In fact, we predicted that H_2GaNH_2 would be planar and have a nonnegligible pi bond energy based on the similarity of the behavior of the Ga–P and Ga–As structures to the corresponding Al–P and Al–As structures (vide infra).

First, the rotational barriers for the monoamides are close to those obtained previously by other groups. McKee (15) and Allen and Fink (16) computed the B–N pi bond to be 32 and 38 kcal/mol, respectively. Fink et al. (17) computed the pi bond strength to be 11.2 kcal/mol for Al–N and 12.8 for Ga–N. Davy and Jaffrey obtained 10.7 kcal/mol for Al–N earlier (18). These values are quite close, although it should be noted that Fink et al. used the energy difference between the planar and perpendicular forms, which exaggerates the difference between Al–N and Ga–N. This also explains why the previously published value of 38 kcal/mol for B–N is higher than the value from the rotational barrier.

A preliminary computation of the rotation barriers for silyl-substituted H_2AlNH_2 was performed in our laboratory, to see if the effects of electropositive (relative to H) groups would have the effect of strengthening the pi bond. These were computed at the B3LYP/6-31G* level of theory (19). The absolute energies are −881.039016 for $H_2AlN(SiH_3)_2$ and −881.013040 for $(SiH_3)_2AlNH_2$. This shows that the silyl group stabilizes the compound more when attached to nitrogen. However, the barrier to rotation for $H_2AlN(SiH_3)_2$ (at 5.9 kcal/mol) is actually lower than in H_2AlNH_2. This is not the case when silyl groups replace hydrogen in silicon compounds; in that case pi bonds are strengthened. When silyl substitution is on the aluminum, the rotation barrier is unaffected (11.6 kcal/mol with no vibration correction).

Finally, we note that Muller also performed calculations on Me_2AlNH_2, finding a rotation barrier slightly less than in H_2AlNH_2 (at 9.7 kcal/mol) (20).

The transition state for rotation about the Group 13–N bond shows that the pi bond energy is significant, but not nearly as large as for ethene (except in the case of H_2BNH_2, which has a pi bond energy of 30 kcal/mol, ethene has a 65 kcal/mol pi energy). The transition-state structures are all of the same type

TABLE 1   Absolute Energies (atomic units) and Relative Energies (kcal/mol) for Ethene Analogs Having N as Group 15 Elements[a]

	B–N	Al–N	Ga–N
SCF/6-31G**			
Planar	−81.499210	−298.705968	−1978.014931
	(0.0)	(0.0)	(0.0)
Perpendicular	−81.446139	−298.688487	−1977.994129
	(30.5)	(9.7)	(11.7)
Transition state	−81.452544		−1977.995435
	(27.3)		(11.3)
MP2/6-31G**			
Planar	−81.765481	−298.949746	−1978.274816
	(0.0)	(0.0)	(0.0)
Perpendicular	−81.704846	−298.929494	−1978.249167
	(35.3)	(11.4)	(14.7)
Transition state	−81.713399	−298.929643	−1978.252197
	(30.6)	(11.6)	(13.3)
MP2/CEP-121G**			
Planar	−15.010717	−14.213364	−269.809824
	(0.0)	(0.0)	(0.0)
Perpendicular	−14.951296	−14.193298	−269.785224
	(34.6)	(11.3)	(14.0)
Transition state	−14.959744	−14.193902	−269.789012
	(29.9)	(11.2)	(12.2)
CCSD(T)[b]			
Planar	−15.044599	−14.246036	−269.803575
	(0.0)	(0.0)	(0.0)
Perpendicular	−14.984415	−14.225250	−269.778913
	(35.1)	(11.7)	(14.1)
Transition state	−14.993436	−14.225985	−269.782843
	(30.0)	(11.6)	(12.1)

[a] The relative energies are in parentheses and have zero-point vibrational energy included.
[b] With the CEP-121G** basis set, at a MP2/CEP-121G** optimized geometry.

(Fig. 3) and are essentially the linkage of a borane, alane, or gallane group with a pyramidal amide. The C_{2v} twist structure has two imaginary frequencies and is hence is of less importance. The difference in energy between the C_{2v} twist and the transition state for H_2BNH_2 is close to the inversion barrier of 5 kcal/mol in ammonia. The inversion of nitrogen for the Ga analog is quite small, and it is almost nonexistent for Al. As a matter of fact, at the HF/6-31G** level of theory the twist structure goes to the perpendicular one, and zero-point energy

TABLE 2   Geometrical Parameters for Group 13-N Ethene Analogs, (bond distances in, Å bond angles in degrees)

Planar	B-N	Al-N	Ga-N
SCF/6-31G**	1.388	1.770	1.821
MP2/6-31G**	1.393	1.780	1.825
MP2/CEP-121G**	1.401	1.783	1.812
	HBH	HAlH	HGaH
HF/6-31G**	121.1	123.1	124.1
MP2/6-31G**	122.0	124.1	126.1
MP2/CEP-121G**	122.5	124.6	126.4
	HNH	HNH	HNH
HF/6-31G**	113.9	110.3	111.2
MP2/6-31G**	114.0	110.4	111.4
MP2/CEP-121G**	114.1	110.3	111.7
	B-H	Al-H	Ga-H
HF/6-31G**	1.193	1.581	1.573
MP2/6-31G**	1.189	1.576	1.564
MP2/CEP-121G**	1.193	1.577	1.537
	N-H	N-H	N-H
HF/6-31G**	0.994	0.997	0.996
MP2/6-31G**	1.005	1.008	1.007
MP2/CEP-121G**	1.009	1.012	1.011
Perpendicular	B-N	Al-N	Ga-N
SCF/6-31G**	1.455	1.792	1.853
MP2/6-31G**	1.458	1.801	1.857
MP2/CEP-121G**	1.467	1.805	1.848
	HBH	HAlH	HGaH
HF/6-31G**	116.6	117.5	119.6
MP2/6-31G**	116.5	117.2	119.7
MP2/CEP-121G**	117.2	118.1	120.9
	HNH	HNH	HNH
HF/6-31G**	114.1	109.4	110.8
MP2/6-31G**	114.0	109.4	110.9
MP2/CEP-121G**	113.9	109.3	111.3
	B-H	Al-H	Ga-H
HF/6-31G**	1.201	1.585	1.580
MP2/6-31G**	1.199	1.582	1.574
MP2/CEP-121G**	1.203	1.583	1.548
	N-H	N-H	N-H
HF/6-31G**	0.993	0.997	0.996
MP2/6-31G**	1.002	1.007	1.006
MP2/CEP-121G**	1.007	1.011	1.009

TABLE 2   Continued

Transition state	B–N	Al–N	Ga–N
SCF/6-31G**	1.469	1.792	1.873
MP2/6-31G**	1.476	1.809	1.890
MP2/CEP-121G**	1.483	1.819	1.878
	H_tBN	H_tAlN	H_tGaN
HF/6-31G**	120.0	121.2	118.4
MP2/6-31G**	119.5	119.7	117.1
MP2/CEP-121G**	119.4	118.6	121.5
	H_cBN	H_cAlN	H_cGaN
HF/6-31G**	122.2	121.2	121.7
MP2/6-31G**	122.4	122.8	122.4
MP2/CEP-121G**	122.1	122.9	121.7
	$B–H_t$	$Al–H_t$	$Ga–H_t$
HF/6-31G**	1.193	1.585	1.575
MP2/6-31G**	1.191	1.579	1.568
MP2/CEP-121G**	1.195	1.579	1.542
	$B–H_c$	$Al–H_c$	$Ga–H_c$
HF/6-31G**	1.200	1.585	1.583
MP2/6-31G**	1.198	1.584	1.577
MP2/CEP-121G**	1.203	1.586	1.552
	HNH	HNH	HNH
SCF/6-31G**	104.2	109.4	106.9
MP2/6-31G**	101.8	107.9	104.7
MP2/CEP-121G**	102.1	106.9	104.6
	N–H	N–H	N–H
SCF/6-31G**	1.006	0.997	1.001
MP2/6-31G**	1.020	1.009	1.017
MP2/CEP-121G**	1.024	1.016	1.020
	BNH	AlNH	GaNH
SCF/6-31G**	110.3	125.3	116.7
MP2/6-31G**	107.7	121.9	112.5
MP2/CEP-121G**	107.8	119.0	111.7

The subscripts t and c refer to trans and cis of H with respect to the NH$_2$ group.

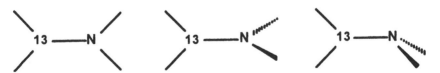

FIGURE 3   Minimum, perpendicular, and transition-state structure types for ethene analogs containing nitrogen as the Group 15 element.

makes the two structures nearly the same energy at correlated levels of theory. Another trend to note is that there is little difference between Al and Ga, except that the Ga compound is predicted to have a stronger pi bond. This is certainly contrary to what would be expected on the basis of orbital size match. The energetics are very similar for all levels of theory, whether a RECP is used or not or whether MP2 or CCSD(T) is used. The HF relative energies are only in error by 5 kcal/mol for H_2BNH_2 and by around 2 kcal/mol for H_2AlNH_2 and H_2GaNH_2.

Examination of the geometrical parameters in Table 2 shows that the Ga–N bonds are only slightly longer than Al–N and that the Ga–H bonds have similar distances to the Al–H bonds. The use of an RECP made little difference, except that the MP2/CEP-121G** Ga–H distance was consistently ~0.03 Å shorter than MP2/6-31G** and a similar trend of ~0.01 shorter Ga–N distances with the RECP. The Al–N and Ga–N bond distances compare favorably with the shortest experimental monoamide values (1.784 and 1.847 Å, respectively (17). Fink et al. also used the Hay–Wadt effective core potential (21) to optimize H_2GaNH_2, getting 1.794 Å for the Ga–N distance (17). This is 0.02 Å shorter than with the CEP-121G** basis.

Comparing the planar to the perpendicular structure, the breaking of the pi bond leads to increase of the 13–N bond, decrease of the H–13–H angles, slightly longer 13–H bonds, and little change in the HNH and N–H values. Certainly the 13–N distance is expected to be longer due to breaking the pi bond; however, other effects may also contribute. The sharper H–13–H angles mean that the hybrid orbitals on the Group 13 atom that form bonds with H have more p character, and therefore the remaining sp^2 hybrid has less p character. This will make the 13–N bond shorter than it otherwise would be and the 13–H bonds longer, as observed (Bent's rule) (22). Davy and Jaffrey (18) suggest that negative hyperconjugation of the lone pair on N with the metal–H σ bonds could also account for some of the M–H distance increase in the twisted isomers. The reason that a hybrid orbital has more p character is that the substituent is more electronegative. When the pi bond is broken, both electrons go to N, making it more negatively charged. These effects are not present in ethene, so the result of breaking the ethene pi bond results in a much larger bond distance change (0.2 Å). When the 13–N bond is twisted, there is no longer the driving force that makes the NH_2 group coplanar with the Group 13 atom, so the tendency to pyramidalize and form a classic lone pair will be stronger. This is quite weak for nitrogen, as the energy differences between the perpendicular and transition-state structures demonstrate. The reason that pyramidalization is more weakly favored than in ammonia is that in the perpendicular structure the b_1 symmetry N–H bonding MO can conjugate with the empty p orbital of boron. Apparently this interaction is strong for Al.

This is reflected in the trends in Mulliken charges based on the SCF/CEP-121G** density (Table 3). Twisting the H_2BNH_2 molecule results in −0.16 elec-

TABLE 3   Mulliken Charges at SCF/CEP-121G** Level of Theory

	Planar			Perpendicular			Transition state		
	B	Al	Ga	B	Al	Ga	B	Al	Ga
Group 13 atom	.20	.65	.89	.38	.70	.97	.32	.68	.91
N	−.57	−.83	−.97	−.73	−.86	−1.02	−.61	−.82	−.93
H bonded to 13	−.09	−.18	−.23	−.11	−.19	−.25	−.09	−.19	−.24
H bonded to N	.27	.27	.28	.28	.27	.28	.24	.26	.25

The H charges of the H bonded to the Group 13 atom are averages.

tron density transferred to N, twisting H_2AlNH_2 leads to -0.03 change, and torsion of H_2GaNH_2 leads to $-.05$ electrons transferred.

3.2. Monophosphides and Monoarsenides

Naturally, the most studied of the H_2MEH_2 compounds with E = P or As is the smallest one, H_2BPH_2. The predicted pi bond strength reported by Allen and Fink of 40 kcal/mol was based on comparing the planar structure to the perpendicular structure. This is a maximum value that depends upon being able to realize a planar phosphorus in the perpendicular form, which is doubtful. They noted that the minimum of H_2BPH_2 is not planar but is pyramidalized about the phosphorus. Figure 4 shows the type of structure that we will call trans bent, in analogy to the Group 14–Group 14 compounds, even though the Group 13 moiety remains nearly planar.

The primary structural determinant is therefore seen to be the identity of the Group 15 element. Nitrogen is different from P and As in its tendency to pyramidalize, as measured by inversion barriers and by bond angles. The inversion barrier in ammonia is 5 kcal/mol, and it is 30 kcal/mol in phosphine and arsine. The inversion of the trans bent form (which goes through the planar form)

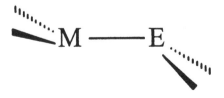

FIGURE 4   Global minimum structure for H_2MEH_2, where E = P or As, M = B, Al, or Ga.

is facilitated by the conjugation in the planar form. However, when the molecule is twisted, there is no such stabilization, and the twist bent geometry is much lower in energy than the perpendicular one. This energy difference is very similar to the phosphine or arsine inversion barriers.

From Table 4, it is clear that the lowest-energy structure is trans bent. The next in terms of energy is the transition state, corresponding to a low rotation barrier—on the order of that for ethane in the molecules that have no first-row atoms. It would be fair to say the latter exhibit very little pi bonding. In the H_2BPH_2 case, the energy difference between the planar geometry and the minimum is half that of the other cases, which is consistent with earlier findings (17,23). The planar structure is a transition state for interconversion of the trans bent forms via inversion. The arsenic has a small but noticeable difference from phosphorus in the energetics—arsenic favors pyramidalization 1–2 kcal/mol more than phosphorus. Boron clearly favors pi bonding a little more, about 6 kcal/mol in H_2BPH_2 and 3 kcal/mol in H_2BAsH_2.

The structure trends support the energetic trends. The geometric parameters in Table 5 show the slightly increased pyramidalization at As compared to P, and the slightly better ability of B to pi bond with P and As (manifested by increases in H–E–H and H–E–B angles).

Recall from the monoamides that the RECP predicted slightly shorter bond distances to Al and 0.01–0.03-Å shorter bonds to gallium. The data needed to compare the results from MP2/6-31G** to MP2/CEP-121G** are omitted, for brevity, so we shall summarize them in text. The central M–E bond length is *longer* by 0.02–0.03 Å with the RECP basis set. This means that the Al–As and Ga–As bond distances are actually closer together than a casual reading of Table 5 would imply. The RECP basis P–H distance is about 0.02 Å longer, whereas the As–H is about the same with either basis. The reader is probably wondering by now why the MP2/CEP-121G** results were not given for H_2GaAsH_2. This is because the predicted forces, for reasons unclear to the authors, were extremely large. In the one case where the optimization algorithm found a point with zero forces, the geometry was absurd.

Before leaving the ethene analogs, we will note that the pi bond strength due to the dative interaction of a lone pair to the empty Group 13 orbital does not depend on electronegativity in an obvious way. Simple reasoning would say that the electronegativity difference between Al and N is much smaller than between B and P, so the pi bond strength will be lower. It may be due to the fact that the increased charge induction due to electronegativity allows for more pi back-donation. This would also explain why putting silyl groups on nitrogen decreases the rotation barrier. The superior ability of Al and N to conjugate is also seen when the benzene analogs are compared: alumazene is planar whereas borophosphazene is puckered (24,25).

TABLE 4　Absolute MP2/CEP-121G** Energies (atomic units) and Relative Energies (kcal/mol) for Ethene Analogs Having P or As as a Group 15 Element[a]

MP2/CEP-121G**	B–P	B–As	Al–P	Al–As	Ga–P	Ga–As[b]
Planar	−11.552240	−11.158336	−10.776749	−10.390671	−266.386098	−4155.703795
	(5.6)	(9.0)	(10.0)	(11.9)	(11.8)	(12.9)
Trans bent	−11.561777	−11.173006	−10.792948	−10.410106	−266.405260	−4155.758107
	(0.0)	(0.0)	(0.0)	(0.0)	(0.0)	(0.0)
Perpendicular	−11.491250	−11.096806	−10.751363	−10.362659	−266.356085	−4155.703795
	(42.7)	(46.8)	(25.4)	(29.1)	(30.0)	(33.3)
Transition state	−11.546702	−11.160809	−10.787983	−10.405894	−266.399179	−4155.752402
	(8.6)	(7.2)	(2.6)	(2.3)	(3.3)	(3.3)

[a] The relative energies are in parentheses and have zero-point vibrational energy included.
[b] MP2/6-31G**.

TABLE 5 MP2/CEP-121G** Geometrical Parameters for Group 13–P and 13–As Ethene Analogs (bond distances in Å, bond angles in degrees)[a]

Planar	B-P	Al-P	Ga-P	B-As	Al-As	Ga-As[b]
	1.808	2.241	2.215	1.887	2.308	2.268
	HBH	HAlH	HGaH	HBH	HAlH	HGaH
	124.7	128.1	129.2	125.5	129.2	128.9
	HPH	HPH	HPH	HAsH	HAsH	HAsH
	109.4	107.3	107.8	108.2	106.1	107.3
	B-H	Al-H	Ga-H	B-H	Al-H	Ga-H
	1.188	1.571	1.529	1.187	1.570	1.557
	P-H	P-H	P-H	As-H	As-H	As-H
	1.395	1.398	1.397	1.468	1.472	1.470
Trans bent	B-P	Al-P	Ga-P	B-As	Al-As	Ga-As[a]
	1.893	2.338	2.309	2.002	2.441	2.380
	HBH	HAlH	HGaH	HBH	HAlH	HGaH
	120.7	121.5	122.0	120.5	120.7	120.4
	HPH	HPH	HPH	HAsH	HAsH	HAsH
	99.7	95.8	96.0	97.8	94.6	93.6
	B-H	Al-H	Ga-H	B-H	Al-H	Ga-H
	1.191	1.581	1.543	1.191	1.582	1.569
	P-H	P-H	P-H	As-H	As-H	As-H
	1.412	1.421	1.419	1.497	1.508	1.499
	HBP	HAlP	HGaP	HBAs	HAlAs	HGaAs
	119.3	119.1	118.9	119.4	119.5	119.6
	HPB	HPAl	HPGa	HAsB	HAsAl	HAsGa
	102.9	95.1	96.6	100.5	92.5	95.6
Perpendicular	B-P	Al-P	Ga-P	B-As	Al-As	Ga-As[b]
	1.964	2.306	2.305	2.047	2.381	2.354
	HBH	HAlH	HGaH	HBH	HAlH	HGaH
	120.0	119.6	122.9	120.7	119.5	121.0
	HPH	HPH	HPH	HAsH	HAsH	HAsH
	115.0	109.0	110.3	114.3	108.4	110.0
	B-H	Al-H	Ga-H	B-H	Al-H	Ga-H
	1.190	1.576	1.540	1.189	1.576	1.567
	P-H	P-H	P-H	As-H	As-H	As-H
	1.388	1.395	1.394	1.461	1.469	1.465
Transition state	B-P	Al-P	Ga-P	B-As	Al-As	Ga-As[b]
	1.974	2.364	2.346	2.079	2.465	2.416
	H_tBP	H_tAlP	H_tGaP	H_tBAs	H_tAlAs	H_tGaAs
	120.1	119.9	119.3	119.9	120.5	118.6
	H_cBP	H_cAlP	H_cGaP	H_cBAs	H_cAlAs	H_cGaAs
	120.6	120.5	120.0	120.7	120.6	122.2
	$B-H_t$	$Al-H_t$	$Ga-H_t$	$B-H_t$	$Al-H_t$	$Ga-H_t$

TABLE 5   Continued

Transition state	1.192	1.581	1.545	1.191	1.582	1.571
(Cont'd)	B-H$_c$	Al-H$_c$	Ga-H$_c$	B-H$_c$	Al-H$_c$	Ga-H$_c$
	1.192	1.583	1.546	1.191	1.584	1.571
	HPH	HPH	HPH	HAsH	HAsH	HAsH
	91.4	92.2	92.2	91.1	91.7	90.4
	P-H	P-H	P-H	As-H	As-H	As-H
	1.427	1.427	1.427	1.514	1.515	1.506
	BPH	AlPH	GaPH	BAsH	AlAsH	GaAsH
	92.0	90.3	91.4	91.6	88.3	91.8

[a] The subscripts *t* and *c* refer to trans and cis of H with respect to the PH$_2$ or AsH$_2$ group.
[b] MP2/6-31G**.

3.3. Hydrogenation and Dimerization

Davy and Jaffrey computed the hydrogenation energy of H$_2$AlNH$_2$ as 62 kcal/mol (18). Assuming the pi energy is 12 kcal/mol, an Al–H bond energy of 83 kcal/mol is predicted (using the bond energies given earlier and N–H bond energy of 92 kcal/mol). This is not unreasonable, especially given the radical change in σ bonding, but is probably too high.

Dimerization of the ethene analogs has been studied by only two groups. Ni et al. studied the combinations of the heavier analogs, using M = In, Ga, and Al, E = P, As (26). Included was an evaluation of the M–H bond energies and the M–E bond energies. The Al–H bond energy they computed was 66 kcal/mol, much more in line with other aluminum–X bonds. The other study was of the dimerization of H$_2$AlNH$_2$, performed by Hamilton and Shaikh (27). This was chosen because the dimer was known (as was the tetramer of the acetylene analog). A search for a transition state was attempted, with none being found. This is consistent with the simplified bonding model given in Figure 2. The energy of dimerization is 66 kcal/mol. This is a little over two times the energy from an Al–N bond in an adduct, such as 26 kcal/mol in ammonia alane (28). The resulting compound is symmetric—all of the Al–N bonds are equivalent. Haaland estimates the Al–N bond energy as 67 kcal/mol for a tetravalent Al atom (7). With the Al–N bond energy of the trivalent Al compound H$_2$AlNH$_2$, one expects the dimerization energy to be the energy of the four tetravalent Al–N bonds in the dimer plus ring strain minus the two trivalent Al–N bonds plus pi bond energy. The resulting equation is $-66 = -4(67) + \text{strain} + 2(84) + 2(12)$. The resulting strain energy is quite reasonable, 10 kcal/mol. If the bond energy of the trivalent Al–N bond includes the pi bonding, the strain will still be a reasonable 34 kcal/mol, comparable to cyclopropane (5).

4. CONCLUSIONS

The strength of the pi bond in H_2MEH_2 ethene analogs is strongest when $E = N$. This is related to the tendency of N to be flatter, as evidenced by inversion barriers and bond angles in this and previous work. The next strongest factor is whether $M = B$ or heavier elements, with boron giving better pi bonds. The compounds have essentially unchanged M centers (which remain nearly planar, no matter how pyramidal the E center is). The bonding can profitably be visualized as a dative sharing of a lone pair with an empty orbital—this explains the weak pi-bonded systems extremely well, for the pyramidal E group has obvious lone pair character, just like the parent phosphine or arsine.

The molecules with $E = N$ are planar. The molecules with $E = P$ or As are nonplanar, with significantly acute angles. For $E = P$ or As, the planar form is the transition state for inversion, which interconverts the nonplanar minima. The twisting about the M–E bond when $E = P$ or As has rotational barriers on the order of ethane, particularly if the M is not boron.

Further work on qualitative understanding of pi bonding in these systems is possible by examining the *electronic* effects of substituents, since most experimental work has focused by necessity on the steric protection requirements. Determining of ring strain in the cage and ring compounds, and improved M–E bond enthalpies, will allow for an independent check on the pi bond energies obtained by computing rotation barriers. Finally, care must be used when using RECPs, which can manifest itself in clearly erroneous results. Of course, that is good advice for any experimental or theoretical procedure.

REFERENCES

1. AG Brook, MA Brook. Adv Organomet Chem 39:71, 1996.
2. PJ Brothers, PP Power. Adv Organomet Chem 39:1, 1996.
3. TP Hamilton, P Pulay. J Chem Phys 84:5728, 1986.
4. LG Wade. Organic Chemistry. 2nd ed. Englewood Cliffs, NJ: Prentice Hall, 1991.
5. FA Carey, RJ Sundberg. Advanced Organic Chemistry. 2nd ed. New York: Plenum Press, 1984.
6. DR Lide, HV Kehiaian. CRC Handbook of Thermophysical and Thermochemical Data. Boca Raton, FL: CRC Press, 1994.
7. GA Robinson, ed. Coordination Chemistry of Aluminum. New York: VCH, 1993.
8. DE Goldberg, DH Harris, MF Lappert. J Chem Soc, Chem Comm 480, 1974.
9. W Cherry, N Epiotis, WT Borden. Acc Chem Res 10:167, 1977.
10. G Trinquier, J-P Malrieu. J Am Chem Soc 109:5303, 1987.
11. RS Grev. Adv in Organomet Chem 33:125, 1991.
12. (a) WJ Stevens, H Basch, M Krauss, J Chem Phys 81:6026, 1984; (b) WJ Stevens, M Krauss, H Basch, PG Jasien. Can J Chem 70:612, 1992; (c) TR Cundari, WJ Stevens. J Chem Phys 98:5555, 1993.

13. K Raghavachari, GW Trucks, JA Pople, ES Replogle. Chem Phys Lett 158:207, 1989.
14. GAUSSIAN 98. MJ Frisch, GW Trucks, HB Schlegel, GE Scuseria, MA Robb, JR Cheeseman, VG Zakrzewski, JA Montgomery, RE Stratmann, JC Burant, S Dapprich, JM Millam, AD Daniels, KN Kudin, MC Strain, O Farkas, J Tomasi, V Barone, M Cossi, R Cammi, B Mennucci, C Pomelli, C Adamo, S Clifford, J Ochterski, GA Petersson, PY Ayala, Q Cui, K Morokuma, DK Malik, AD Rabuck, K Raghavachari, JB Foresman, J Cioslowski, JV Ortiz, BB Stefanov, G Liu, A Liashenko, P Piskorz, I Komaromi, R Gomperts, RL Martin, DJ Fox, T Keith, MA Al-Laham, CY Peng, A Nanayakkara, C Gonzalez, M Challacombe, PMW Gill, BG Johnson, W Chen, MW Wong, JL Andres, M Head-Gordon, ES Replogle, JA Pople. Pittsburgh, PA: Gaussian, Inc., 1998.
15. M McKee. J Phys Chem 96:5380, 1992.
16. TL Allen, WH Fink. Inorg Chem 32:4230, 1993.
17. WH Fink, PP Power, TL Allen. Inorg Chem 36:1431, 1997.
18. RD Davy, KL Jaffrey. J Phys Chem 98:8930, 1994.
19. (a) AD Becke. J Chem Phys 98:5648, 1993; (b) C Lee, W Yang, RG Parr. Phys Rev B 37:785, 1988.
20. J Muller. J Am Chem Soc 118:6370, 1996.
21. (a) JP Hay, WR Wadt. J Chem Phys 82:270, 1985; (b) JP Hay, WR Wadt. J Chem Phys 82:284, 1985; (c) JP Hay, WR Wadt. J Chem Phys 82:299, 1985.
22. HA Bent. J Chem Ed. 37:616, 1960.
23. TL Allen, WH Fink. Inorg Chem 32:4230, 1993; MB Coolidge, WT Borden. J Am Chem Soc 112:1704, 1990.
24. KM Waggoner, H Hope, PP Power. Angew Chem Int Ed Engl 27:1699, 1988.
25. PP Power. Angew Chem Int Ed Engl 29:449, 1990.
26. H Ni, DM York, L Bartolotti, RL Wells, W Yang. J Am Chem Soc 118:5732, 1996.
27. TP Hamilton, AW Shaikh. Inorg Chem 36:754, 1997.
28. CMB Marsh, TP Hamilton, Y Xie, HF Schaefer. J Chem Phys 96:5310, 1992.

16

Main Group Half-Sandwich and Full-Sandwich Metallocenes

Ohyun Kwon and Michael L. McKee
Auburn University, Auburn, Alabama

1. INTRODUCTION

The term *metallocene* defines a bis(cyclopentadienyl)metal complex; however, we extend it to cyclopentadienyl (Cp) complexes with main group elements (1). The cyclopentadienyl ligand (Cp) has played a major role in the development of organometallic chemistry since the structure of ferrocene (Cp_2Fe) was identified in 1952 (2) and has continued to be the archetype of cyclic polyene ligands. Cyclopentadienyl is usually present as a pentahapto ligand in complexes with transition metal elements, where it exists formally as an anion (Cp^-) acting as a six-electron donor. Over the last three decades, much effort in experiment and theory has been focused on transition metal metallocenes. In contrast, somewhat less is known about main group metallocenes due to the diversity of unusual structural properties and bonding phenomena between Cp and the main group elements (E). Recently, main group metallocenes have become important because of their structural fluxionality and synthetic utility in organometallic chemistry. However, most studies to date on main group metallocenes have concentrated on the structures of the neutral species in the gas phase or salt forms in the solid state, and only a small number of investigations have focused on syntheses and reactivities (3). Moreover, only a limited number of theoretical studies have focused on main group metallocenes, because the different types of bonding character between the main group element (E) and Cp ligand (due to less involvement of d orbitals compared to the transition metal metallocenes) result in numerous structural possibilities with different bonding patterns. In main group metallocenes, π-type interactions become weaker due to the absence of d orbitals, and deviation from typical pentahapticity (η^5) is often observed. The broad range of electronegativities of main group elements (E) also leads to either ionic or covalent bonding for Cp–E interactions. For example, bonding of Cp to s-block elements (Groups 1, 2) shows highly ionic character, while the p-block elements (Groups 13, 14), which have electronegativities similar to that of transition metals, exhibit large covalent bonding character to the Cp ligand.

This chapter will provide a review of previous theoretical studies of main group metallocene compounds and a guide for theoretical calculations of main group metallocenes. Owing to the tremendous progress of computational hardware and software in the last decade, computational chemistry has rapidly expanded to various fields of chemistry and now plays an important role as a real partner in most chemical research where experiment and theory are complementary tools to each other. This is because the accuracy of computed equilibrium geometries, energetics, and other molecular properties, such as vibrational frequencies and NMR chemical shifts, can often be comparable or even superior to experimental data. Therefore, a general overview of the application of computational chemistry to the main group metallocenes should be helpful for those who want to model these systems theoretically. Additionally, many of the methods and approaches discussed should be generally useful for quantum modeling for main group compounds.

The review part of this chapter will be limited to the calculation of metallocenes of Groups 1 and 2 (s-block elements) and Groups 13 and 14 (p-block elements). There have been good reviews of synthetic procedures, structural characterization, and other experimental features for main group metallocenes (4–6). We will restrict our discussion to simple main group metallocenes that have only one Cp ring bonded (half-sandwich) or two Cp rings bonded (full-sandwich) to the main group element, although there are various metallocenes that include more than two Cp ligands or contain substituted Cp ligands such as pentamethylcyclopentadienyl (abbreviated as Cp*) and trimethylsilylcyclopentadienyl ligands. However, we will include a few examples of the effect the bulky ligand Cp* has on the structure and electronic properties of main group metallocenes. Finally, some examples of calculations for main group metallocene will be presented in order to show fundamental differences between various computational techniques and theoretical methods.

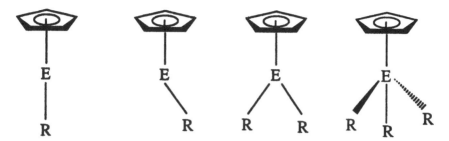

FIGURE 1   Possible structures of half-sandwich metallocenes. (R is either an electron lone pair or an auxiliary substituent.)

2. STRUCTURAL FEATURES OF MAIN GROUP METALLOCENES

The structural classification of metallocene compounds can be divided into half-sandwich complexes and full-sandwich complexes with respect to the number of Cp ligands bonded to the main group element (E). Half-sandwich metallocenes exhibit the possible structural forms in Figure 1. The parent structure of a half-sandwich metallocene is assumed to have C_{5v} symmetry. These half-sandwich metallocenes can be bonded to auxiliary ligands or possess an electron lone pair. It is known that stable half-sandwich metallocenes exist for most main group elements with covalent bonded substituents or ligands, depending on the valence and the atomic radius of main group elements (5a). Half-sandwich metallocenes exist in a monomeric form in the gas phase, but a polymeric arrangement or highly symmetric cluster form occurs in the solid state. A full-sandwich metallocene generally has two Cp ligands, in which two Cp rings are either parallel or nonparallel (bent), as shown in Figure 2. The parallel structure can have staggered

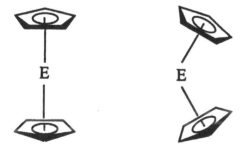

FIGURE 2   Possible structures of full-sandwich metallocenes.

D_{5d} or eclipsed D_{5h} conformations, while the bent structure can have C_{2v} or C_s symmetry. In typical metallocenes, all of the C–C bond distances are equal and the rings are parallel. However, there are several cyclopentadienyl compounds in which the rings are tilted with respect to one another. Full-sandwich metallocenes of heavier elements of Group 2, such as Cp_2Ca and Cp_2Ba, have bent structures, which can be explained by the fact that larger metal cations increase core polarizability and decrease ligand–ligand repulsions, which results in less linear rigidity, or even bent, structures. One or more auxiliary substituents, such as a solvent molecule, can also lead to a bent conformation of full-sandwich metallocenes. Structure determinations of the simplest Group 14 full-sandwich compounds, such as Cp_2Ge and Cp_2Pb, have shown that they adopt a bent conformation in monomeric form due to the lone pair repulsion in Ge(II) and Pb(II) (5).

Some geometrical parameters must be defined in order to compare various structural patterns of metallocenes, as shown in Figure 3, that is, the angle α between the Cp ring planes and the angle β between Cp(centroid)–E–Cp(centroid) and the distances of C(Cp)–E and Cp(Centroid)–E. For example, parallel magnesocene (Cp_2Mg) shows D_{5d} symmetry and has α = 0° and β = 180° (7), while nonparallel stannocene (Cp_2Sn) has α = 47° and β = 146° (8). For bond distances in Cp_2Mg, X-ray diffraction shows 2.304(8) Å for C(Cp)–Mg and 1.977(8) Å for Cp(centroid)–Mg (7).

Since main group metallocenes show weak proclivity toward directional bonding between the Cp ring and the central atom due to the lack of available d orbitals for π-type interaction, different hapticities (referred to as *ring slippage*) from η^1 to η^4 (Fig. 4) rather than prototype η^5 can often be observed. If a metallocene has two or more equidistant C(Cp)–E bonds, it can be classified as a π complex with hapticities of η^2–η^5, while a metallocene with an η^1–Cp has a σ type of interaction between C(Cp)–E (9). These different hapticities are due to

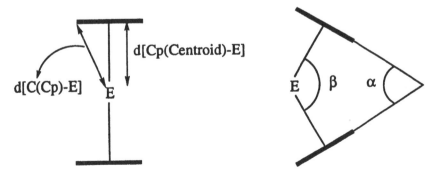

FIGURE 3   Geometrical descriptions of the metallocenes.

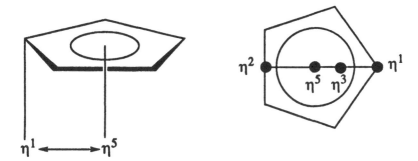

Figure 4 Cp–E bonding arrangements with respect to different hapticities.

the nature of the main group elements, such as electronegativity and ionic radius, other substituents bonded to the main group element, and the substituted Cp ligand effect (10). However, the comparison of energetics among different hapto-metallocenes shows very small energy differences (4–8 kJ/mol) (11), which makes it difficult to determine a certain hapticity for main group metallocenes.

3. MOLECULAR ORBITAL INTERACTIONS OF CYCLOPENTADIENYL RINGS AND MAIN GROUP ELEMENTS (E)

In this section, we will state simply the general qualitative molecular orbital interactions between Cp and main group element for main group metallocenes. While these schemes have been presented and discussed elsewhere (5), we include a simple MO interaction diagram for a half-sandwich complex in Figure 5. Unlike ferrocene, where the bonding interaction results from the interaction of the a_1 and e_1 orbitals of Cp with the d_{z^2}, d_{xz}, and d_{yz} orbitals of iron, in main group metallocenes the comparable interaction involves the s, p_x, and p_y orbitals of the main group atom (Fig. 5).

Due to the broad range of electronegativities of the main group elements, the bonding in metallocenes can vary from being strongly ionic to being mainly covalent. The largest contributing factor to Cp–E interaction is the energy of the atomic orbitals of the main group element relative to the highest occupied molecular orbital (HOMO) of the Cp ligand. In the case of the half-sandwich metallocene, if the main group element's valence orbitals are much higher in energy than the degenerate occupied e_1 orbitals, this should make it easy to transfer electrons from the main group element to the Cp ligand, generating ionic bonding of Cp–E. For CpLi, the first ionization energy of Li (5.4 eV) (12) is much lower than that of Cp (8.4 eV) (13). On the other hand, Group 13, 14 elements have

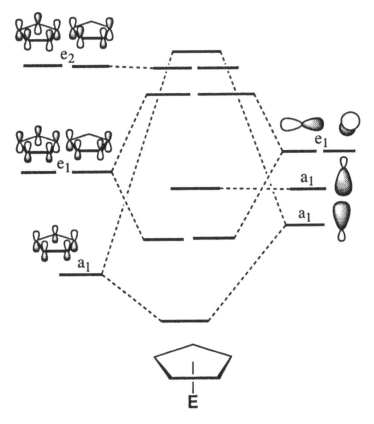

Figure 5 MO diagram for the half-sandwich main group metallocene (C_{5v} symmetry).

valence orbitals comparable in energy with that of the Cp's HOMO, which gives rise to much larger covalent bonding of Cp–E. For CpMg$^+$, the first ionization energy of Mg (7.6 eV) (12) is very compatible with that of Cp (8.4 eV) (13). In this complex, the direction of electron transfer is best described as being from the Cp ligand to the main group element. For full-sandwich metallocenes, we can split the metallocene compound into Cp$_2$ and E units under either D_{5d} or D_{5h} symmetry, much like the ferrocene MO interaction example (12). However, the absence of available d orbitals in most main group elements leads to weaker π bonding with the degenerate e_1 orbitals of Cp ligand due to poorer overlap of the p_x and p_y main group orbitals with the ligand compared to overlap with d_{xz} and d_{yz} orbitals. Since the degeneracy of these e_1 orbitals easily breaks down by the second order Jahn–Teller effect, D_{5d} or D_{5h} parallel conformations tend to distort to lower-symmetry C_{2v} or C_s bent structures for some main group metallocenes (14,15).

4. RESULTS AND DISCUSSIONS OF PREVIOUS COMPUTATIONAL STUDIES OF MAIN GROUP METALLOCENES

Tables 1 and 2 show available experimental and calculated geometrical parameters for known s-block metallocenes of main group elements in Groups 1 and 2, while Table 3 gives the same information for p-block elements in Groups 13 and 14. As mentioned before, main group metallocenes can be classified into two categories: s-block metallocenes and p-block metallocenes, depending on the main group element. The s-block metallocenes (Groups 1, 2) are considered to have mainly ionic bonding, while most p-block metallocenes (Groups 13, 14) show covalent interaction of Cp–E. However, there is not enough data from high-level theoretical calculations to have a quantitative understanding of the factors involving covalent or ionic bonding between the Cp ligand and the main group element.

4.1. s-Block Metallocenes

LiCp and Cp_2Li^- are the simplest metallocenes and have been of considerable interest in terms of the interaction between lithium and the π electron system.

TABLE 1 Experimental and Calculated Geometrical Parameters for Group 1 Metallocenes

	C(Cp)–E (Å)	Cp(centroid)–E (Å)	β (∠Cp–E–Cp) (°)
LiCp		2.06(XD) [16]	
		1.700(MNDO) [17]	
		1.957(PM3) [17]	
		1.79(SCF) [18]	
Cp_2Li^-		2.008(XD) [19]	Linear (expt.; calc.)
		1.974(MNDO) [17]	
		2.034(PM3) [17]	
		2.015(B3LYP) [20]	
Cp_2Na^-	2.630(XD) [22]	2.366(XD) [22]	Linear (expt.; calc.)
	2.600(VWN) [23]	2.461(VWN) [23]	
	2.700(BP) [23]	2.566(BP) [23]	
Cp_2K^-	2.800(VWN) [23]	2.671(VWN) [23]	Linear (VWN, BP)
	2.820(BP) [23]	2.692(BP) [23]	

ED = electron diffraction; XD = X-ray diffraction; VWN, BP, B3LYP = DFT level.
Source: Reference numbers appear in brackets.

TABLE 2   Experimental and Calculated Geometrical Parameters for Group 2 Metallocenes

	C(Cp)–E (Å)	Cp(centroid)–E (Å)	β (∠Cp–E–Cp) (°)
CpBeH	1.920(XD) [24]	1.49(XD) [24]	
	1.991(MNDO) [24]	1.557(MNDO) [24]	
	1.976(SCF) [24]	1.563(SCF) [24]	
Cp₂Be[a]	1.94(XD) [25]	1.52(XD) [25]	
	1.624(SCF) [24]	1.466(SCF) [24]	
MgCp⁺	2.251(MP2) [26]		
Cp₂Mg	2.339(ED) [27]	2.008(ED) [27]	Linear (exptl.; calc.)
	2.304(XD) [7]	1.977(XD) [7]	
	2.376(SCF) [28]	2.050(SCF) [28]	
	2.270(VWN) [23]	1.930(VWN) [23]	
	2.360(BP) [23]	2.040(BP) [23]	
	2.357(B3LYP) [28]	2.022(B3LYP) [28]	
Cp₂Ca	2.80(XD) [29]		119(XD) [29]
	2.714(SCF) [30]	2.428(SCF) [29]	Linear (SCF, MP2) [30]
	2.611(MP2) [30]	2.321(MP2) [30]	150(VWN, BP) [23]
	2.540(VWN) [23]	2.240(VWN) [23]	149.6 (B3LYP) [28]
	2.580(BP) [23]	2.290(BP) [23]	
	2.613(B3LYP) [28]	2.317(B3LYP) [28]	
Cp₂Sr	2.883(SCF) [30]	2.623(SCF) [30]	Linear (SCF, MP2) [30]
	2.801(MP2) [30]	2.533(MP2) [30]	145 (VWN, BP) [23]
	2.600(VWN) [23]	2.310(VWN) [23]	
	2.680(BP) [23]	2.440(BP) [23]	
Cp₂Ba	3.083(SCF) [30]	2.842(SCF) [30]	Linear (SCF) [30]
	2.976(MP2) [30]	2.725(MP2) [30]	142.7(MP2) [30]

[a] Cp₂Be is a C_s slipped sandwich structure.
ED = electron diffraction; XD = X-ray diffraction; VWN, BP, B3LYP = DFT level.
Source: Reference numbers appear in brackets.

Many calculations on LiCp have been done at various levels (17,18,42,43–46). Using a double-zeta basis set, Schaefer and coworkers (42) found that LiCp had a pentahapto structure with a Cp–Li distance of 1.79 Å, with the CH bonds bent slightly out of the Cp plane away from the Li atom. Jemmis and Schleyer (18) optimized the Cp–Li distance to 1.79Å at the HF/3-21G level. Waterman et al. (44) reinvestigated the out-of-plane bending of the CH bonds and concluded (in agreement with Schaefer) that the interaction was mainly Coulombic. Jemmis and Schleyer (18) explained the bonding of CpLi as an aromatic six-membered nido polyhedron in terms of "aromaticity in three dimensions." High-level ab initio calculations with larger basis sets reproduced the experimental Cp–Li dis-

TABLE 3 Experimental and Calculated Geometrical Parameters for Groups 13 and 14 Metallocenes

	C(Cp)–E (Å)	Cp(centroid)–E (Å)	β ($\angle$Cp–E–Cp) (°)
BCp	2.136(SCF) [18]	1.764(SCF) [18]	
AlCp	2.468(MP2) [31]	2.037(MP2) [31]	
Cp_2Al^+	2.204(SCF) [32]	1.843(SCF) [32]	Linear
	2.174(MP2) [32]		
GaCp	2.420(MP2) [33]	2.096(MP2) [33]	
InCp	2.62(ED) [5b]	2.32(ED) [5b]	
	2.688(SCF) [34]		
TlCp	2.705(ED) [5b]	2.41(ED) [5b]	
	2.832(SCF) [34]		
Cp_2Tl		2.66(XD) [35]	155(XD) [35]
	3.01–3.27(SCF) [35]	2.85,2.90(SCF) [35]	144(SCF) [35]
CCp^+	1.822(SCF) [36]	1.357(SCF) [36]	
Cp_2C	η^3–Cp_2C is the most stable conformer [37]		
$SiCp^+$	2.126(SCF) [36]	1.745(SCF) [36]	
Cp_2Si	η^3–Cp_2Si is the most stable conformer [37]		
$GeCp^+$	2.32(SCF) [38]	1.99(SCF) [38]	
Cp_2Ge		2.23(XD) [5b]	152(XD) [5b]
		2.34[SCF] [39]	Linear (SCF) [39]
$SnCp^+$	2.474(SCF) [34]		
Cp_2Sn	2.70(XD) [8]	2.42(XD) [5b,8]	146(XD) [5b,8]
	2.56–3.09(SCF) [40]		
Cp_2Pb	2.78(ED) [41a]	2.55–2.82(ED) [41a]	135 $\pm$ 15(ED) [41a]
	2.76(XD) [41b]	2.50(XD) [41b]	
	2.71–2.97(SCF) [40]		

ED = electron diffraction; XD = X-ray diffraction.
Source: Reference numbers appear in brackets.

tance, while the semiempirical MNDO method overestimated the C(Cp)–Li strength (17,45). Kwon and Kwon (20) showed that density functional theory (B3LYP/6-31G*) results for the Cp_2Li^- anion were far superior to Hartree–Fock (HF) or semiempirical in terms of Cp–Li distance and CH bending (which is inward toward the Li atom in contrast to the case of LiCp). The staggered Cp_2Li^- anion was predicted (DFT) to be more stable than the eclipsed form by 0.26 kJ/ mol, which is much less than in Cp_2Fe (~5 kJ/mol).

Bridgeman (23) calculated Cp_2Na^-, Cp_2K^-, and Cp_2Rb^- using DFT methods (BP and VWN) and showed that calculated geometries were in reasonable agreement with the available experimental data. The alkali metal metallocenes were predicted to have parallel equilibrium structures, and Mulliken orbital population analysis indicated that d orbitals of heavier alkali metals contributed largely to π-type interactions in Cp_2K^- and Cp_2Rb^-.

Although a number of beryllium compounds containing only one Cp ligand have been synthesized, there is no evidence of the simple $BeCp^+$ cationic compound to our knowledge. Theoretical calculations at different levels have been done for BeCpH (24). It has been shown that CpBeH prefers C_{5v} symmetry and η^5-type bonding with a large covalent character. The direction of the calculated dipole moment, which has a negative end toward the BeH group, contradicts an ionic bonding view of Cp^-BeH^+ and implies considerable electron donation from the Cp ligand to the Be atom. The structure of Cp_2Be is unusual and still somewhat uncertain. Both X-ray and electron diffraction data have been interpreted in terms of a "slipped-sandwich" complex with a C_s symmetry in which one Cp ring is pentahapto and the other is either weakly π bonded or perhaps even ρ bonded (see Fig. 6) (25,47). However, a previous SCF calculation (24) indicated that the lowest-energy form has one η^5-Cp ring and one η^1-Cp ring bonded to the Be atom, which differs from the experimental structures (see Fig. 6). Recently, MP2 and DFT (B3LYP) calculations on Cp_2Be revealed that DFT reproduced the experimental C_s slipped-sandwich structure as the lowest conformer, while MP2 failed (48). When comparing experiment and theory, there is still disagreement about the equilibrium geometry of Cp_2Be. Cotton and Wilkinson postulated that the radius of the Be atom is so small that even at the closest distance of

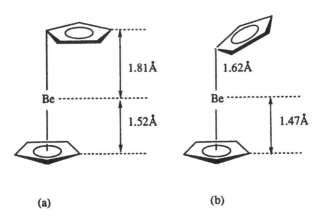

(a)   (b)

FIGURE 6 Experimental and calculated structures of Cp_2Be: (a) solid-state structure, (b) optimized structure at the SCF level.

the two parallel Cp rings, the Be atom cannot be bonded to both Cp ligands simultaneously (49).

There is no available structural characterization of the $MgCp^+$ cation due to its instability in the gas phase. Recently, a combination of tandem mass spectrometry and high-level ab initio calculations (26) revealed that the η^5 bonded structure with C_{5v} symmetry is the lowest conformer at the MP2/6-31G** level. With a natural bond orbital (NBO) analysis (50) of the Cp–Mg interactions, the authors (26) obtained a natural charge of 1.8 for Mg, which implies ionic bonding between the Mg^{2+} cation and a Cp^- anion. Cp_2Mg has a staggered parallel conformation in the solid state but is reported to be eclipsed in the gas phase (5a). However, the rotational barrier is computed to be only 0.13 kJ/mol at the HF level (51). A D_{5d} symmetry structure of Cp_2Mg has been identified by previous theoretical calculations (23,28,51) in which DFT produces a better optimized geometry of Cp_2Mg than the HF method. However, the nature of the bonding between the Cp ligand and Mg atom (ionic or covalent) is still controversial. Because the ionic chemistry of Mg^{2+} is quite similar to that of Li^+, one might expect that the bonding of Cp_2Mg is comparable to that of Cp_2Li^- anion. However, the Mulliken analysis at the HF level of theory revealed an atomic charge of 1.39 for Mg (51), while the Mulliken charge for Mg at the DFT level predicted 0.66, which would suggest somewhat covalent character in metal–ligand bonding (23).

Heavier alkaline earth metallocenes such as Cp_2Ca, Cp_2Sr, and Cp_2Ba have bent polymeric structures due to the d orbital contribution from the Ca, Sr, and Ba atoms (5a). However, with sufficiently bulky substituents on the Cp ring, the heavier Group 2 metallocenes can also be isolated in a monomeric form (5a). Previous calculations (23,30) showed that they have strong ionic bonding character and structurally nonrigid systems. The energy barrier from the bent to the parallel form for Cp_2Ca was computed to be about 10 kJ/mol at DFT levels (B3LYP, BP, VWN) (23,28). Mulliken population analysis at the DFT level of theory revealed that the greater π bonding in Cp_2Ca and Cp_2Sr was influenced by metal d orbitals, while the d orbital populations of Mg were found to be negligible in Cp_2Mg (23).

4.2. P-Block Metallocenes

Due to the inert-pair effect, Group 13 elements sometimes produce low-valent monomeric metallocenes as well as high-valent oligomeric metallocene clusters and polymeric metallocene chains. Half-sandwich metallocenes of Group 13 elements can be found in the gas phase, but exist as clusters and polymeric forms in the solid state. Schleyer calculated the CpE, E = Be, B, C, and N, molecules at the SCF level and found the shortest C–C ring bond length in CpB (18). The

full-sandwich boron metallocene, $Cp*_2B^+$, was prepared by Jutzi, who characterized its structure as $[(\eta^5-Cp*)(\eta^1-Cp*)B]^+$ by NMR spectroscopy (52). Boron
metallocene compounds are predicted to be highly fluxional involving $\eta^1-\eta^5$ rearrangement of the Cp rings (1), but there is still a lack of theoretical results.
    The characterization of AlCp* crystals revealed that solid AlCp* forms a
tetramer, Al_4Cp*_4. Ab initio MP2 calculations of the model compounds AlCp
and Al_4Cp_4 have been done, and the calculated C(Cp)–Al distance (2.368 Å)
(31) is in quite good agreement with the C(Cp)–Al distance (2.39 Å) of AlCp
in the gas phase (53). In addition, the tetramerization energy was calculated to
be -243 kJ/mol, which is smaller than that (-569 kJ/mol) of the ideal molecule
Al_4H_4 (53), which implies that there are weaker intermolecular interactions between the AlCp units in the Al_4Cp_4 cluster. Cp ring slippage paths in $\eta^2-CpAlH_2$
and $\eta^2-CpAlMe_2$ have been calculated at various levels of theory, and it has
been shown that there is a shallow potential energy surface connecting the ring
hapticities in these compounds (11). The calculated energies of the η^2 to η^5 structures were within 4–8 kJ/mol of each other, which indicates that the movement
of the Al atom over the Cp ring plane is basically fluxional. Although the Cp_2Al^+
cation has been synthesized, it has not yet been structurally characterized (55).
Ab initio MP2 calculations (32) have shown that the calculated C(Cp)–Al distance (2.174 Å) is quite close to the experimental distance of 2.137–2.184 Å
from $Cp*_2Al^+$ cation X-ray diffraction data (32). The Cp_2Al^+ cation is expected
to have a typical D_{5d} pentahapto-metallocene based on the known structure of
$[\eta^5-Cp*_2Al]^+$ and model calculations. Mulliken population analysis shows that
the contribution of $3d$ orbitals is smaller than $3s$ and $3p$ orbitals, which indicates
that d orbitals are not important in the bonding of the Cp_2Al^+ cation. The net
charge on Al(1.14) and the Cp ligand (-0.07) signifies strong covalent interactions if we consider Cp_2Al^+ system as $Al^{3+}(Cp^-)_2$. Half-sandwich metallocenes
for Ga, In, and Tl are predicted to have the same bonding pattern as AlCp with
C_{5v} symmetry. An MP2 optimization (33) of the C(Cp)–Ga distance in GaCp
(2.420 Å) is in good agreement with the C(Cp)–Ga distance in the gas-phase
structure of GaCp* (56). The gas-phase structures of InCp and TlCp indicate that
they have a half-sandwich η^5 structure (5b). Calculated distances of C(Cp)–E
from SCF calculations compare satisfactorily to the experiment in the gas phase
(34). In addition, calculated Mulliken charges for the In and Tl atoms were both
0.5, indicating that the bonding between Cp and In and Tl can be expected to
be covalent rather than ionic. The calculation also suggests that the influence of
d orbitals can be negligible for TlCp. The full-sandwich compound of Tl was
characterized by X-ray crystallography and studied with ab initio methods (35).
The most stable structure of Cp_2Tl^- anion has a bent conformation in the solid
state. Electron-correlated MP2, MP3, MP4 single-point energy calculations at the
HF optimized geometry indicated that the bent geometry is slightly more stable

than the D_{5d} conformer (at the MP4 level, the difference in energy is 3.4 kJ/mol), which suggests that the geometry of the Cp_2Tl^- anion is highly flexible (35). The Group 14 metallocenes are the most widely known series of isolated main group sandwich compounds. Schleyer calculated the model compounds η^5-CpC^+ and η^5-CpSi^+ and found that the pentahapto carbon compound is less stable than the single-bonded fulvenyl cation or the phenyl cation (36). On the other hand, η^5-CpSi^+ has been predicted to be the most stable one among possible isomers. A mass spectrometry study of Group 14 half-sandwich metallocenes has been made and their atomization energies were calculated (57). It was concluded that *nido*-cluster is the most stable structure and that the stability of these metallocenes increases from $CpPb^+$ to CpC^+.

Experimentally known Group 14 full-sandwich metallocenes adopt a bent conformation with a subsequent difference in C(Cp)–E distances. The first Group 14 full-sandwich compound is the plumbocene, Cp_2Pb, which has a zigzag polymeric form with bridging Cp rings in the solid state (58,41b). However, Cp_2Pb has been found to be monomeric in the gas phase, with an angle between Cp rings of 45 ± 15° and a mean C(Cp)–Pb distance of 2.78 Å (41a). The lightest Group 14 full-sandwich metallocene, $Cp*_2Si$, adopts a highly symmetric D_{5d} structure in the crystal (59). A second conformer, which was also prepared by Jutzi, shows a bent angle (25.3°) between Cp rings and broad ranges of C(Cp)–Si distances from 2.324 Å to 2.541 Å (59). Cp_2Ge has been characterized as a bent structure with an angle between Cp rings of 152° in the solid state (60). All Ge metallocenes having substituted Cp's have been reported to have a bent conformation. However, decaphenylstannocene (phenyl-substituted Cp_2Sn) has been found to be linear (rings parallel) (61), while Cp_2Sn and $Cp*_2Sn$ have bent conformations (41,62). Actually, most Group 14 full-sandwich metallocenes have bent structures, which is supported by the fact that the driving force responsible for the bending is the nonbonding electron pair (62).

Schoeller et al. have calculated Cp_2C and Cp_2Si at the B3LYP level and found that the ideal D_{5d} Cp_2C structure is 672 kJ/mol higher in energy than a C_{2v} symmetry structure with weak (η^3–Cp)–C bonding (37). In fact, the lowest-energy conformer of Cp_2C corresponds to phenylcyclopentadiene resulting from carbene rearrangement. In comparison, although Cp_2Si adopts a lowest-energy conformation of C_2 symmetry, the energy required to change from D_{5d} to C_2 symmetry structure is only 17 kJ/mol at the B3LYP level (37). In addition, an NBO analysis (50) shows that there is no d orbital contribution in Cp–E bonding in Cp_2C and Cp_2Si. However, SCF calculations by Lee and Rice demonstrated that d orbital participation is larger in D_{5d} or C_{2v} conformers than in C_s or C_2 conformers for the Cp_2Si system (63), thus explaining why the potential surface for ring slippage of the Cp_2Si system is flatter than that of the Cp_2C system, which has no available low-lying d orbitals.

An ab initio SCF calculation of Cp_2Ge calculated a linear structure (38) rather the experimentally reported bent structure (5b,60). However, it was pointed out that the lone pair electrons on Ge were stabilized in a bent geometry and that the $3d$ electrons of Ge were of secondary importance in bonding between Cp and Ge (39). Recent SCF calculations of Cp_2Sn and Cp_2Pb showed both metallocenes have C_{2v} eclipsed structures as the lowest-energy conformer (40). However, the calculated C(Cp)–E distances are quite different from the experimental values (see Table 3) (5b,8,41).

5. PRACTICAL STRATEGY FOR THE CALCULATIONS OF MAIN GROUP METALLOCENES

This section introduces the available computational methods and explains how to treat main group metallocenes effectively using computational methods.

5.1. Relevant Computational Methods

The scope of this section is restricted to semiempirical and ab initio quantum mechanical methods. Semiempirical methods have been used in many active areas of current research over more than two decades. The most popular methods are modified neglect of differential overlap (MNDO) (64), Austin method 1 (AM1) (65), and parameterization method 3 (PM3) (66). These methods are based on the approximation of "neglect of diatomic differential overlap" (NDDO) (67). Both AM1 and PM3 methods are improved parameterizations of the MNDO method. All three methods consider only the valence electrons and are currently implemented in several computational packages, such as MOPAC2000 (68) and Gaussian 98 (69). Although semiempirical methods are generally less accurate than high-level ab initio methods for energetics, semiempirical optimized geometries might be good initial guesses for the higher-level ab initio optimizations.

Ab initio methods are currently widely used in computational chemistry. At a suitable level, they can provide accurate equilibrium geometries, reliable energetics, and other molecular properties, such as vibrational frequencies and NMR chemical shifts, that are in accord with experimental results. The accuracy of ab initio methods highly depends on the size of basis sets and the levels of electron correlation, such as Hartree–Fock (HF), MP2, or coupled clusters methods (70). However, the cost in computer time becomes enormous as the size of basis sets increases and more electron correlation is applied. For most main group systems, some sort of correction for electron correlation effects needs to be included. However, for traditional computational methods the demand for computer CPU time and disk space goes up very steeply with the amount of electron correlation (70). Recently, density functional theory (DFT) (71) has emerged as a successful method for predicting various molecular properties, such as molecular geometries and spectroscopic properties, often giving results of a quality comparable or even better than second-order Möller–Plesset (MP2) for a computational cost that is substantially less than that of traditional correlation techniques (72). For instance, recent theoretical calculations on s block metallocenes has shown that DFT-reproduced equilibrium geometries and energetics of main group metallocenes compared well to experiments (23,28).

5.2. Illustrative Examples with Model Calculations

Constructing Input Data

The first step in computing a metallocene compound is determination of the molecular symmetry, since symmetry conservation may be the easiest way to reduce computer time and facilitate analyzing the molecular orbitals of the metallocene. There are two kinds of input construction methods: one is the Z-matrix (internal coordinate) type, the other is the Cartesian coordinate type. The Z-matrix coordinate input method makes it easier to constrain higher symmetries, such as D_{5d} or C_{2v} symmetry, as well as controlling the Cp(centroid)–E distance using dummy atoms. For half-sandwich metallocenes, C_{5v} symmetry can be applied to η^5-CpE and C_s symmetries to other hapticities. Full-sandwich compounds have more possibilities, which are D_{5d}, D_{5h}, C_{2v}, C_s, and C_2 symmetry structures, as suggested by Lee and Rice (63). There are six models for these given symmetry constraints, as shown in Figure 7. Structure **a** is a D_{5d} structure in which two Cp rings are parallel and staggered, and structure **b** is a D_{5h} structure where the Cp rings are parallel and eclipsed. Some metallocenes, such as Cp_2Li and Cp_2Mg, are known to have parallel structures. Structure **c** (''point to edge,'' C_s symmetry) can be formed from the D_{5d} parallel structure (**a**) when the Cp rings are allowed to bend. Structure **d** (''point to point,'' C_{2v} symmetry) can be formed from D_{5h} parallel structure (**b**) when the two Cp rings are allowed to bend. Structure **e** (''edge to edge,'' C_{2v} symmetry) is characterized by two Cp rings approaching closely via one C–C bond of each Cp ring. Finally, structure **f** (C_2 symmetry) has two equivalent but nonparallel Cp rings. Since some metallocenes have very small activation barriers between different conformers, it is advisable that all possible conformers be considered for geometry optimization. In addition, vibrational frequencies should be calculated in order to determine whether the structure is a minimum (0 imaginary frequencies), transition state (1 imaginary frequency), or higher stationary point. Indeed, it is possible that the global minimum has no symmetry and that full geometry optimization in C_1 symmetry is required.

Choosing a Theoretical Level and Basis Set

Semiempirical methods are generally much faster than ab initio methods; hence it is suggested to carry out exploratory calculations at this level first. However, each semiempirical method has different parameterizations, and some methods do not have parameters for all main group elements. In this regard, PM3 is proba-

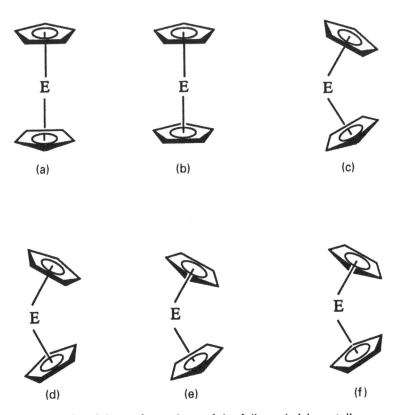

FIGURE 7 Possible conformations of the full-sandwich metallocene.

bly the most complete (in MOPAC2000 there are parameters for all Group 13 and Group 14 elements through the fourth period as well as Li, Be, Mg). MNDO or MNDO/d is parameterized for nearly as many elements; however, some theoretical calculations have pointed out that MNDO overestimates the C(Cp)–E bond strength (17,45).

In ab initio calculations, there are many different possible choices of theoretical levels and basis sets. Since HF calculations fail to reproduce the bent structure and C(Cp)–E bond distances in alkaline earth and Group 14 metallocenes (30,39,40), it is probably safe to assume that electron correlation must be included in geometry optimizations. Thus, electron correlation is required to reproduce the donor–acceptor nature of the main group–Cp interaction as well as the structural consequence of the main group lone pairs. For instance, previous DFT calculations (which include electron correlation) predicted reasonable C(Cp)–E distances and the bent structures for heavier alkaline-earth metallocenes (23) unlike the previous HF calculations (30). Lee and Rice also emphasized that electron correlation plays an important role in determining the angle between two Cp rings and relative energetics among possible conformers for the Cp_2Si system (63). In addition, it has been reported that calculated C(Cp)–E distances for s block metallocenes are highly dependent on the correlation effects (20,23,28,51,73) However, from a comparison of the different electron correlation methods, it has been found that the MP2 method might overestimate the stability of higher-symmetry species (20,63). For example, the MP2 method favors the higher-symmetry D_{5d} conformer over a C_{2v} conformer for Cp_2Si (63). Also, recent MP2 calculations on Cp_2Be showed that the lowest-energy structure has D_{5d} symmetry rather than C_s symmetry, which is inconsistent with experiment (48). From a comparison of the calculated ring slippage potential energy surfaces for $CpAlMe_2$ and $Cp^*\ AlMe_2$ at the HF and MP2 levels, it was found that MP2 surface was more shallow than the HF one (11). MP2 also predicts the wrong symmetry structure for the Cp_2Li anion. While experiment (supported by HF and DFT calculations) found the D_{5d} symmetry structure to be more stable than the D_{5h} structure, MP2 calculations gave the opposite stability ordering (20). Even the bending of CH bonds in Cp rings is not described correctly by MP2 (MP2: outward bending from Li atom; DFT: inward bending to Li atom). Since d orbital correlation effects are important in determining equilibrium geometries of heavier alkaline earth metallocenes (74), electron correlation methods are required to account qualitatively for the structures of these systems. A significant underestimation of Cp–E–Cp bending in heavier alkaline earth metallocenes has been found at the HF level of theory (30). Since a major factor in determining the Cp–E–Cp angle for these structures is the degree of s–d hybridization, it is likely that this hybridization is underestimated at the HF level.

It is known that gradient-corrected functionals, such as BLYP (nonhybrid) and B3LYP (hybrid), are almost always superior to the local exchange functionals, such as VWN, in terms of energy and geometry. In addition, among the gradient-corrected functionals, it is also known that the choice of exchange functional is more important than the choice of correlation functional for DFT calculations (48). For instance, in a comparison of different exchange functionals for Cp_2Mg, the optimized Cp–Mg distance by the B3LYP (hybrid) functional was superior to that of BP (nonhybrid) functional (see Table 2). Likewise, for Cp_2Be, the BP86 (nonhybrid) and B3P86 (hybrid) methods incorrectly predicted the D_{5d} conformer to be the lowest-energy form, while the B3LYP (hybrid) method produced the correct trend (48).

Basis sets are an important ingredient in the calculations of main group metallocenes, since there is a large difference in the required degree of flexibility due to the broad range of electronegativities found in main group elements. In general, larger basis sets produce more accurate electronic descriptions and yield more reliable energies and geometries, but they require much more computer time. Thus, choosing an appropriate basis set is one of the important factors in the calculation of metallocenes. A good comprehensive review of basis sets is provided in the chapter by Feller and Davidson in *Reviews in Computational Chemistry* (75). Since the nature of C(Cp)–E bonding ranges from highly ionic to covalent, small basis sets, such as STO-3G and 3-21 G, may result in significant BSSE (basis set superposition error). Previous calculations indicate that at least double-zeta plus polarization basis sets, such as 6-31G*, should be applied to main group metallocene calculations. For example, Waterman and Streitweiser found a considerable effect of the basis set on the CH bending angle of CpLi at the HF level (44). In an HF and DFT study of Cp_2Li, Kwon and Kwon (20) found that adding polarization functions to the basis set improved relative energies and bond distances in conformers and that adding diffuse functions to the basis set had an effect on the CH bending angle. Basis set requirements for atoms beyond the third row of the periodic table are different from the elements with lower atomic number. For heavier metals, such as Ba and Pb, relativistic effects become important, and core electrons must be treated differently from valence electrons. The effective core potential (ECP) can be used to handle the core electrons of these heavier atoms effectively. The ECPs and associated basis sets of Hay–Wadt (76), Stevens and coworkers (77), and Stuttgart–Dresden ECP (78) are widely used and already implemented in many computational chemistry packages. A good review of using ECPs is provided in two chapters in *Reviews in Computational Chemistry*, vol. 8 (79). The main advantage of using ECPs plus valence basis sets rather than all-electron basis sets is that the former consider relativistic effects effectively and require less computer time. But it should be noted that these larger basis sets should be accompanied by electron-correlated methods in order to obtain reliable results. It is also noted that the addition of the polarization function for the Cp ligand is required in order to get a correct bent conformation for Cp_2Ca (28).

Population Analysis Methods

It is easy to understand the nature of bonding between Cp ligand and main group elements using available population analysis methods. Among them, Mulliken population analysis and natural population analysis (NPA) (50) have been widely used to calculate orbital and bonding populations and partial atomic charges of main group metallocenes. It should be noted that the values from these population analyses do not have physical meanings but rather give an interpretation. Mulliken population analysis is strongly basis set dependent. Because of this, the Mulliken population analysis fails to produce meaningful results when the basis set includes diffuse functions. Mulliken population analysis overestimates the electron density on alkali metals and thus underestimates the polar character of the carbon–metal bond (80). In contrast, NPA gives a more realistic picture almost independent of the basis set. Natural population analysis indicates about 80–90% ionic character for the C–Li bond in general for organolithium compounds (80).

Another way to produce atomic charges is electrostatic charge analysis, which assigns point charges to fit the computed electrostatic potential at points on the van der Waals surface. This is also less basis set dependent and gives a good description of partial charges for molecules with polar interactions. CHelpG (81) and Merz–Kollman–Singh (MKS) (82) charge analyses are well-known electrostatic charge methods and can be compared with Mulliken and NPA charge methods.

Vibrational Frequencies and Nuclear Magnetic Resonance Chemical Shift Calculations

Since experimental observations may be difficult or even impossible for some main group metallocenes, theoretical calculations of physical properties may provide an important source of information. Vibrational frequencies play an important role in characterizing the potential energy surface (PES) and can be used to identify minima among the possible structural alternatives. Thus, calculated normal vibrational modes can be used to characterize stationary points on the PES in order to distinguish local minima, which have all real frequencies, from saddle points, which have one imaginary frequency. It must be noted that frequency calculations are valid only at stationary points on the potential energy surface, which means that frequency calculations must be done on optimized structures. Also, a frequency calculation must use the same theoretical level and basis set as the optimized geometry (otherwise the structure would not correspond to a stationary point). Vibrational frequencies can also be used to make zero-point corrections (ZPCs), which are necessary to make accurate predictions when only small energy differences separate different hapticities.

Nuclear magnetic resonance is a powerful tool for characterizing main group metallocenes, since it can often reveal information about fluxionality among different isomers. However, it has been known that due to the nonrigid structure and the free rotation of Cp rings with respect to the principal C_5 axis of main group metallocenes, NMR chemical shifts are often difficult to measure and interpret when there are no available experimental data for comparison. Thus, accurate calculation of absolute chemical shieldings (relative to the calculated absolute chemical shielding of a known standard) can be compared with experimental spectra to help elucidate the exact hapticities of metallocenes. Many ab initio and DFT approaches to the NMR chemical shift calculations have become available in the last decade, supported by the improvement of computer hardware and program algorithms.

Among known efficient techniques for calculating NMR chemical shifts, the individual gauge for localized orbital (IGLO) (83) and gauge-including atomic orbital (GIAO) (84,85) methods have been widely used during the last decade. The most important factor is to obtain reliable equilibrium geometries at the electron-correlated levels, since calculated chemical shifts are highly dependent on the geometrical environments. Since it is known that the HF method is not always reliable, electron correlation methods might be preferable for computing chemical shifts. MP2 or DFT methods seem to be adequate for this purpose, especially DFT, which is more efficient in terms of computer time and disk space than MP2 in calculating chemical shifts. Usually, it is recommended to use gradient-corrected or hybrid functionals (such as B3LYP) for the DFT calculation. It should also be noted that basis sets at least as large as 6-31G* must be used in computing chemical shifts.

Substituted Cyclopentadienyl Ligand Effects on the Metallocene

The sterically bulky pentamethylcyclopentadienyl (Cp*) ligand has successfully been used to synthesize metallocenes, where the simpler cyclopentadienyl group led to polymerization (86). Thus, many Cp_2^*E systems have been characterized experimentally in monomeric form, such as Cp_2^*Ca and Cp_2^*Si. From ab initio calculations of Cp_2Ca and Cp_2^*Ca, it has been shown that adding ten methyl groups on the Cp ligands reduced the Cp(centroid)–Ca distance by only 0.01 Å (28,73). Therefore, the often-simplifying approximation of replacing Cp* with Cp in theoretical calculations is not expected to introduce a large discrepancy between theory and experiment. Also, B3LYP calculations on Cp_2Si and Cp_2^*Si suggest that methyl substitutions on Cp ligand do not cause any significant influence on the electronic and geometrical structures of metallocenes. (37)

Model Calculations

This section provides some model calculations of main group metallocenes to compare different theoretical methods. Optimized geometrical parameters for the Cp_2Li anion at various levels of theory are shown in Table 4. The superiority of the DFT method over HF and semiempirical methods is most striking in the prediction of geometrical parameters. Semiempirical methods overestimate the CH bending angle as well as the C–C distances in Cp rings compared to the DFT and experimental results. The calculated distances in Cp– Li at B3LYP/6-31G* and B3LYP/6-31G** levels are very close to experiment. In a comparison of different basis sets (Table 4), it can be seen that including diffuse functions in the basis set results in longer a C–Li bond distance, and adding p functions on hydrogen atoms has little influence on the Cp_2Li^- anion geometry.

Table 5 represents the optimized geometrical parameters of Cp_2Mg at various computational levels. The PM3 and HF methods predict slightly longer C–Mg distances compared to experiment and B3LYP/6-31G*. Similar to the Cp_2Li^- anion, including diffuse functions in the basis set results in larger C–Mg bond distances, while adding p functions on hydrogen atoms has little influence on the geometry of Cp_2Mg.

TABLE 4 Optimized Geometrical Parameters for Cp_2Li^- Anion at Various Theoretical Levels (distance in Å, angle in degrees)

Theoretical level	C(Cp)–Li	Cp–Li	C–C	C–H	C–H bending angle[a]
MNDO/d	2.323	1.974	1.439	1.086	6.50
PM3	2.367	2.033	1.424	1.089	5.48
HF/6-31G*	2.390	2.069	1.406	1.076	0.06
HF/6-31+G*	2.399	2.079	1.408	1.076	−0.49
HF/6-31G**	2.390	2.070	1.406	1.076	0.13
B3LYP/6-31G*	2.348	2.015	1.417	1.086	−0.27
B3LYP/6-31+G*	2.356	2.024	1.419	1.086	−1.06
B3LYP/6-31G**	2.349	2.016	1.417	1.085	−0.17
Exptl. data[b]		2.008	1.362		

[a] The negative value of CH bending angle indicates inward bending.
[b] X-ray diffraction data.

TABLE 5 Optimized Geometrical Parameters for Cp_2Mg at Various Theoretical Levels (distance in Å, angle in degrees)

Theoretical level	C(Cp)–Mg	Cp–Mg	C–C	C–H	C–H bending angle[a]
PM3	2.407	2.079	1.425	1.091	6.51
HF/6-31G*	2.383	2.058	1.411	1.073	1.62
HF/6-31+G*	2.385	2.061	1.412	1.073	1.47
HF/6-31G**	2.382	2.058	1.411	1.073	1.61
B3LYP/6-31G*	2.364	2.030	1.424	1.084	1.29
B3LYP/6-31+G*	2.366	2.032	1.425	1.084	0.97
B3LYP/6-31G**	2.363	2.030	1.423	1.083	1.28
Exptl. data	2.339[b]	2.008[b]			
	2.304[c]	1.977[c]			

[a] The negative value of CH bending angle indicates inward bending.
[b] Electron diffraction data.
[c] X-ray diffraction data.

TABLE 6   Optimized Distances of Cp–E for LiCp, LiCp*, MgCp, [MgCp*]⁺, AlCp, and AlCp* at Various Theoretical Levels (distance in Å, angle in degrees)

	PM3	HF/6-31G*	B3LYP/6-31G*
LiCp	1.956	1.765	1.730
LiCp*	2.007	1.748	1.717
MgCp⁺	2.039	1.908	1.881
MgCp*⁺	2.058	1.888	1.864
AlCp	2.173	2.050	2.065
AlCp*	2.197	2.008	2.022

The effects caused by methyl substitution (Cp → Cp*) on the main structural features of metallocenes are shown in Table 6. Comparison of calculated Cp–E distance between Cp-containing half-sandwich metallocenes and Cp*-containing half-sandwich metallocenes suggests that methyl substitutions on the Cp rings has little affect on Cp–E distances. This fact is also supported by previous calculations of Cp_2Ca, in which optimized geometrical parameters are in good agreement with the experimental data on Cp_2^*Ca (28).

One of the interesting features for main group metallocenes is electron distribution between main group elements and Cp rings. We have evaluated atomic charges and bond populations on the basis of some useful partitioning schemes for the total electron density distributions, such as Mulliken population analysis, NPA, CHelpG, and MKS, and results are shown in Tables 7 and 8. Calculated atomic charges of different elements, such as Li, Mg, and Al, shows

TABLE 7   Calculated Atomic Charges of Some Half-Sandwich and Full-Sandwich Metallocenes at the B3LYP/6-31G*//B3LYP/6-31G*

		Mulliken	NPA	CHelpG	MKS
LiCp	Q(Li)	0.154	0.902	0.439	0.451
LiCp*	Q(Li)	0.176	0.916	0.499	0.540
Cp_2Li^-	Q(Li)	0.034	0.906	0.306	0.252
MgCp⁺	Q(Mg)	0.679	1.746	1.025	1.073
MgCp*⁺	Q(Mg)	0.615	1.717	1.060	1.174
Cp_2Mg	Q(Mg)	0.310	1.757	0.367	0.401
AlCp	Q(Al)	0.155	0.629	−0.069	0.011
AlCp*	Q(Al)	0.112	0.657	−0.169	−0.074
Cp_2Al^+	Q(Al)	0.531	1.817	0.233	0.209

TABLE 8   Calculated Mulliken Bond Populations and Wiberg Bond Indices of C(Cp)–E for Some Half-Sandwich and Full-Sandwich Metallocenes at the B3LYP/6-31G*// B3LYP/6-31G*

	Mulliken bond population	Wiberg bond index from NPA
LiCp	0.097	0.038
LiCp*	0.104	0.032
Cp_2Li^-	0.058	0.017
$MgCp^+$	0.129	0.069
$MgCp^{*+}$	0.128	0.099
Cp_2Mg	0.102	0.045
AlCp	0.036	0.138
AlCp*	0.035	0.123
Cp_2Al^+	0.115	0.197

positive charges, as expected. The values of electrostatic charges are between those of Mulliken and NPA charges. Mulliken charges overestimate the polarity of main group elements, while NPA and electrostatic charges give a reasonable interpretation of the electron distribution of main group elements. Mulliken bond populations for metallocenes of Li and Mg show that there is a certain covalent interaction between C(Cp) and E, which is somewhat controversial given that these metallocenes have a large amount of ionic bonding character between C(Cp) and E. On the other hand, the Wiberg bond indices (WBI), which are included in the NBO analysis, shows a reasonable description of C(Cp)–E bonding. Thus, the C(Cp)–Al bonding is predicted to be stronger than C(Cp)–Li and C(Cp)–Mg bonding, which explains why Group 13 metallocenes have larger amount of covalent bond character between carbon and the central atom. When a Cp ligand is replaced by Cp*, the Mulliken bond populations and WBI are changed very little, which implies that methyl substitution on the Cp ring should not significantly influence the nature of bonding of metallocenes.

6. CONCLUSIONS

While metallocenes are often thought to be entirely in the domain of transition metal chemistry, we have shown that main group chemists have a legitimate claim as well. The lack of d orbital participation in the metal–ligand bonding may result in less thermodynamic stability but does not preclude the possibility of novel structural motifs or future "real-world" applications. Computational chemistry is expected to play a very active role in the development of this field, calculating properties of known compounds and predicting properties of unknown compounds. Even at the present stage of computer software and hardware, it can be expected that the calculation of main group metallocene potential energy surfaces will give the experimentalist clues for the successful synthesis of new compounds. We hope that this chapter provides an introduction and motivation to continued explorations in this field. Happy hunting!

7. RECENT DEVELOPMENTS

A series of Group 14 metallocenes with substituted Cp rings $[C_5Me_4(SiMe_2Bu^t)]$ was recently reported (87). Single-crystal X-ray structural analysis of each metallocene (E = Ge, Sn, Pb) showed that the mixed alkyl- and silyl-substituted Cp rings are parallel for all three metallocenes (in contrast to most Group 14 with unsubstituted Cp rings). Theoretical calculations at the DFT level indicated that the preference of parallel Cp rings over bent ones is due to the SiR_3 substituent on the Cp rings, which lowers the a_{1g}^* orbital (stereochemically active lone pair) below the e_{1u} orbital.

REFERENCES

1. C Elschenbroich, A Salzer. Organometallics. Weinheim: VCH Verlagsgesellschaft, 1989.
2. G Wilkinson, M Roseblum, MC Whiting, RB Woodward. J Am Chem Soc 74:2125–2126, 1952.
3. P Jutzi. J Organomet Chem 400:1–17, 1990.
4. MA Beswick, JS Palmer, DS Wright. Chem Soc Rev 27:225–232, 1998.
5. (a) P Jutzi, N Burford. In: A Togni, RL Halterman, eds. Metallocenes. New York: Wiley-VCH, 1998, pp 3–54. (b) P Jutzi, N Burford. Chem Rev 99:969–990, 1999.
6. DJ Burkey, TP Hanusa. Comments Inorg Chem 17:41–77, 1995.
7. W Bünder, E Weiss. J Organomet Chem 92:1–6, 1975.
8. SG Baxter, AH Cowley, JG Lasch, M Lattman, WP Sharum, CA Stewart. J Am Chem Soc 104:4064–4069, 1982.
9. NT Anh, M Elian, R Hoffmann. J Am Chem Soc 100:110–116, 1978.
10. P Jutzi. Chem Rev 86:983–996, 1986.
11. JD Fisher, PHM Budzelaar, PJ Shapiro, RJ Staples, GPA Yap, AL Rheingold. Organometallics 16:871–879, 1997.
12. JE Huheey, EA Keiter, RL Keiter. Inorganic Chemistry. 4th ed. New York: HarperCollins College, 1993.
13. FP Floss, JC Traeger. J Am Chem Soc 97:1579–1580, 1975.
14. ESJ Robles, AM Ellis, TA Miller. J Phys Chem 96:8791–8801, 1992.
15. WW Schoeller, O Friedrich, A Sundermann, A Rozhenko. Organometallics 18:2099–2106, 1999.

16. H Chen, P Jutzi, W Leffers, MM Olmstead, PP Power. Organometallics 10:1282–1286, 1991.
17. LM Pratt, IM Khan. J Comput Chem 16:1067–1080, 1995.
18. ED Jemmis, PvR Schleyer. J Am Chem Soc 104:4781–4788, 1982.
19. S Harder, MH Prosenc. Angew Chem Int Ed Engl 33:1744–1746, 1994.
20. O Kwon, Y Kown. J Mol Struct (Theochem) 401:133–139, 1997.
21. J Wessel, E Lork, R Mews. Angew Chem Int Ed Engl 34:2376–2378, 1995.
22. S Harder, MH Prosenc, U Rief. Organometallics 15:118–122, 1996.
23. AJ Bridgeman. J Chem Soc Dalton Trans 2887–2893, 1997.
24. ED Jemmis, S Alexandratos, PvR Schleyer, A Streitwieser Jr, HF Schaefer III. J Am Chem Soc 100:5695–5700, 1978.
25. C Wong, TY Lee, S Lee. Acta Crystallogr B28:1662–1665, 1972.
26. J Berthelot, A Luna, J Tortajada. J Phys Chem A 102:6025–6034, 1998.
27. A Haaland, J Lusztyk, J Brunvoll, KB Starowieyski. J Organomet Chem 85:279–285, 1975.
28. I Bytheway, PLA Popelier, RJ Gillespie. Can J Chem 74:1059–1071, 1996.
29. R Zerger, G Stucky. J Organomet Chem 80:7–17, 1974.
30. M Kaupp, PvR Schleyer, M Dolg, H Stoll. J Am Chem Soc 114:8202–8208, 1992.
31. J Gauss, U Schneider, R Ahlrichs, C Dohmeier, H Schnöckel. J Am Chem Soc 115:2402–2408, 1993.
32. C Dohmeier, H Schnöckel, C Robl, U Schneider, R Ahlrichs. Angew Chem Int Ed Engl 32:1655–1657, 1993.
33. D Loos, H Schnoeckel, J Gauss, U Schneider. Angew Chem Int Ed Engl 31:1362–1364, 1992.
34. E Canadell, O Eisenstein, J Rubio. Organometallics 3:759–764, 1984.
35. DR Armstrong, R Herbst-Irmer, A Kuhn, D. Moncrieff, MA Paver, CA Russell, D Stalke, A Steiner, DS Wright. Angew Chem Int Ed Engl 32:1774–1776, 1993.
36. K Krogh-Jesperson, J Chandrasekhar, PvR Schleyer. J Org Chem 45:1608–1614, 1980.
37. WW Schoeller, O Friedrich, A Sundermann, A Rozhenko. Organometallics 18:2099–2106, 1999 (Correction: 18: 3554, 1999).
38. A Haaland, B Schilling. Acta Chem Scand A38:217–222, 1984.
39. J Almoef, L Fernholt, K Faegri, Jr, A Haaland, BR Schilling, R Seip, K Taugbol. Acta Chem Scand A37:131–140, 1983.
40. DR Armstrong, MA Beswick, NL Cromhout, CN Harmer, D Moncrieff, CA Russell, PR Raithby, A Steiner, AEH Wheatley, DS Wright. Organometallics 17:3176–3181, 1998.
41. (a) A Almenningen, A Haaland, T Motzfeldt. J Organomet Chem 7:97–104, 1967.
    (b) JS Overby, TP Hanusa, VG Young Jr. Inorg Chem 37:163–165, 1998.
42. S Alexandratos, A Streitwieser Jr, HF Schaefer III. J Am Chem Soc 78:7959–7962, 1976.
43. M Lattman, AH Cowley. Inorg Chem 23:241–247, 1984.
44. KC Watermann, A Streitwieser. J Am Chem Soc 106:3138–3140, 1984.
45. LA Paquette, W Bauer, MR Sivik, M Bühl, M Feigel, PvR Schleyer. J Am Chem Soc 112:8776–8789, 1990.
46. R Blom, K Faegri, Jr, T Midtgaard. J Am Chem Soc 113:3230–3235, 1991.

47. A Almenningen, A Haaland, J Lusztyk. J Organomet Chem 170:271–284, 1979.
48. LW Mire, SD Wheeler, E Wagenseller, DS Marynick. Inorg Chem, 37:3099–3106, 1998.
49. FA Cotton, G Wilkinson. Advanced Inorganic Chemistry. 3rd ed. New York, Wiley-Interscience, 1972.
50. JP Foster, F Weinhold. J Am Chem Soc 102:7211–7218, 1980.
51. K Faegri, Jr, J Almoef, HP Lüthi. J Organomet Chem 249:303–313, 1983.
52. P Jutzi, A Seufert. J Organomet Chem 161:C5–C7, 1978.
53. A Haaland, K-G Martinsen, SA Shlykov, HV Volden, C Dohmeier, H Schnöckel. Organometallics 14:3116–3119, 1995.
54. C Dohmeier, D Loos, H Schnöckel. Angew Chem Int Ed Engl 35:129–149, 1996.
55. M Bochmann, DM Dawson. Angew Chem Int Ed Engl 35:2226–2228, 1996.
56. A Haaland, K-G Martinsen, HV Volden, D Loos, H Schnöckel. Acta Chem Scand 48:172–174, 1994.
57. YS Nekrasov, VF Sizoi, DV Zagorevskii, YA Borisov. J Organomet Chem 205: 157–160, 1981.
58. C Pannattoni, G Bombieri, U Croatto. Acta Crystallogr 21, 823–826, 1966.
59. P Jutzi, D Kanne, C Krüger. Angew Chem, Int Ed Engl 25:164, 1986.
60. M Grenz, E Hahn, W-W du Mont, J Pickardt. Angew Chem, Int Ed Engl 23:61–63, 1984.
61. MJ Heeg, C Janiak, JJ Zuckerman. J Am Chem Soc 106:4259–4261, 1984.
62. P Jutzi, F Kohl, P Hofmann, C Krüger, Y-H Tsay. Chem Ber 113:757–769, 1980.
63. TJ Lee, JE Rice. J Am Chem Soc 111:2011–2017, 1989.
64. MJS Dewar, WJ Thiel. J Am Chem Soc 99:4899–4907, 1977.
65. MJS Dewar, EG Zoebisch, EF Healy, JJP Stewart. J Am Chem Soc 107:3902–3909, 1985.
66. JJP Stewart. J Comput Chem 10:209–220, 1989.
67. JA Pople, DP Santry, GA Segal. J Chem Phys 43:S129–S135, 1965.
68. JJP Stewart. "MOPAC2000," Fujitsu Limited, Tokyo, Japan, 1999.
69. Gaussian 98 (Revision A4). MJ Frisch, GW Trucks, HB Schlegel, GE Scuseria, MA Robb, JR Cheeseman, VG Zakrzewski, JA Montgomery, Jr, RE Stratmann, JC Burant, S Dapprich, JM Millam, AD Daniels, KN Kudin, MC Strain, O Farkas, J Tomasi, V Barone, M Cossi, R Cammi, B Mennucci, C Pomelli, C Adamo, S Clifford, J Ochterski, GA Petersson, PY Ayala, Q Cui, K Morokuma, DK Malick, AD Rabuck, K Raghavachari, JB Foresman, J Cioslowski, JV Ortiz, BB Stefanov, G Liu, A Liashenko, P Piskorz, I Komaromi, R Gomperts, RL Martin, DJ Fox, T Keith, MA Al-Laham, CY Peng, A Nanayakkara, C Gonzalez, M Challacombe, PMW Gill, B Johnson, W Chen, MW Wong, JL Andres, M Head-Gordon, ES Replogle, JA Pople. Gaussian, Inc, Pittsburgh PA, 1998.
70. WJ Hehre, L Radom, PvR Schleyer, JA Pople. Ab Initio Molecular Orbital Theory. New York: Wiley, 1986.
71. RG Parr, W Yang. Density Functional Theory of Atoms and Molecules. Oxford: Oxford University Press, 1989.
72. AA El-Azhary, HU Suter. J Phys Chem 100:15056–15063, 1996.
73. R Blom, K Faegri Jr, HV Volden. Organometallics 9:372–379, 1990.
74. TP Hanusa. Chem Rev 93:1023–1036, 1993.

75. D Feller, ER Davidson. In: KB Lipkowitz, DB Boyd, eds. Reviews in Computational Chemistry. New York: VCH, 1996, pp 1–45.
76. PJ Hay, WR Wadt. J Chem Phys 82:270–310, 1985.
77. WJ Stevens, H Basch, M Krauss. J Chem Phys 81:6026–6033, 1984.
78. M Dolg, U Wedig, H Stoll, H Preuss. J Chem Phys 86:866–872, 1987.
79. (a) G Frenking, I. Antes, M. Böhme, S. Dapprich, AW Ehlers, V Jonas, A Neuhaus, M Otto, R Stegmann, A Veldkamp, SF Vyboishchikov. In: KB Lipkowitz, DB Boyd, eds. Reviews in Computational Chemistry. New York: VCH, 1996, pp 63–144. (b) TR Cundari, MT Benson, ML Lutz, SO Sommerer. In: KB Lipkowitz, DB Boyd, eds. Reviews in Computational Chemistry. New York: VCH, 1996, pp 145–202.
80. C Lambert, PvR Schleyer. Angew Chem, Int Ed Engl 33:1129–1140, 1994.
81. CM Breneman, KB Wiberg. J Comput Chem 11:361–373, 1990.
82. BH Besler, KM Merz Jr, PA Kollman. J Comput Chem 11:431–439, 1990.
83. W Kutzelnigg. Isr J Chem 19:193–200, 1980.
84. HF Hameka. Mol Phys 1:203–215, 1958.
85. R Ditchfield. Mol Phys 27:789–807, 1974.
86. R Blom, J Boersma, PHM Budzelaar, B Fischer, A Haaland, HV Volden, J Weidlein. J Acta Chem Scand A40:113–120, 1986.
87. SP Constantine, H Cox, PB Hitchcock, GA Lawless. Organometallics 19:317–326, 2000.

Index

425

Printed in the United States
by Baker & Taylor Publisher Services